Oncology of the Nervous System

Cancer Treatment and Research

WILLIAM L. MCGUIRE, *series editor*

Livingston, R.B., ed., Lung Cancer 1. 1981. ISBN 90-247-2394-9.
Bennett Humphrey, G., Dehner, Louis P., Grindey, Gerald B., and Acton, Ronald T., eds., Pediatric Oncology 1. 1981. ISBN 90-247-2408-2.
DeCosse, Jerome J., and Sherlock, Paul, eds., Gastrointestinal Cancer 1. 1981. ISBN 90-247-2461-9.
Bennett, John M., ed., Lymphomas 1, including Hodgkin's Disease. 1981. ISBN 90-247-2479-1.
Bloomfield, C.D., ed., Adult Leukemias 1. 1982. ISBN 90-247-2478-3.
Paulson, David F., ed., Genitourinary Cancer 1. 1982. ISBN 90-247-2480-5.
Muggia, F.M., ed., Cancer Chemotherapy 1. ISBN 90-247-2713-8.
Bennett Humphrey, G., and Grindey, Gerald, B., eds., Pancreatic Tumors in Children. ISBN 90-247-2702-2.
Costanzi, John J., ed., Malignant Melanoma 1. ISBN 90-247-2706-5.
Griffiths, C.T., and Fuller, A.F., eds., Gynecologic Oncology. ISBN 0-89838-555-5.
Greco, Anthony F., ed., Biology and Management of Lung Cancer. ISBN 0-89838-554-7.

Oncology of the Nervous System

Edited by

MICHAEL D. WALKER

National Institutes of Health
Bethesda, Maryland, U.S.A.

1983 **MARTINUS NIJHOFF PUBLISHERS**
a member of the KLUWER ACADEMIC PUBLISHERS GROUP
BOSTON / THE HAGUE / DORDRECHT / LANCASTER

Distributors

for the United States and Canada: Kluwer Boston, Inc., 190 Old Derby Street, Hingham, MA 02043, USA
for all other countries: Kluwer Academic Publishers Group, Distribution Center, P.O.Box 322, 3300 AH Dordrecht, The Netherlands

Library of Congress Cataloging in Publication Data

Main entry under title:

Oncology of the nervous system.

   (Cancer treatment and research ; v. 12)
   1. Brain--Tumors. 2. Nervous system--Tumors.
I. Walker, Michael D. II. Series. [DNLM:
1. Brain neoplasms. 2. Spinal cord neoplasms.
3. Brain neoplasms--Therapy. 4. Spinal cord neoplasms
--Therapy. W1 CA693 v.12 / WL 358 O58]
RC280.B7O53 1983   616.99'28   82-18824
ISBN 0-89838-567-9

ISBN 0-89838-567-9

Copyright

© 1983 by Martinus Nijhoff Publishers, Boston.

All rights reserved. No part of this publication may be reproduced, stored in a retrieval system, or transmitted in any form or by any means, mechanical, photocopying, recording, or otherwise, without the prior written permission of the publisher
Martinus Nijhoff Publishers, 190 Old Derby Street, Hingham, MA 02043, USA.

PRINTED IN THE NETHERLANDS

To JB who was a pioneer in the field.

Contents

Foreword ix

Preface xi

List of Contributors xv

1. The Epidemiology of Central Nervous System Tumors
   BRUCE S. SCHOENBERG   1

2. A Review of Animal Brain Tumor Models that have been used for Therapeutic Studies
   S. CLIFFORD SCHOLD Jr. and DARELL D. BIGNER   31

3. Chemotherapy of Brain Tumors – Basic Concepts
   WILLIAM R. SHAPIRO and THOMAS N. BYRNE   65

4. The Role of Intracranial Surgery for the Treatment of Malignant Gliomas
   JOSEPH RANSOHOFF   101

5. Surgical Management of Endocrine-Active Pituitary Adenomas
   CHARLES B. WILSON   117

6. Immunologic Considerations of Patients with Brain Tumors
   M.S. MAHALEY Jr. and G. YANCEY GILLESPIE   151

7. Corticosteroids: Their Effect on Primary and Metastatic Brain Tumors
   ROBERT G. SELKER   167

8. Applied Radiophysics for Neuro-Oncology
   HARRY R. KATZ and ROBERT L. GOODMAN   193

9. Radiotherapy of Adult Primary Cerebral Neoplasms
   GLENN E. SHELINE   223

10. Tumor Cell and Host Response Parameters in Designing Brain Tumor Therapy
    PAUL L. KORNBLITH, BARRY H. SMITH and MAURICE K. GATELY   247

11. Metabolic Therapy of Malignant Gliomas
    JAMES T. ROBERTSON, E. STANFIELD ROGERS, W.L. BANKS and HAROLD F. YOUNG   273

12. Pathologic Effects of Chemotherapy
    KURT JELLINGER   285

13. The Management of Brain Metastases
    J. GREGORY CAIRNCROSS and JEROME B. POSNER   341

14. Prognostic Factors for Malignant Glioma
    DAVID P. BYAR, SYLVAN B. GREEN and THOMAS A. STRIKE   379

15. The Challenge to Oncology of the Nervous System
    MICHAEL D. WALKER   397

Index   405

Cancer Treatment and Research

FOREWORD

Where do you begin to look for a recent, authoritative article on the diagnosis of management of a particular malignancy? The few general oncology textbooks are generally out of date. Single papers in specialized journals are informative but seldom comprehensive; these are more often preliminary reports on a very limited number of patients. Certain general journals frequently publish good in-depth reviews of cancer topics, and published symposium lectures are often the best overviews available. Unfortunately, these reviews and supplements appear sporadically, and the reader can never be sure when a topic of special interest will be covered.

Cancer Treatment and Research is a series of authoritative volumes which aim to meet this need. It is an attempt to establish a critical mass of oncology literature covering virtually all oncology topics, revised frequently to keep the coverage up to date, easily available on a single library shelf or by a single personal subscription.

We have approached the problem in the following fashion. First, by dividing the oncology literature into specific subdivisions such as lung cancer, genitourinary cancer, pediatric oncology, etc. Second, by asking eminent authorities in each of these areas to edit a volume on the specific topic on an annual or biannual basis. Each topic and tumor type is covered in a volume appearing frequently and predictably, discussing current diagnosis, staging, markers, all forms of treatment modalities, basic biology, and more.

In Cancer Treatment and Research, we have an outstanding group of editors, each having made a major commitment to bring to this new series the very best literature in his or her field. Martinus Nijhoff Publishers has made an equally major commitment to the rapid publication of high quality books, and worldwide distribution.

Where can you go to find quickly a recent authoritative article on any major oncology problem? We hope that Cancer Treatment and Research provides an answer.

WILLIAM L. MCGUIRE
Series Editor

Preface

MICHAEL D. WALKER

Oncology of the nervous system has gradually been coming of age through the development and establishment of the field of Neuro-Oncology. In the past, physicians tended to consider the involvement of the brain or spinal cord with cancer as often being the point at which they drew back and let nature take its course for there was 'nothing more that could be done'. A few investigators saw early the importance to cancer chemotherapy of pharmacologic 'oddities' such as the blood-brain barrier and the brain. Zubrod and Rall were among the first to point this out in the practical context of the pharmacodynamics of treatment for the cancer patient with CNS involvement. Over the course of time, increasing interest in neoplastic disease of the brain has been evident at the presentations in national meetings as well as articles in scientific journals. In fact, the monograph entitled *Modern Concepts in Brain Tumor Therapy: Laboratory and Clinical Investigation* was the eighth most frequently sited and ordered monograph in 1978.

This volume is a series of independent but interrelated chapters devoted to specifically defined as well as to controversial areas of research in CNS neoplasm. Because such information such as epidemiology, immunology and radiophysics appears in diverse journals, it is the intent of this volume to bring together some of these specific areas of importance to the entire field of neuro-oncology under one cover and for reference purposes.

From the therapeutic point-of-view, the treatment of nervous system tumors has been approached independently by the traditional disciplines (surgery, radiotherapy, chemotherapy, etc.). At one time in the not too distant past, surgery was the *only* therapeutic maneuver utilized for any intracranial neoplasm and only then when it had reached a large enough size to provide clear neurologic localization. As surgical procedures become refined, illustrated by the chapters of both Ransohoff and Wilson, they have become polarized into those operations where a discrete encapsulated independent lesion with sharp borders and little evidence of invasion (such as the majority of pituitary adenomas) can be removed, while those which are parenchymal, invasive and mixed (such as gliomas) cannot. Surgical boldness has justifiably increased over the years as control of the patient and his environment has become a reality.

Well controlled neuroanesthesiology, fluid management and replacement, blood pressure regulation and the control of intracerebral pressure by use of corticosteroids have all made procedures much less risky. The use of the operating microscope through which precise dissection and absolute control over hemostasis and the surgical field is obtained has extended the vision of the surgeon to smaller and smaller tumors in more discrete locations. Despite the ability to manage the environment surrounding brain tumors, those of the astrocytic series which are inherently locally invasive in vital tissues cannot be resected.

During the course of the last decade, corticosteroids have become the most frequently utilized therapy for the primary management of brain tumors. These drugs are recognized not only for the dramatic effect which they have on the symptomatic patient harboring intracranial neoplasm but also for their effect on diagnosis and therapeutic approaches. Within 24 to 48 hours of starting a patient on corticosteroids, dramatic improvement in the level of consciousness, aphasia, hemiparesis and sensory loss may all be demonstrated. Such loss of symptomatology is accompanied by a decrease in intracranial pressure as cerebral edema abates. 'Medical decompression' with steroid provide the clinician with the necessary time to carry out more detailed examinations, restore the patient to appropriate fluid electrolyte and metabolic normality, and plan a careful surgical attack. Rarely, is a patient rushed to the operating room for an emergency decompression under suboptimal conditions and with less than all the diagnostic information that one could have at hand. From the point-of-view of comfort of the patient (and the physician) there is no more remarkable substance than corticosteroids. But they are only temporary and do not interfere with tumor growth or replication from a practical point-of-view.

The development and widespread use of CT Scanning has also had a major impact on the treatment of central nervous system tumors. All of the previously diagnostic procedures, such as pneumoencephalography, arteriography and ventriculography often yielded less than precise information; all are uncomfortable and are not without hazard. The radionuclide scan which preceded the development of CT Scanning has extremely poor resolution and thus tumors often had to be of large size in order to be visualized. CT Scanning now permits the definition of areas only a few millimeters in diameter, and because it is a non-invasive method with extremely little risk to the patient, it is used more frequently early in the course of the disease and with greater assurance. Although a histologic diagnosis cannot be made on CT Scanning, there are observations of tumor size, shape, density, effect and involvement with surrounding brain which allow the CT Scan interpreter to make a well-informed estimate of the histologic appearance of the tumor. In addition to its technical advantages this instrument has also added in an obvious and immediately measurable way, to the comfort of the patient and the assuredness with which the physician can approach therapeutic intervention.

Brain tumor research is in the process of shifting as the immediate short-term needs to control the disease and save the patient's life by decompression have been met by corticosteroids and more certain and anatomically accurate diagnosis have been achieved by the CT Scan. Thus, the identification and removal of acoustic tumors, pituitary adenomas, meningiomas and host of other highly localized tumors with minimal defect for the patient is a reality.

On the other hand, the most frequently occurring tumors of the astrocytic series with their locally invasive properties may be more clearly defined and temporarily controlled but nevertheless are lethal in the outcome. In order to change this, both radiotherapy and chemotherapy have been employed with the hope of effecting better control. Although both are effective they are limited in their value. Attention has been turned toward understanding the disease process more fully. Controlled clinical trials have been established in order to determine the efficacy of putitative therapy. Parallel with these studies, there has been a progressive realization that the future lies in a better understanding of the metabolic and biochemical events which distinguish a tumor from normal tissue. In addition, the milieu in which the tumor exists and the very special properties of the blood-brain barrier which modulate fluid exchange and the egress of substances of modest molecular size into the extracellular space are all of importance.

ACKNOWLEDGMENTS

The Editor would like to acknowledge the foresightedness of Martinus Nijhoff Publishers for identifying the field of Neuro-Oncology as one of importance and for pursuing the development of this volume. In particular, our thanks to Dr. William L. McGuire the Series Editor and Jeffrey K. Smith, Publisher – Medical Division for their thoughtful advice, encouragement and insight. Much of the routine effort, such as letter writing and telephone calls which make a volume such as this so successful has been carried out by Mrs. Geraldine Bitango to whom I express my special thanks. Finally, a volume is only as good as the Chapters of which it is comprised. The Contributors have worked hard, prepared excellent illustrations and addressed their assigned topic thoughtfully. The field of Neuro-Oncology is in their debt.

Michael D. Walker

List of Contributors

BANKS, William L., Ph.D., Professor of Biochemistry and Microbiology, Medical College of Virginia, Box 614, Richmond, VA 23298, U.S.A.

BIGNER, Darell D., M.D., Ph.D., Professor of Pathology, Department of Pathology, Duke University Medical Center, Box 3156, Durham, NC 27710, U.S.A.

BYAR, David P., M.D., Chief, Clinical and Diagnostic Trials Section, NCI, Biometry Branch, Landow, Rm. C509, Bethesda, MD 20205, U.S.A.

BYRNE, Thomas N., M.D., Association for Brain Tumor Research Fellow, Department of Neurology, Memorial Sloan-Kettering Cancer Center, New York, NY 10021, U.S.A.

CAIRNCROSS, J. Gregory, M.D., Department of Clinical Neurosciences, Victoria Hospital, South Street, London, Ontario, Canada

GATELY, Maurice K., M.D., Ph.D., Staff Fellow, NINCDS, SN Bldg. 9, Rm. 1W 120, Bethesda, MD 20205, U.S.A.

GILLESPIE, G. Yancy, III, Ph.D., Research Assistant Professor, Department of Surgery, Division of North Carolina Medical Center, 148 Clinical Science Bldg. 229H, Chapel Hill, NC 27514, U.S.A.

GOODMAN, Robert L., M.D., Professor and Chairman, Department of Radiation Therapy, American Oncologic Hospital, Central and Shelmire Avenues, Philadelphia, PA 19111, U.S.A.

GREEN, Sylvan B., M.D., Medical Researcher, Clinical and Diagnostic Trials Section, NCI, Biometry Branch, Landow Bldg., Room 5C09, Bethesda, MD 20205, U.S.A.

JELLINGER, Prof. Dr. Kurt, Ludwig Boltzmann – Institute of Clinical, Neurobiology, Lainz-Hospital, 1 Wolkersbergenstrasse, A-1130 Vienna, Austria

KATZ, Harry, M.D., Assistant Professor of Radiation Therapy, University of Pennsylvania School of Medicine, 3400 Spruce Street, Philadelphia, PA 19104, U.S.A.

KORNBLITH, Paul L., M.D., Chief, Surgical Neurology Branch, NINCDS, Building 10A, 3E68, Bethesda, MD 20205, U.S.A.

MAHALEY, M. Stephen, Jr., M.D., Ph.D., Professor and Chief of Neurological Surgery, Department of Surgery, Division of Neurological Surgery, University of North Carolina Medical Center, 148 Clinical Science Bldg., 229H, Chapel Hill, NC 27514, U.S.A.

POSNER, Jerome B., M.D., Head, Department of Neurology, Memorial Sloan-Kettering Cancer Center, 1275 York Avenue, New York, NY 10021, U.S.A.

RANSOHOFF, Joseph, II, M.D., Professor and Chairman, Department of Neurosurgery, New York University Medical Center, 550 First Avenue, New York, NY 10016, U.S.A.

ROBERTSON, James T., M.D., Professor and Chairman, Department of Neurosurgery, 956 Court Street, Memphis, TN 38163, U.S.A.

ROGERS, E. Stanfield, M.D., Professor Biochemistry and Microbiology, University of Tennessee, 800 Madison Avenue, Memphis, TN 38163, U.S.A.

SCHOENBERG, Dr. Bruce S., Chief, Section on Neuroepidemiology, NINCDS, IRP, Federal Bldg., Rm. 904, Bethesda, MD 20205, U.S.A.

SCHOLD, S. Clifford, Jr., M.D., Assistant Professor, Division of Neurology, Duke University Medical Center, Laboratory of Neuro-Oncology, P.O. Box 2905, Durham, NC 27710, U.S.A.

SELKER, Robert G., M.D., Professor of Neurosurgery, Department of Neurosurgery, University of Pittsburgh, Montefiore Hospital, 3459 Fifth Avenue, Pittsburgh, PA 15213, U.S.A.

SHAPIRO, William R., M.D., Attending Neurologist and Head, Cotzias Laboratory of Neuro-Oncology, Department of Neurology, Memorial Sloan-Kettering Cancer Center, 1275 York Avenue, New York, NY 10021, U.S.A.

SHELINE, Glenn E., Ph.D., M.D., Professor & Vice-Chairman, Division of Radiation Oncology, University of California, San Francisco, 330 – Moffitt Hospital, San Francisco, CA 94143, U.S.A.

SMITH Barry H., M.D., Ph.D., Deputy Chief Surgical Neurology, NINCDS, Bldg. 10, Rm. 3E68, Bethesda, MD 20205, U.S.A.

STRIKE, Thomas A., Ph.D., Special Assistant, Division of Cancer Treatment, NCI, Landow Bldg., Rm. 4C33, Bethesda, MD 20205, U.S.A.

WALKER, Michael D., M.D., National Institutes of Health, NINCDS, STP, Federal 8A-08, Bethesda, MD 20205, U.S.A.

WILSON, Charles B., M.D., Professor and Chairman, Department of Neurological Surgery, University of California, SF, The Editorial Office, 350 Parnassus, Suite 807, 786 M/School of Medicine, San Francisco, CA 94143, U.S.A.

YOUNG, Harold F., M.D., Co-Director, Division of Neurosurgery, Medical College of Virginia, Box 631, Richmond, VA 23298, U.S.A.

1. The Epidemiology of Central Nervous System Tumors

BRUCE S. SCHOENBERG

INTRODUCTION

An essential part of any epidemiologic investigation is defining and classifying the disease or diseases under study. For central nervous system (CNS) neoplasms, several different terms have been applied to the same tumor types, thus resulting in a somewhat cumbersome nomenclature. This chapter follows the classification scheme for CNS neoplasms as outlined in Table 1. It is adapted from the systems of nomenclature suggested by Kernohan and Sayre [1] and Rubinstein [2], and is based on the presumed cell type of origin of the tumor. Because of their close anatomic proximity, pituitary gland and craniopharyngeal duct tumors are included with CNS neoplasms. The optic nerve and retina are derived from the primary cerebral vesicle, and tumors of these sites will also be discussed. The term *astrocytoma* designates grade 1 and 2 gliomas of the astrocytic series, while the term *glioblastoma* refers to grade 3 and 4 gliomas of this series.

CNS neoplasms can produce symptoms directly by causing irritation or destruction of adjacent neural tissue, and indirectly by increasing intracranial or intraspinal pressure as the tumor mass expands within a fixed volume. The same histologic type of tumor may produce very different clinical symptoms, depending on its anatomic location. It may be difficult to determine the malignancy of a CNS tumor solely by its clinical presentation. A histologically benign, slow-growing tumor may produce marked and devastating effects in a relatively short time because of its critical location, whereas a histologically malignant neoplasm may not produce overt symptoms for several months. Because of this problem, the terms "benign" and "malignant" are often not used in epidemiologic studies of CNS neoplasms.

The neoplasms mentioned above rarely metastasize to sites outside the CNS, although seeding within the CNS is not uncommon. On the other hand, the brain is a frequent metastatic site for tumors originating elsewhere in the body. Bronchogenic carcinoma in males and mammary carcinoma in females are the two most common primary sites of cerebral metastases [2, 4]. Other tumors that frequently metastasize to the brain are malignant melanoma, and carcinoma of the kidney and gastrointestinal tract [2, 5]. The skull and spine are also common

Table 1. Classification of central nervous system neoplasms

Tumors of neuroglial origin (gliomas)
    Astrocytic series
        Astrocytoma (grades 1 and 2)
        Glioblastoma (grades 3 and 4)
    Oligodendroglioma
    Ependymoma
Tumors of neuronal cells and primitive bipotential precursors
    Medulloblastoma
    Ganglioneuroma
    Ganglioglioma
    Neuroblastoma
Tumors of mesodermal tissues
    Meningioma
    Sarcoma
Tumors of nerve roots (neurilemoma is used here as a general term to refer to nerve sheath tumors)
    Neurofibroma
    Schwannoma
Tumors of lymphoreticular system
    Reticulum cell sarcoma – microglioma
Tumors of blood vessel origin
    Hemangioblastoma
Tumors of pituitary gland
    Chromophobe adenoma
    Acidophilic adenoma
    Basophilic adenoma
Tumors of choroid plexus
    Choroid plexus papilloma
Tumors of pineal region
    Germ cell origin
        Germinoma (pinealoma)
Tumors of pineal parenchyma
    Pineocytoma
    Pineoblastoma
Tumors of maldevelopmental origin
    Teratoma
    Craniopharyngioma

Reproduced with permission from Schoenberg, 1977 [3].

metastatic sites for primary cancers of the breast, prostate, thyroid, kidney, and lung [5]. Cancer of the nasopharynx or nasal sinuses may spread intracranially by direct extension. Bony metastatic deposits may likewise produce neurologic abnormalities by affecting underlying neural tissue. When dealing with an intracranial mass in a patient with a known malignancy outside the nervous system, it is essential to microscopically examine the intracranial lesion in order to distinguish primary intracranial neoplasms from metastatic lesions.

DESCRIPTIVE STUDIES

Mortality data are readily available for primary nervous system neoplasms, since many countries have standardized operational methods for collecting and publishing such information. Despite the relative ease and low cost of obtaining these data, there are several very basic problems. One is accuracy and completeness of reporting. Most readers are familiar with the rather haphazard way in which the death certificate is often filled out. Furthermore, mortality tabulations for most countries are based only on the underlying cause of death, thus eliminating from consideration diseases that are listed as the immediate or contributing cause of death. It may be difficult to choose one underlying cause of death when, actually, several diseases are present and contribute to the patient's demise. Finally, not all patients with a given disorder die of that particular disorder. In spite of these drawbacks, mortality data have provided valuable information.

A comparison of international death rates for CNS neoplasms during the 1950's revealed a consistent rate of around 5 deaths per 100,000 population per year [6]. Massey *et al.* [7] recently updated these data for the years 1967 through 1974. For males, the average annual age-adjusted death rate was 7.50 deaths per 100,000 per year. The corresponding rate for females was 5.85 deaths per 100,000 per year. This male excess is noted consistently in available reports. The increasing rate over time is thought to represent more complete case ascertainment rather than a true rise in the frequency of these neoplasms. Age-specific mortality rates for primary CNS neoplasms in the United States and other countries demonstrate a small peak in childhood, followed by a higher and sharper peak in adult life [6, 8]. Moreover, this is also valid when U.S. age-specific death rates are analyzed separately by race (white versus nonwhite) and sex. Whites have higher rates than nonwhites, and males have higher rates than females [8].

MORBIDITY

Early reports dealing with the frequency of neoplasms affecting the nervous system were based on the experience of individual physicians or on proportional rates of hospital admissions. Recently, investigations described tabulations from well-defined population groups in order to minimize the problems of selection bias usually inherent in such investigations. Studies derived from tumor registry data or information based on special case-finding surveys have yielded more accurate measures of morbidity [8 – 23].

Primary nervous system neoplasm incidence data were analyzed for 62 population groups worldwide [24]. Information was available for various years between 1956 and 1967. With few exceptions, males had higher rates than females. The median age-adjusted incidence rate (adjusted to the 1950 U.S. population) was between 4 and 5 cases per 100,000 per year; 90% of the age-adjusted rates fell within the range of 1 to 8 cases per 100,000 per year. These rates, however, probably underestimate the true rates. Caution must be exercised when comparing figures from these various data resources. For example, some registries record information only for CNS tumors specified as malignant, while others restrict themselves to tumors of the brain.

Despite variations among the different data resources in reporting and diagnostic practices, a general pattern of age-specific incidence emerged: a small peak in childhood, followed by a higher peak, reaching a maximum between ages 60 and 80, and then a decline after those ages. The shape of the age-specific incidence curve resembles that of the age-specific mortality curve.

Schoenberg et al. [22] recently analyzed data for primary intracranial neoplasms from the Connecticut Tumor Registry as a model of the large population-based registries. A total of 3,210 primary tumors diagnosed between 1935 and 1964 in Connecticut residents was the basis for this study. Slightly over 75% of the neoplasms were microscopically confirmed. Only those tumors with such confirmation were then histologically classified. There is good evidence that the various histologic types of brain tumors reveal sufficiently distinct epidemiologic patterns to be considered as separate diseases. Table 2 presents the age-adjusted incidence rates by histologic type, anatomic site, and sex, together

Table 2. Primary intracranial tumors, Connecticut, 1935 – 1964

Tumor	Incidence rate*		Average age
	Male	Female	
Glioblastoma	2.07	1.51	50
Meningioma	0.49	0.70	52
Astrocytoma	0.52	0.34	38
Medulloblastoma	0.19	0.15	12
Neurilemoma	0.08	0.07	50
Hemangioma**	0.08	0.08	35
Ependymoma	0.08	0.07	27
All tumors of:			
Brain and meninges	4.91	4.04	46
Pituitary	0.43	0.46	45
Pineal	0.04	0.02	28

*Cases per 100,000 per year. These rates are age-adjusted using the direct method and using the 1950 U.S. population as the standard.
**In data from the Connecticut Tumor Registry, tumors of blood vessel origin are termed 'hemangiomas'; no coding distinction is made between 'hemangioma' and 'hemangioblastoma'.
Reproduced with permission from Schoenberg et al. [22].

Table 3. Histologically confirmed primary intracranial neoplasms: frequency distribution, Connecticut, 1935 – 1964

Children (0 – 14 years)			Adults ($\geq$ 15 years)		
Type	No.	%	Type	No.	%
Medulloblastoma	74	24.2	Glioblastoma	1,105	52.1
Astrocytoma	63	20.6	Meningioma	389	18.4
Glioblastoma	62	20.3	Astrocytoma	214	10.1
Ependymoma	20	6.5	Chromophobe adenoma	96	4.5
Craniopharyngioma	17	5.6	Neurilemoma	46	2.2
Meningioma	14	4.6	Hemangioma	41	1.9
Hemangioma	9	2.9	Craniopharyngioma	30	1.4
Neuroblastoma	8	2.6	Medulloblastoma	27	1.3
Teratoma	6	2.0	Ependymoma	27	1.3
Pinealoma	6	2.0	Acidophilic adenoma	26	1.2
Sarcoma	5	1.6	Oligodendroglioma	22	1.0
Oligodendroglioma	2	0.7	Sarcoma	19	0.9
Neurilemoma	2	0.7	Pinealoma	6	0.3
Others, specified	11	3.6	Others, specified	26	1.2
Others, unspecified	7	2.3	Others, unspecified	45	2.1
Total	306	100.2	Total	2,119	99.9

Reproduced with permission from Schoenberg *et al.* [25].

with the average age of occurrence. Note that meningioma is the only relatively common primary intracranial neoplasm with a higher incidence in females.

In children, the relative frequencies of these various tumors differ markedly from their frequencies in adults [25] (Table 3). While medulloblastoma tops the list in children, making up approximately 24% of the confirmed tumors, this tumor type is not among the most common intracranial neoplasms in adults. Astrocytoma is second in children, accounting for 21%, whereas it is third in adults at 10%. Glioblastoma, ranking third in children at 20%, constitutes more than half of the histologically confirmed intracranial tumors in adults. Craniopharyngioma accounts for 5.6% of confirmed childhood tumors and only 1.4% of the total in adults. Chromophobe adenoma, while common among adults, is relatively rare in children. Meningioma, which ranks second in adults (accounting for 18%), is relatively infrequent in childhood. Hemangioma has similar relative rankings in both children and adults.

Figure 1 presents the age-specific incidence rates for the more common primary neoplasms of the brain and cranial meninges. The pattern for glioblastoma holds for all brain tumors: a small childhood peak and the higher adult peak. The curve for astrocytoma also has an early peak, but has a much flatter and smaller rise for the older age groups. The curve for meningiomas shows, with some fluctuations, an increase in incidence in going from the

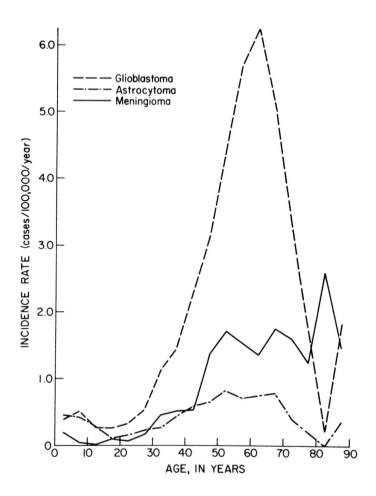

Figure 1. Average annual age-specific incidence rates by histologic type for the more common primary neoplasms of the brain and cranial meninges, Connecticut, 1935–1964. (Reproduced with permission from Schoenberg et al. [22]).

younger to the older age groups.

The less common tumors are illustrated in Figure 2. Medulloblastoma peaks in childhood. Ependymoma has two peaks, one for the younger ages and a second around age 55. Neurilemoma has one peak around age 65. The curve for hemangioma rises slightly between ages 40 and 65, followed by a decline beyond that point.

One notable exception to the pattern for all tumors of the brain and cranial meninges is provided by the tabulations for Rochester, Minnesota, where after a

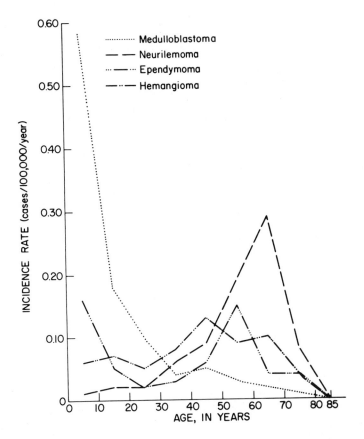

Figure 2. Average annual age-specific incidence rates by histologic type for the less common primary neoplasms of the brain, Connecticut, 1935–1964. (Reproduced with permission from Schoenberg *et al.* [22]).

small childhood peak, there is a sustained increase in incidence with increasing age [26]. Furthermore, the Rochester data reveal higher incidence rates for brain tumors than are reported from other sources. Finally, the most common intracranial neoplasm in the Rochester series was meningioma, as contrasted with glioblastoma in most other investigations, including Connecticut. These discrepancies were elucidated by comparing the Rochester figures with those for Connecticut [26]. The majority of cases first diagnosed at autopsy in Rochester (as opposed to Connecticut) account, in large part, for the different age-specific incidence curves, particularly among the elderly. This also explains the higher percentage of meningiomas. These neoplasms account for a larger percentage of tumors only diagnosed after death as compared to tumors discovered before death. The results of comparisons between Connecticut and Rochester suggest that (a) substantial numbers of asymptomatic tumors are missed in older individuals, and (b) as larger proportions of the population die and come to autopsy, the age-specific incidence pattern will more closely resemble the Rochester curve.

Incidence rates for all pituitary tumors increase with increasing age. This is also true for chromophobe adenoma, the most common histologically confirmed tumor of the pituitary gland [22]. Most studies of the incidence of pituitary tumors report rates of approximately 1 case/100,000/year [8]. Data for Olmsted County, Minnesota indicate, since 1970, an increase in the number of pituitary adenomas diagnosed in women of childbearing age. The rates for men and for older women have remained stable since 1935, however. Using a case-control study, the investigators explored a possible relationship between the use of oral contraceptives and the rise in the incidence of pituitary adenomas among women of childbearing age, but found no association. They attributed the increase to improvements in diagnostic techniques, and the fact that the discovery of such lesions originates from complaints of galactorrhea, amenorrhea, or infertility [27].

Based on data from the Connecticut Tumor Registry the incidence rate for retinoblastoma has been estimated at about 1 case/1,000,000/year, while the comparable figure for optic gliomas is approximately 1.1 cases/10,000,000/year [28]. Tumors of the peripheral nervous system are much more common, with a reported incidence rate of about 1.5/100,000/year [20].

Mortality tabulations demonstrate an increased risk of dying from nervous system neoplasms in whites as compared to blacks. This finding led to a number of morbidity studies which explored racial differentials in brain tumor incidence. Several investigations of large case series of CNS neoplasms have implied that the morbidity experience among blacks is different than that among whites. For example, meningiomas and pituitary tumors appear to occur relatively more frequently in blacks than in whites [29]. These findings are confirmed in both U.S. and African series [29, 30], and have been verified in a population-based study of residents in the Washington, D.C. metropolitan area [31].

SURVIVAL

Survival data for patients with primary CNS neoplasms are difficult to interpret, especially when attempting to determine changes over time. Survival is a function of tumor type, tumor grade, anatomic location(s) of the neoplasm, the patient's age, the presence of other disease, the responsiveness of the tumor to radiation or chemotherapy, the surgeon's skill, post-operative care, etc. Perhaps the most comprehensive tabulations on survival of brain tumor patients are available from the National Cancer Institute [127], and are based on information

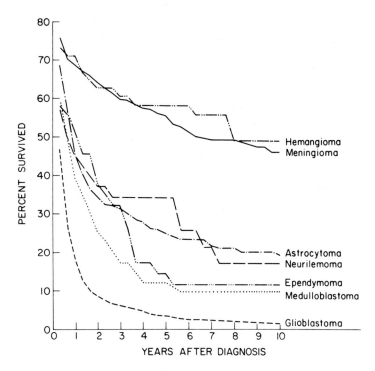

Figure 3. Survival by histologic type for patients with primary neoplasms of the brain and cranial meninges, Connecticut, 1935–1964. (Reproduced with permission from Schoenberg [24]).

from the California and Connecticut Tumor Registries, the Charity Hospital of Louisiana, and the State University of Iowa Hospital. Data from 1950 through 1973 reflect a gradual improvement in survival over time. Whether this change indicates earlier diagnosis, reduction in surgical mortality, improved methods of treatment, or some other factors remains to be elucidated. Despite the multiple interactions of these various factors, the effect of tumor type on survival is striking as demonstrated by data from the Connecticut Tumor Registry (Figure 3). Individuals with hemangioma or meningioma have the longest survival, while those with glioblastoma have the shortest. With the exception of patients with hemangioma and meningioma (showing a relatively linear decline in survival over time), patients with other forms of primary brain tumor generally show a relatively rapid decrement within the first two years after diagnosis, after which the slope of the survival curve changes and the decline is more gradual.

GENETIC ASPECTS

Except for the known familial occurrence of bilateral retinoblastoma or chemodectoma (glomus tumors) revealing an autosomal dominant pattern of inheritance in certain instances, and the rare phenomenon of familial glioma [24, 32], no genetic factor has been found to influence the incidence of primary nervous system tumors taken as a whole. Table 4 outlines the currently recognized role of genetic factors in the occurrence of nervous system neoplasms.

The phakomatoses included in this list are associated with nervous system neoplasms as detailed in Table 5. Hamartomas of the heart, kidney, and eye are associated with tuberous sclerosis. The most common intracranial neoplasms found in patients with this condition are astrocytomas, thought to arise from the foci of subependymal astrocytes that characterize tuberous sclerosis [2].

Other less common forms of primary intracranial neoplasms found with tuberous sclerosis include glioblastoma [46], ependymoma [47], and ganglioneuroma [48]. Despite the well-documented associations between tuberous sclerosis and primary intracranial tumors, such neoplasms are relatively rare when large series of tuberous sclerosis patients are examined [2].

Three forms of von Recklinghausen's neurofibromatosis are recognized: a peripheral form, a central form, and a visceral form [2]. The peripheral form is

Table 4. Genetic factors and tumors of the nervous system*

Tumor or condition	Comment
Retinoblastoma	Often shows an autosomal dominant inheritance with incomplete penetrance [8, 33]
Glomus tumor	Occasionally familial with autosomal dominant inheritance pattern and incomplete penetrance [8, 33]
Von Recklinghausen's disease	Often shows an autosomal dominant pattern [8, 33]
Tuberous sclerosis**	Often shows an autosomal dominant pattern [8, 33]
Von Hippel-Lindau disease	Some cases show an autosomal dominant pattern of inheritance [8, 33]
Glioma	Some authors believe that a genetic factor is involved [34 – 38], while others have reported no evidence for such a factor [39, 40]
Medulloblastoma, meningioma	No evidence of a genetic pattern presented [41], although there have been isolated case reports of familial aggregation [42 – 45]

*Although the Sturge-Weber syndrome, on rare occasions, shows an autosomal dominant pattern of inheritance, the cerebral capillary-venous malformation characteristic of this condition is not considered to be neoplastic. The syndrome is therefore not included in the table.

**Since astrocytic glioma associated with tuberous sclerosis appears to originate from the subependymal lesions characteristic of this condition, it is included in the above listing.

Reproduced with permission from Schoenberg et al. [32].

Table 5. Nervous system neoplasms and the phakomatoses

Type of phakomatosis	Nervous system tumors
Tuberous sclerosis	Astrocytic glioma; ependymoma; ganglioneuroma; hamartoma of eye
Von Recklinghausen's neurofibromatosis	
Peripheral	Multiple peripheral and subcutaneous nerve sheath tumors
Central	Multiple CNS tumors arising from nervous system parenchyma, meninges, and cranial and spinal nerve roots
Visceral	Ganglioneuromas and nerve sheath tumors involving viscera and autonomic nervous system; visceral neoplasms of neural crest origin
Von Hippel-Lindau disease	Hemangioblastoma (often multiple); angiomatosis of retina; pheochromocytoma
Neurocutaneous melanosis	Malignant leptomeningeal melanoma; peripheral neurofibromas

Reproduced with permission from Schoenberg [3].

characterized by the presence of café-au-lait skin pigmentation and large numbers of tumors involving the nerve sheaths of the peripheral nervous system. The central form of the disease consists of multiple tumors (most frequently schwannomas) of the cranial and spinal nerve roots. The most common site for these tumors involves the eighth cranial nerve, and they are often bilateral [49, 50]. Multiple meningiomas and multiple ependymomas are also seen in this condition [2]. Astrocytic neoplasms of the central nervous system (including optic nerve gliomas) [51], and the retina [52] have been reported. The occurrence of bilateral acoustic schwannomas or other multiple primary neoplasms [53] of the CNS should alert the clinician to the possible presence of von Recklinghausen's disease. Harmartomatous malformations involving neuroglia, meningeal cells, blood vessels, and Schwann cells have also been reported in central neurofibromatosis [2], as has the occurrence of syringomyelia [53, 54].

The visceral form of von Recklinghausen's disease is characterized by neoplasms of neural crest origin, such as neuroblastoma, adrenal ganglioneuroma, and pheochromocytoma, in addition to nerve sheath tumors and ganglioneuromas of the viscera and autonomic nervous system [2].

Patients with von Hippel-Lindau disease have hemangioblastomas (often multiple) of the CNS together with angiomatosis of the retina. The cerebellum is the most common site for hemangioblastoma. The condition also frequently includes pancreatic and kidney cysts and renal carcinoma [2]. Patients with this disease are also at increased risk for pheochromocytoma [55, 56] and ependymoma [57]. Occurrences of both syringomyelia [54] and erythrocythemia [58] are well-documented in von Hippel-Lindau disease.

Diffuse proliferation of melanocytes throughout the meninges together with large cutaneous nevi characterize neurocutaneous melanosis [2, 59, 60]. Multiple

Table 6. Multiple primary tumors and genetic syndromes; index primary neoplasms of the nervous system

Condition	Nervous system neoplasm	Non-nervous system neoplasm	References
Wermer's syndrome (multiple endocrine adenomatosis I)	Anterior pituitary	Parathyroid; pancreatic islet cells; thyroid; adrenal cortex; carcinoid tumor (intestine, bronchus)	Johnson *et al.* [61]
Sipple's syndrome (multiple endocrine adenomatosis II)	Pheochromocytoma; neurofibroma (multiple); submucosal neuromas	Medullary thyroid carcinoma; parathyroid neoplasm	Mulvihill [62]; Schimke *et al.* [63]
Turcot's syndrome	Brain tumor	Polyposis coli	Baughman *et al.* [64]; Turcot *et al.* [65]
Nevoid basal cell carcinoma syndrome	Medulloblastoma	Basal cell carcinomas; ovarian tumor	Gorlin *et al.* [66]; Graham *et al.* [67]; Hermans *et al.* [68]; Herzberg and Wiskemann [69]; Jackson and Gardere [70]; Meerkotter and Shear [71]; Moynahar [72]; Neblett *et al.* [73]; Stout [74]; Strong [75]
Cowden's disease (multiple hamartoma syndrome)	Meningioma	Lip and mouth papillomas; breast cancer; thyroid adenoma and carcinoma; lipoma; polyps; bone and liver cysts	Mulvihill [62]

Reproduced with permission from Schoenberg [3].

peripheral neurofibromas are also commonly seen in this syndrome [2].

In addition to the phakomatoses, nervous system neoplasms are associated with other multiple primary tumors as part of several genetic syndromes as outlined in Table 6. Details of these various conditions have been reviewed by Schoenberg [3].

Brain tumors have also been reported to occur in patients with ataxia-telangiectasia or in family members of these patients, as have lymphoreticular tumors, leukemia, epithelial tumors, and stomach cancer [62, 76, 77].

Table 7. Multiplicity of primary nervous system neoplasms

Tumor	References
Chemodectoma	Anderson [82]
Ependymoma	Russell and Rubinstein [78]
Ganglioneuroma	McFarland and Sappington [83]; Stout [74]
Glioblastoma	Batzdorff and Malamud [84]; Borovich *et al.* [85]; Courville [86]; Moertel *et al.* [87]; Scherer [88]
Hemangioblastoma	Raney and Courville [41]; Russell and Rubinstein [78]
Meningioma	Mufson and Davidoff [89]
Neuroblastoma	Russell and Rubinstein [78]
Neurofibroma	Rubinstein [2]
Optic nerve glioma	Davis [51]
Pheochromocytoma	Cahil [90]; Cragg [91]; Russell and Rubinstein [78]; Symington and Goodall [92]
Retinoblastoma	Anderson [82]; Waldmann *et al.* [58]
Schwannoma	Gardner and Turner [49]; Moyes [50]
Teratoma	Ingraham and Bailey [93]; Russell and Rubinstein [78]

Reproduced with permission from Schoenberg [3].

MULTICENTRIC NERVOUS SYSTEM NEOPLASMS

In some cases nervous system tumors may have a multicentric origin (Table 7). Multifocal occurrences of tumors of the nervous system may signal the presence of a hereditary syndrome (e.g., bilateral retinoblastoma) or the presence of one of the phakomatoses (described earlier).

There have also been reports of different types of CNS neoplasms occurring together in the same patient. Thus, ganglioneuroma may be associated with primary tumors of the autonomic nervous system (neuroblastoma and pheochromocytoma) [78]. Meningioma has been reported to occur together with glioma in a single patient [79–81]. The associations outlined in Tables 5, 6, and 7, although representing relatively rare phenomena, are important to the clinician when identifying individuals at high risk for nervous system neoplasms.

ANALYTIC STUDIES

Numerous epidemiologic and pathologic investigations have been carried out to explain the role of various putative etiologic factors in the pathogenesis of brain tumors. Much experimental work in the area of oncogenic viruses and chemical carcinogenesis has been reported. Those individuals proposing a viral etiology for nervous system neoplasms support their hypothesis by the following: (a) induction of nervous system tumors in experimental animals or cell cultures following administration of particular viruses, and (b) association of virus-like particles with human brain tumors. A variety of neoplasms, including glioblastoma, medulloblastoma, neuroblastoma, ependymoma, pineocytoma, and intracranial sarcoma arise after inoculation of viruses. Tumor induction depends on several factors, e.g., the host animal, age of the animal, site of inoculation, virus type, etc. These studies are well-summarized by Bigner [94] in a 1978 review. Despite the intriguing reports of virus-like particles with human nervous system neoplasms, "none of the above studies have presented any conclusive evidence for tumor virus particles, genomes, or antigens in human brain tumors." [94]

With regard to chemical carcinogenesis, early workers were able to induce brain tumors *only* following intracranial administration of certain chemical agents into experimental animals. Since 1965, however, more than 30 compounds have been identified that result in a high incidence of nervous system tumors after *systemic* administration. The most effective of these substances include the N-nitrosamides, dialkyl-aryltriazenes, azo, azoxy, and hydrazo compounds and a polycyclic aromatic hydrocarbon (7, 12-dimethylbenz-(a) anthracene) [95].

These experimental observations have been complemented by an increasing number of analytic epidemiologic investigations, summarized in Table 8. Unfortunately, the majority of these studies treat all brain tumors together as a single group. This is contrary to the evidence of descriptive epidemiologic data which suggest that each histologic type of nervous system neoplasm is a separate disease, with specific incidence and survival patterns. It would, therefore, appear likely that they would have distinct risk factors which could be obscured in studies considering all brain tumors as a single entity.

The first category in Table 8 reviews the relationship between trauma to the head and subsequent brain tumor. There have been several case reports of brain tumors, particularly meningiomas, occurring at sites of previous trauma resulting from injury or surgical procedures [96, 97]. Other authors found no such association [98]. A case-control study of risk factors for meningiomas in women found a statistically significant association with head injury [99]. However, an earlier case-control study which explored risk factors for all brain tumors discovered no association with head trauma [100]. Using a nonconcurrent prospective approach, Annegers *et al.* [101] examined the subsequent brain tumor experience of Rochester, Minnesota residents who had suffered head trauma. The number of brain tumors developing in this cohort was not significantly different from the expected number of neoplasms.

Studies of prenatal X-ray exposure and subsequent brain tumors have yielded conflicting results [100, 102].

In 1974, Modan *et al.* [103] described a study based on 11,000 children who had received radiation to treat tinea capitis. This cohort was followed retrospectively from 12 to 23 years. When the tumor experience of this group was com-

Table 8. Epidemiologic studies of demographic or possible etiologic factors associated with brain tumors

Factor	References	Association with brain tumor*	Comment
I. Trauma			
Head injury	Reynolds [96]; Tanaka et al. [97]; Parker and Kernohan [98]; Preston-Martin [99]	± +	Case reports have related anecdotal evidence of an association One case-control study found an association between head injury and meningioma in women
	Choi et al. [100]	–	A second case-control study of all brain tumors found no association with head injury.
	Annegers et al. [101]	–	A noncurrent prospective investigation of individuals with head injury revealed no increase in the incidence of subsequent brain tumors
II. X-ray Exposure			
Prenatal X-ray exposure	MacMahon [102] Choi et al. [100]	+ –	
X-ray therapy to scalp in childhood	Modan et al. [103]; Shore et al. [104]	+	Increased risk of brain tumors for individuals receiving X-ray therapy for tinea capitis during childhood
Medical/dental x-rays	Preston-Martin [99]	+	Case-control study revealed association between medical or dental x-rays and meningioma in women

Table 8. (Continued)

Factor	References	Association with brain tumor*	Comment
III. Other Exogenous Exposures			
A. Non-Occupational			
Smoking	Choi et al. [100]	−	Case-control study of women with meningioma
	Preston-Martin [99]	−	
Alcohol consumption	Choi et al. [100]	−	
Consumption of meats with high sodium nitrite levels	Preston-Martin [99]	+	Case-control study revealed association between consumption of meats with high sodium nitrite concentrations and meningioma in women
Lead exposure	Schreier et al. [105]	+	Two case reports of children with elevated urinary lead levels who subsequently developed astrocytoma
Insecticide exposure	Gold et al. [106]	±	Reported association (not statistically significant) between brain tumors in children and their exposure to insecticides
Living on farms or exposure to farm animals	Gold et al. [106]	+	Statistically significant association between brain tumors in children and a history of having lived on a farm or a history of exposure to farm animals
Exposure to sick pets	Gold et al. [106]	±	Reported association (not statistically significant) between brain tumors in children and a history of exposure to sick pets
Barbiturates	Gold et al. [107]	+	Association between barbiturate use and brain tumors in children
Oral contraceptives	Annegers et al. [27]	−	Study of women with pituitary adenoma

Table 8. (Continued)

Factor	References	Association with brain tumor*	Comment
B. Occupational			
Work in rubber manufacturing industries	Lamperth-Seiler [108]; Mancuso [109]	+	
Vinyl chloride exposure	Waxweiler et al. [110]	+	Reported excess of glioblastoma multiforme among vinyl chloride workers
IV. Miscellaneous Host Factors			
Birth order	Choi et al. [100]	−	Reported association (not statistically significant) between being first born and brain tumors in children
	Gold et al. [106]	±	
Birth weight	Gold et al. [106]	±	Reported association (not statistically significant) between birth weight of 8 lbs. or greater and brain tumors in children
ABO blood groups	Mayr et al. [111]	+	Association between blood group O and pituitary tumor patients
	Buckwalter et al. [112]	+	Association between blood group A and male intracranial tumor patients
	Selverstone and Cooper [113];	+	Association between blood group A and patients with glioma
	Yates and Pearce [114]	+	
	Choi et al. [100];	−	
	Garcia et al. [115]	−	
Religion	Gold et al. [106]	±	Reported negative association (not statistically significant) between being Jewish and brain tumors in children
Tonsillectomy	Gold et al. [106]	±	Reported negative association (not statistically significant) between brain tumors in children and a history of tonsillectomy

Table 8. (Continued)

Factor	References	Association with brain tumor*	Comment
Renal transplant patients on immunosuppressive therapy	Kersey and Spector [77]	+	Increased risk of reticulum cell sarcoma of brain
V. Presence of Other Diseases			
Diabetes mellitus	Paton and Petch [116]	+	A case report suggesting an association
	Aronson and Aronson [117]	1/+	Negative association between gliomas and diabetes mellitus
Toxoplasmosis	Schuman et al. [118]	+	
Multiple sclerosis	Reagan and Freiman [119]	+	Review of several cases of astrocytomas arising in multiple sclerosis plaques
Breast cancer	Schoenberg et al. [120]	+	Associated significantly with occurrence of meningioma
Osteosarcoma	Jensen and Miller [121] Kitchin and Ellsworth [122]	+	Associated with bilateral retinoblastoma
VI. Family Characteristics and History			
Maternal age at birth of index case	Choi et al. [100]	−	Study of brain tumors in children
Paternal age at birth of index case	Gold et al. [106]	−	Study of brain tumors in children
Prenatal illness in mother of index case	Gold et al. [106]	−	
	Choi et al. [100]	−	
Abortions and complications of delivery in mothers of patients	Choi et al. [100]	±	

Table 8. (Continued)

Factor	References	Association with brain tumor*	Comment
Oral contraceptive use by mother prior to index birth	Gold et al. [106]	−	Study of brain tumors in children
Maternal smoking history prior to index pregnancy	Gold et al. [106]	−	Study of brain tumors in children
Use of hormones by mother during index pregnancy	Gold et al. [106]	−	Study of brain tumors in children
Maternal mean age of menarche	Gold et al. [106]	−	Study of brain tumors in children
Occupational chemical exposure of parents	Gold et al. [106]	−	Study of brain tumors in children
Epilepsy in siblings	Gold et al. [106]	+	Statistically significant association between brain tumors in children and epilepsy in a sibling
Epilepsy or stroke in mother of case	Gold et al. [106]	±	Reported association (not statistically significant) between brain tumors in children and epilepsy or stroke in their mothers
Family history of congenital disease	Gold et al. [106]	−	Study of brain tumors in children
Family history of cancer	Preston-Martin [99]	−	Study of meningiomas in women
	Gold et al. [106]	−	Study of brain tumors in children

*Legend: + positive association; 1/+ negative association; − no association; ± results equivocal.
Modified and reproduced with permission from Schoenberg [26].

pared to the experience of two matched control groups, the children who had received scalp irradiation showed an excess of both benign and malignant primary brain tumors. Although the authors of this report stated that meningioma accounted for part of the excess brain tumor risk, they did not mention the other histologic types involved. A study by Shore *et al.* [104] supports these results. In this investigation the number of tumors developing among 2,215 children who received radiation therapy for tinea capitis was compared to the number of tumors developing among a matched control group of children with tinea capitis who had not received any X-ray therapy for their condition. Again, there was an excess incidence of brain tumors in the cohort receiving scalp irradiation. The primary intracranial neoplasms developing in the irradiated group were: 3 gliomas, 2 meningiomas, and 1 nerve sheath tumor. Therefore, one particular histologic type was not responsible for the excess risk of primary brain tumors following X-ray therapy to the scalp.

Preston-Martin [99] in a case-control study of women with meningioma found a statistically significant association with medical or ddental X-rays. This same investigator observed a relationship between meningioma in women and head trauma. In this study it was not possible to distinguish the possible etiologic role contributed by head trauma as opposed to head X-rays, which are commonly used to evaluate the patient who has sustained a head injury.

Factors such as smoking or alcohol consumption are not significantly associated with brain tumors [99, 100]. Preston-Martin [99] discovered a statistically significant association between meningiomas in women and a history of consuming meats with a high concentration of sodium nitrite. Schreier *et al.* [105] reported on 28 institutionalized children in whom urinary lead was measured. Three individuals were found to have increased lead levels. Two of these children developed astrocytic gliomas 2 and 4 years after the urine was tested. Although this report is largely anecdotal, it is supported by laboratory studies in which rats on a diet with relatively large quantities of lead acetate subsequently developed gliomas [123].

Gold *et al.* [106] conducted a case-control study of several putative etiologic factors in childhood brain tumors. Because of the small number of cases, several of their findings were not statistically significant, although the investigators considered them to have potential biologic significance. Included in this category was the reported association between childhood brain tumors and exposure to either insecticides or sick pets. One relationship with statistical significance was between childhood brain tumors and living on a farm, or a history of exposure to farm animals. This same group of investigators found a statistically significant association between childhood brain tumors and the patient's prior use of barbiturates [107].

Unfortunately, it was not possible to rule out with certainty whether some of this barbiturate use reflected treatment of seizure activity resulting from a brain tumor which was not yet detected clinically.

Annegers *et al.* [27] carried out a case-control study to determine the possible role of oral contraceptives in the recent increase in the incidence of pituitary adenomas in women of childbearing age. No relationship was found, however.

A few occupational studies have demonstrated that workers in rubber manufacturing industries are at increased risk for developing brain tumors [108, 109]. These results require further confirmation and more precise definition of the factors responsible for any increased risk. Waxweiler *et al.* [110], in a study of workers exposed to vinyl chloride, observed an excess of glioblastoma multiforme.

Although one case-control study of the possible association of brain tumors and birth order found no relationship [100], a later study restricted to childhood brain tumors detected relationships with being first born and having a higher birth weight [106]. Neither of these associations was statistically significant.

There are different and often conflicting conclusions in investigations relating ABO blood groups and brain tumors [100, 111–115].

In their study of childhood brain tumors, Gold *et al.* [106] reported a negative association with being Jewish or having had a tonsillectomy. Neither of these findings was statistically significant.

Renal transplant patients receiving immunosuppressive therapy have a markedly increased risk of reticulum cell sarcoma of the brain, an otherwise relatively uncommon tumor. It is generally believed that the high relative frequency of this malignancy is the result of immunosuppression, rather than an oncogenic effect of the techniques used to suppress the immune system [77].

The relationship of other diseases to the occurrence of brain tumors has been investigated. Whereas one report [116] suggested a relationship between diabetes mellitus and brain tumor, a second paper concluded that diabetic patients were at decreased risk of brain tumor [117].

Although evidence of toxoplasmosis was found to be linked with brain tumors in one investigation [118], this has not been substantiated by others. There have been several reports of astrocytomas in multiple sclerosis plaques [119], but this association has yet to be documented in well-controlled studies.

Analyses of the association between CNS neoplasms and primary tumors of other sites have, with two exceptions, not revealed any definite patterns. Although there have been numerous case reports of nervous system neoplasms occurring together with other primary cancers, most of these studies failed to determine whether or not such an association of multiple primaries is a chance phenomenon [120]. The two exceptions to this have been the reported relationships between (a) meningioma and breast cancer, and (b) retinoblastoma and osteosarcoma.

A significant association between breast cancer and meningioma in women was discovered in an analysis of data from the Connecticut Tumor Registry [120]. Although it is yet to be confirmed by other investigations, this result raises interesting etiologic possibilities when considered together with other epidemiologic features of meningioma. First, meningioma is the only common intracranial neoplasm with a higher incidence in females [22]; second, the abrupt appearance or enlargement of this tumor during pregnancy has been reported [124]. Thus, the association between breast cancer and meningioma may be related to hormonal factors. Moreover, this hypothesis is supported by the recent finding of estrogen receptor protein in intracranial meningiomas. In some female patients with meningioma, the level of estrogen receptor protein approached the level found in hormonally sensitive breast tissue [125].

A nationwide investigation of 1,623 children hospitalized with retinoblastoma revealed 30 patients with second primary cancer [121]. In 11 such cases the second tumor could not be attributed to radiotherapy for the retinoblastoma. Of the 11 cases, three of the second primaries were osteosarcomas. A second study dealing with 1,130 retinoblastoma patients reported seven additional bone tumors in areas far removed from the radiation field [122]. The excess of second primary tumors, mainly osteosarcomas, has been limited to patients with bilateral disease. Since a familial pattern is usually apparent in cases of bilateral retinoblastoma (and only in a small percentage of cases with unilateral tumor), a genetic mechanism may be involved in the development of both retinoblastoma and osteosarcoma.

Among the many factors examined in their case-control investigation of childhood brain tumors, Gold *et al.* [106] found no association with the following items; maternal or paternal age at the birth of the index case, maternal oral contraceptive use prior to the index birth, maternal smoking history prior to the index pregnancy, maternal use of hormones during the index pregnancy, maternal mean age of menarche, parental occupational exposure to chemicals, and a family history of congenital diseases or cancer. A statistically significant association was discovered for the presence of epilepsy in siblings; although a relationship was found with the presence of epilepsy or stroke in the mothers of cases, it was not statistically significant. In the study of Choi *et al.* [100], there was no association between brain tumors and either maternal age at the birth of the index case or a history of maternal prenatal illness during the index pregnancy. The results with regard to maternal abortion history or history of complicated delivery were equivocal. Preston-Martin's 1978 case-control investigation concerning meningiomas in women found no relationship with a family history of cancer [99].

CONCLUSIONS

Descriptive epidemiologic investigations of primary nervous system neoplasms in well-defined populations have demonstrated that despite our diagnostic sophistication, current incidence rates probably underestimate the magnitude of this health problem. The various histologic types of brain tumors reveal sufficiently distinct epidemiologic patterns to be classified as separate diseases. It may be misleading to consider primary nervous system neoplasms as a single group, since the overall picture simply reflects (a) these distinctive individual patterns, and (b) the relative frequency of diagnosed cases of these histologically separate neoplasms in the population under study. Epidemiologic analysis of the histologic and clinical features of these tumors may lead to new systems of classification and may provide the clinician with more useful indices of prognosis and response to theapy [126].

The rare familial aggregation of certain nervous system neoplasms and their association with conditions demonstrating a defined pattern of inheritance identify the high-risk patient. Such identification is of immediate clinical value in alerting the physician to the likely presence of these tumors. Further analytic epidemiologic investigations are required to confirm the associations suggested in the few previous studies and to identify other currently unknown associations. The initial case-control studies present a number of methodologic problems. The rationale and need for restricting such investigations to specific histologic tumor types have already been discussed.

Future studies must address the problem of selective recall bias. Thus, the family of a brain tumor patient may actively search for past events in an attempt to explain this devastating illness. They are also more familiar with neurologic terms and symptoms, and more likely to recognize similar disorders in other family members. This is usually not so with healthy controls. Such differences between cases and controls may themselves cause an artifactual association. In an attempt to minimize the problem of selective recall bias, Gold *et al.* [106] used two sets of controls: (a) healthy children, and (b) children with other forms of cancer. Presumably, families of children with other forms of cancer would be equally motivated to remember past events.

Unfortunately, many of their statistically significant associations were valid only when brain tumor cases were compared to healthy controls. This raises some doubts as to whether such relationships resulted from selective recall bias. Future studies must use more refined epidemiologic techniques and require input from clinical neurologists and neurosurgeons. More recent investigations have focused on potential etiologic factors within the occupational environment [128 – 131], while others have utilized a more general exploratory approach to identify risk factors [132]. It is hoped that the more detailed evaluation of patients at high risk for brain tumors will lead to a better understanding of the mechanisms involved in oncogenesis within the nervous system.

REFERENCES

1. Kernohan JW, Sayre GP: Tumors of the central nervous system. Fascicle 35, Atlas of tumor pathology. Armed Forces Institute of Pathology, Washington, DC, 1952.
2. Rubinstein LJ: Tumors of the central nervous system. In: Atlas of tumor pathology, Second Series, Fascicle 6. Washington, DC, Armed Forces Institute of Pathology, 1972.
3. Schoenberg BS: Multiple primary neoplasms and the nervous system. Cancer 40:1961 – 1967, 1977.
4. Richards P, McKissock W: Intracranial metastases. Br Med J 1:15 – 18, 1963.
5. Escourolle R, Poirier J: Manual of basic neuropathology, translated by LJ Rubinstein. WB Saunders, Philadelphia, 1973.
6. Goldberg ID, Kurland LT: Mortality in 33 countries from diseases of the nervous system. World Neurol 3:444 – 465, 1962.
7. Massey EW, Schoenberg DG, Schoenberg BS: Mortality in 33 countries from diseases of the nervous system. Submitted for publication, 1983.
8. Kurtzke JF, Kurland LT: The epidemiology of neurologic disease. In: Clinical neurology. Baker AB, Baker LH (eds), New York, Harper and Row, 1971, Chap. 48, pp 4 – 9.
9. Barker DJP, Weller RO, Garfield JS: Epidemiology of primary tumours of the brain and spinal cord: A regional survey of southern England. J Neurol Neurosurg Psychiatry 39:290 – 296, 1976.
10. Biometry Branch, National Cancer Institute: In: Third National Cancer Survey; Incidence Data. National Cancer Institute Monograph 41. DHEW Publication No. (NIH) 75-787, Bethesda, MD, National Institutes of Health, 1975, pp 17, 21, 25.
11. Brewis M, Poskanzer DC, Rolland C, Miller H: Neurological disease in an English city. Acta Neurol Scand (Suppl. 24) 42:21, 23, 41 – 46, 1966.
12. Clemmesen J: Statistical studies in the aetiology of malignant neoplasms. I. Review and results. Acta Pathol Microbiol Scand (Suppl 174) Part 1:422 – 424, 538 – 539, 542 – 543, 1965.
13. Clemmesen J: Statistical studies in the aetiology of malignant neoplasms. II. Basic tables: Denmark 1943 – 1957. Acta Pathol Microbiol Scand (Suppl 174) Part 2:3, 8 – 9, 28 – 29, 68 – 69, 122 – 123, 194 – 195, 226 – 227, 252 – 253, 294 – 295, 308 – 309, 1965.
14. Clemmesen J: Statistical studies in the aetiology of malignant neoplasms. III. Testis cancer: Basic tables, Denmark 1958 – 1962. Acta Pathol Microbiol Scand (Suppl 209) pp LXV, 3, 14 – 15, 36 – 39, 44 – 47, 58 – 59, 91 – 92, 130 – 133, 142 – 143, 1969.
15. Cohen A, Modan B: Some epidemiologic aspects of neoplastic diseases in Israeli immigrant population. III. Brain tumors. Cancer 22:1323 – 1328, 1968.
16. Dorn HF, Cutler SJ: Morbidity from cancer in the United States. US Dept. of Public Health Monograph No. 29. Washington, DC, US Government Printing Office, 1955, pp 7 – 8, 11 – 12, 144 – 145, 151 – 159.
17. Gudmundsson KR: A survey of tumours of the central nervous system in Iceland during the 10-year period 1954 – 1963. Acta Neurol Scand 46:538 – 552, 1970.
18. Haenszel W, Marcus SC, Zimmerer EG: Cancer morbidity in urban and rural Iowa. US Dept. of Public Health Monograph No. 37. Washington, DC, US Government Printing Office, 1956, pp 1 – 6, 55, 60, 63, 81.
19. Kurland LT: The frequency of intracranial and intraspinal neoplasms in the resident population of Rochester, Minnesota. J Neurosurg 15:627 – 641, 1958.
20. Leibowitz U, Yablonski M, Alter M: Tumors of the nervous system: Incidence and population selectivity. J Chronic Dis 23:707 – 721, 1971.
21. Percy AK, Elveback LR, Okazaki H, Kurland LT: Neoplasms of the nervous system: Epidemiologic considerations. Neurology (Minneap) 22:40 – 48, 1972.
22. Schoenberg BS, Christine BW, Whisnant JP: The descriptive epidemiology of primary intracranial neoplasms – The Connecticut experience. Am J Epidemiol 104:499 – 510, 1976.

23. Schoenberg BS, Christine BW, Whisnant JP: The resolution of discrepancies in the reported incidence of primary brain tumors. Neurology (Minneap) 28:817–823, 1978.
24. Schoenberg BS: Primary intracranial neoplasms: A study of incidence, epidemiological trends, and the association of these neoplasms with primary malignancies of other sites (thesis). University of Minnesota, Rochester, Minn., 1974.
25. Schoenberg BS, Schoenberg DG, Christine BW, Gomez MR: The epidemiology of primary intracranial neoplasms of childhood: A population study. Mayo Clin Proc 51:51–56, 1976.
26. Schoenberg BS: Epidemiology of primary nervous system neoplasms. In: Advances in neurology Vol. 19: Neurological epidemiology: Principles and clinical applications. Schoenberg BS (ed), New York, Raven Press, 1978, pp 475–493.
27. Annegers JF, Coulam CB, Abboud CF, Laws ER Jr., Kurland LT: Pituitary adenoma in Olmsted County, Minnesota, 1935–1977; a report of an increasing incidence of diagnosis in women of childbearing age. Mayo Clin Proc 53:641–643, 1978.
28. Jordan BD, Christine BW, Schoenberg BS: Descriptive epidemiology of primary malignant eye tumors in Connecticut. Submitted for publication, 1983.
29. Mahalak LW Jr., Schoenberg BS: Differentials in brain tumor occurrence by race: Experience of the University of Mississippi Medical Center. Submitted for publication, 1983.
30. Fan KJ, Kovi J, Earle KM: The ethnic distribution of primary central nervous system tumours: Armed Forces Institute of Pathology, 1958–1970. J Neuropathol Exp Neurol 36:41–49, 1977.
31. Heshmat MY, Kovi J, Simpson C, Kennedy J, Fan KJ: Neoplasms of the central nervous system: Incidence and population selectivity in the Washington, DC Metropolitan area. Cancer 38:2135–2142, 1976.
32. Schoenberg BS, Glista GG, Reagan TJ: The familial occurrence of glioma. Surg Neurol 3:139–145, 1975.
33. Kurland LT, Myrianthopoulos NC, Lessell S: Epidemiologic and genetic considerations of intracranial neoplasms. In: The biology and treatment of intracranial tumors. Fields, WS, Sharkey PC (eds), Springfield, Illinois, CC Thomas, 1962, pp 5–47.
34. Armstrong RM, Hanson CW: Familial gliomas. Neurology (Minneap) 19:1061–1063, 1969.
35. Kjellin K, Muller R, Astrom KE: The occurrence of brain tumors in several members of a family. J Neuropathol Exp Neurol 19:528–537, 1960.
36. Metzel E: Betrachtungen zür Genetik der familiaren Gliome. Acta Genet Med Gemellol (Roma) 13:124–131, 1964.
37. Metzel E, Mohadjer M: Familial incidence of brain tumors. In: Present limits of neurosurgery. Fusek I, Kune J (eds), Prague, Avicenum, Czechoslovak Medical Press, 1974, pp 17–18.
38. Van der Wiel HJ: Inheritance of glioma: The genetic aspects of cerebral glioma and its relation to status dysraphicus. Amsterdam, Elsevier Press, 1959, pp 249–252.
39. Harvald B, Hauge M: On the heredity of glioblastoma. J Natl Cancer Inst 17:289–296, 1956.
40. Hauge M, Harvald B: Genetics in intracranial tumours. Acta Genet 7:573–591, 1957.
41. Raney RB, Courville CB: Multiple hemangioblastomas of the central nervous system. Bull Los Angeles Neurol Soc 2:104–114, 1937.
42. Gaist G, Piazza G: Meningiomas in two members of the same family (with no evidence of neurofibromatosis). J Neurosurg 16:110–113, 1959.
43. Joynt RJ, Perret GE: Meningiomas in a mother and daughter: Cases without evidence of neurofibromatosis. Neurology (Minneap) 11:164–165, 1961.
44. Refsum S, Mohr J: Genetic aspects of neurology. In: Clinical neurology. Baker AB, Baker LH (eds), New York, Harper and Row, 1971, Chap. 47, p 40.
45. Sahar A: Familial occurrence of meningiomas: Case report. J Neurosurg 23:444–445, 1965.
46. Jervis GA: Spongioneuroblastoma and tuberous sclerosis. J Neuropathol Exp Neurol 13:105–116, 1954.
47. Norman RM, Taylor AL: Congenital diverticulum of the left ventricle of the heart in a case of epiloia. J Pathol Bacteriol 50:61–68, 1940.

48. Davis RL, Nelson E: Unilateral ganglioglioma in a tuberosclerotic brain. J Neuropathol Exp Neurol 20:571–581, 1961.
49. Gardner WJ, Turner O: Bilateral acoustic neurofibromas – further clinical and pathological data on hereditary deafness and Recklinghausen's disease. Arch Neurol 44:76–99, 1940.
50. Moyes PD: Familial bilateral acoustic neuroma affecting 14 members from four generations. Case Report. J Neurosurg 29:78–82, 1968.
51. Davis FA: Primary tumors of the optic nerve (a phenomenon of Recklinghausen's disease) – a clinical and pathological study with a report of five cases and a review of the literature. Arch Ophthalmol 23:735–827, 957–1022, 1940.
52. Saran N, Winter FC: Bilateral gliomas of the optic discs associated with neurofibromatosis. Am J Ophthalmol 64:607–612, 1967.
53. Rodriguez HA, Berthrong M: Multiple primary intracranial tumors in von Recklinghausen's neurofibromatosis. Arch Neurol 14:467–475, 1966.
54. Poser CM: The relationship between syringomyelia and neoplasms. Springfield, Illinois, CC Thomas. 1956.
55. Chapman RC, Diaz-Perez R: Pheochromocytoma associated with cerebellar hemangioblastoma. Familial occurrence. JAMA 182:1014–1017, 1962.
56. Nibbelink DW, Peters BH, McCormick WF: On the association of pheochromocytoma and cerebellar hemangioblastoma. Neurology (Minneap) 19:455–460, 1969.
57. Fraumeni JF Jr: Genetic factors. In: Cancer medicine. Holland JF, Frei E III (eds), Philadelphia, Lea and Febiger, 1973, pp 7–15.
58. Waldmann TA, Levin EH, Baldwin M: The association of polycythemia with a cerebellar hemangioblastoma. The production of an erythropoiesis stimulating factor by the tumor. Am J Med 31:318–324, 1961.
59. Hoffman HJ, Freeman A: Primary malignant leptomeningeal melanoma in association with giant hairy nevi. Report of two cases. J Neurosurg 26:62–71, 1967.
60. Slaughter JC, Hardman JM, Kempe LG, Earle KM: Neurocutaneous melanosis and leptomeningeal melanomatosis in children. Arch Path 88:298–304, 1969.
61. Johnson GJ, Summerskill WHJ, Anderson VE, Keating FR Jr: Clinical and genetic investigation of a large kindred with multiple endocrine adenomatosis. N Engl J Med 277:1379–1385, 1967.
62. Mulvihill JJ: Congenital and genetic diseases. In: Persons at high risk of cancer: an approach to cancer etiology and control. Fraumeni JF Jr (ed), New York, Academic Press, 1975, pp 3–38.
63. Schimke RN, Hartmann WH, Prout TE, Rimoin DL: Syndrome of bilateral pheochromocytoma, medullary thyroid carcinoma, and multiple neuromas. N Engl J Med 279:1–7, 1968.
64. Baughman FA Jr, List CF, Williams JR, Muldoon JP, Segarra JM, Vokel JS: The gliomapolyposis syndrome. N Engl J Med 281:1345–1346, 1969.
65. Turcot J, Despres JP, St. Pierre F: Malignant tumors of the central nervous system associated with familial polyposis of the colon – Report of two cases. Dis Colon Rectum 2:465–468, 1959.
66. Gorlin RJ, Vickers RA, Kellen E, Williamson JJ: The multiple basal-cell nevi syndrome. Cancer 18:89–104, 1965
67. Graham JK, McJimsey BA, Hardin JC Jr: Nevoid basal cell carcinoma syndrome. Arch Otolaryngol 87:90–95, 1968.
68. Hermans EH, Grosfeld JCM, Spaas JAJ: The fifth phacomatosis. Dermatologica 130:446–476, 1965.
69. Herzberg JJ, Wiskemann A: Die fünfte Phakomatose. Dermatologica 126:106–123, 1963.
70. Jackson R, Gardere S: Nevoid basal cell carcinoma syndrome. Can Med Assoc J 105:850–862, 1971.
71. Meerkotter VA, Shear M: Multiple primordial cysts associated with bifid rib and ocular defects. Oral Surg 18:498–503, 1964.
72. Moynahan EJ: Multiple basal cell naevus syndrome – Successful treatment of basal cell tumours with 5-fluorouracil. Proc R Soc Med 66:627–628, 1973.

73. Neblett CR, Waltz TA, Anderson DA: Neurological involvement in the nevoid basal cell carcinoma syndrome. J Neurosurg 35:577–584, 1971.
74. Stout AP: Ganglioneuroma of the sympathetic nervous system. Surg Gynecol Obstet 84:101–110, 1947.
75. Strong LC: Genetic and environmental interactions. Cancer 40:1861–1866, 1977.
76. Haerer AF, Jackson JF, Evers CG: Ataxia-telangiectasia with gastric adenocarcinoma. JAMA 210:1884–1897, 1969.
77. Kersey JH, Spector BD: Immune deficiency diseases. In: Persons at high risk of cancer: An approach to cancer etiology and control. Fraumeni JF Jr (ed), New York, Academic Press, 1975, pp 55–67.
78. Russell DS, Rubinstein LJ: Pathology of tumours of the nervous system, 3rd ed. Baltimore, The Williams and Wilkins Co., 1971.
79. Alexander WS: Multiple primary intracranial tumours – Meningioma associated with a glioma – Report of a case. J Neuropathol Exp Neurol 7:81–88, 1948.
80. Feiring EH, Davidoff LM: Two tumors, meningioma and glioblastoma multiforme, in one patient. J Neurosurg 4:282–289, 1947.
81. Kirschbaum WR: Intrasellar meningioma and multiple cerebral glioblastomas. J Neuropathol Exp Neurol 4:370–378, 1945.
82. Anderson DE: Familial susceptibility. In: Persons at high risk of cancer: An approach to cancer etiology and control. Fraumeni JF Jr (ed), New York, Academic Press, 1975, pp 39–55.
83. McFarland J, Sappington SW: A ganglioneuroma in the neck of a child. Am J Pathol 11:429–448, 1935.
84. Batzdorff U, Malamud N: The problem of multicentric gliomas. J Neurosurg 20:122–136, 1963.
85. Borovich B, Mayer M, Gellei B, Peyser E, Yahel M: Multifocal glioma of the brain. Case report. J Neurosurg 45:229–232, 1976.
86. Courville CB: Multiple primary tumors of the brain (review of the literature and report of 21 cases). Am J Cancer 26:703–731, 1936.
87. Moertel CG, Dockerty MB, Baggenstoss AH: Multiple primary malignant neoplasms. III. Tumors of multicentric origin. Cancer 14:238–248, 1961.
88. Scherer HJ: The forms of growth in gliomas and their practical significance. Brain 63:1–35, 1940.
89. Mufson JA, Davidoff LM: Multiple meningiomas (report of two cases). J Neurosurg 1:45–57, 1944.
90. Cahill GF: Pheochromocytomas. JAMA 138:180–186, 1948.
91. Cragg RW: Concurrent tumors of the left carotid body and both Zuckerkandl bodies. Arch Path 18:635–645, 1934.
92. Symington T, Goodall AL: Studies in phaeochromocytoma. I. Pathological aspects. Glasgow Med J 34:75–96, 1953.
93. Ingraham FD, Bailey OT: Cystic teratomas and teratoid tumors of the central nervous system in infancy and childhood. J Neurosurg 3:511–532, 1946.
94. Bigner DD: Role of viruses in the causation of neural neoplasia. In: Biology of brain tumors. Laerum OD, Bigner DD, Rajewsky MF (eds), Geneva, International Union Against Cancer, 1978, pp 85–111.
95. Kleihues P: Chemical carcinogenesis in the nervous system. In: Biology of brain tumors. Laerum OD, Bigner DD, Rajewsky MF (eds), Geneva, International Union Against Cancer, 1978, pp 113–128.
96. Reynolds ES: Trauma as a possible cause of brain tumor. Lancet 2:13–14, 1923.
97. Tanaka J, Garcia JH, Netsky MG, Williams JP: Late appearance of meningioma at the site of partially removed oligodendroglioma. J Neurosurg 43:80–85, 1975.
98. Parker HL, Kernohan JW: The relation of injury and glioma of the brain. JAMA 97:535–540, 1931.

99. Preston-Martin S: Abstract: A case-control study of intracranial meningiomas in women. Am J Epidemiol 108:233–234, 1978.
100. Choi NW, Schuman LM, Gullen WH: Epidemiology of primary central nervous system neoplasms: II. Case-control study. Am J Epidemiol 91:467–485, 1970.
101. Annegers JF, Kurland LT, Grabow JD, Groover RV, Laws ER Jr: Abstract: The incidence of head trauma and subsequent risk of seizures and brain tumors. Neurology (Minneap) 29:578, 1979.
102. McMahon B: Prenatal X-ray exposure and childhood cancer. J Natl Cancer Inst 28:1173–1191, 1962.
103. Modan B, Baidatz D, Mart H, Steinitz R, Levin SG: Radiation-induced head and neck tumours. Lancet 1:277–279, 1974.
104. Shore RE, Albert RE, Pasternack BS: Follow-up study of patients treated by X-ray epilation for tinea capitis: Resurvey of post-treatment illness and mortality experience. Arch Environ Health 31:21–28, 1976.
105. Schreier HA, Sherry N, Shaughnessy E: Lead poisoning and brain tumors in children: a report of 2 cases. Ann Neurol 1:599–600, 1977.
106. Gold E, Gordis L, Tanascia J, Szklo M: Risk factors for brain tumors in children. Am J Epidemiol 109:309–319, 1979.
107. Gold E, Gordis L, Tanascia J, Szklo M: Increased risk of brain tumors in children exposed to barbiturates. J Natl Cancer Inst 61:1031–1034, 1978.
108. Lamperth-Seiler E: Harnweg- und Hirntumoren bei Gummiarbeitern. Schweiz Med Wochenschr 104:1655–1659, 1974.
109. Mancuso TF: Tumors of the central nervous system: Industrial considerations. Acta Unio Internat Contra Cancrum 19:488–489, 1963.
110. Waxweiler RJ, Stringer W, Wagoner JK, Jones J, Falk H, Carter C: Neoplastic risk among workers exposed to vinyl chloride. Ann NY Acad Sci 271:40–48, 1976.
111. Mayr E, Diamond LK, Levine RP, Mayr M: Suspected correlation between blood-group frequency and pituitary adenomas. Science 124:932–934, 1956.
112. Buckwalter JA, Turner JH, Gamber HH, Raterman K, Soper RT, Knowler LA: Psychoses, intracranial neoplasms, and genetics. Arch Neurol Psychiat 81:480–485, 1959.
113. Selverstone B, Cooper DR: Astrocytomas and ABO blood groups. J Neurosurg 18:602–604, 1961.
114. Yates PO, Pearce KM: Recent change in blood-group distribution of astrocytomas. Lancet 1:194–195, 1960.
115. Garcia JH, Okazaki H, Aronson SM: Blood group frequencies and astrocytoma. J Neurosurg 20:397–399, 1963.
116. Paton A, Petch CP: Association of diabetes mellitus with cerebral tumour. Br Med J 1:855–856, 1954.
117. Aronson SM, Aronson BE: Central nervous system in diabetes mellitus: Lowered frequency of certain intracranial neoplasms. Arch Neurol 12:390–398, 1965.
118. Schuman LM, Choi NW, Gullen WH: Relationship of central nervous system neoplasms to toxoplasma gondii infection. Am J Public Health 57:848–856, 1967.
119. Reagan TJ, Freiman IS: Multiple cerebral gliomas in multiple sclerosis. J Neurol Neurosurg Psychiatry 36:523–528, 1973.
120. Schoenberg BS, Christine BW, Whisnant JP: Nervous system neoplasms and primary malignancies of other sites. The unique association between meningiomas and breast cancer. Neurology (Minneap) 25:705–712, 1975.
121. Jensen RD, Miller RW: Retinoblastoma – Epidemiologic characteristics. N Engl J Med 285:307–311, 1971.
122. Kitchin FD, Ellsworth RM: Pleiotropic effects of the gene for retinoblastoma. J Med Genet 11:244–246, 1974.

123. Oyasu R, Battifora HA, Clasen RA, McDonald JH, Hass GM: Induction of cerebral gliomas in rats with dietary lead subacetate and 2-acetylaminofluorene. Cancer Res 30:1248–1261, 1970.
124. Michelson J, New PFI: Brain tumour and pregnancy. J Neurol Neurosurg Psychiatry 32:305–307, 1969.
125. Donnell MS, Meyer GA, Donegan WL: Estrogen-receptor protein in intracranial meningiomas. J Neurosurg 50:499–502, 1979.
126. Winston K, Gilles FH, Leviton A, Fulchiero A: Cerebellar gliomas in children. J Natl Cancer Inst 58:833–838, 1977.
127. End Results Section, Biometry Branch, National Cancer Institute: In: Cancer patient survival, report number 5. DHEW Publication No. (NIH) 77–992, Bethesda, MD, National Institutes of Health, 1976, pp 234–240.
128. Peters, JM, Preston-Martin S, Yu MC: Brain tumors in children and occupational exposure of parents. Science 213:235–237, 1981.
129. Blair A, Hayes HM Jr: Mortality patterns among US veterinarians, 1947–1977: an expanded study. Int J Epidemiol 11:391–397, 1982.
130. Musicco M, Filippini G, Bordo BM, Melotto A, Morello G, Berrino F: Gliomas and occupational exposure to carcinogens: case-control study. Am J Epidemiol 116:782–790, 1982.
131. Selikoff IJ, Hammond EC (eds): Brain tumors in the chemical industry. Ann NY Acad Sci 381:1–364, 1982.
132. Hochberg FH, Cole P, Salcman M, Kadish SP, Glicksman A, Saunders R, Toniolo P, Taylor L: Abstract: Risk factors in glioblastoma development. Neurology (Minneap) 32(2):A75, 1982.

2. A Review of Animal Brain Tumor Models that have been used for Therapeutic Studies

S. CLIFFORD SCHOLD, Jr. and DARELL D. BIGNER

INTRODUCTION

Laboratory investigations of primary human brain tumors have largely depended on the use of animal models. While studies of tumors in animals often lack the precision and elegance of research on the cell biology of individual neoplastic cells, observations on the behavior of *in vivo* models are more readily applied to clinical problems. The proper experimental use of a well-characterized animal brain tumor model can therefore be a powerful intermediate step in the overall attack on human brain tumors.

For experimental purposes, an animal model system ought to mimic the particular aspect of a problem in which one is interested. A number of aspects of human brain tumors are of concern, but none has greater importance currently than the choice of appropriate therapy. Many diverse animal models have been developed to investigate brain tumors, and in this chapter we will review a series of these *in vivo* brain tumor models which have been used for therapeutic studies.

BIOLOGICAL CHARACTERISTICS OF SELECTED ANIMAL BRAIN TUMOR MODELS

Non-neurogenic Intracerebral Tumors

Background. A host of non-neurogenic tumors have been transplanted into the brains of animals and used as models of nervous system tumors. The murine leukemias L1210 and P388 both readily spread to the nervous system after systemic implantation, and both have been directly implanted into the nervous system for use as models of CNS leukemia. Murine solid tumors, such as B16 melanoma, Erlich ascites carcinoma, sarcoma 180, and Walker 256 carcinoma, also grow readily in the nervous system and have been used for therapeutic studies. These tumors have the advantages of being well characterized when growing systemically and of having known therapeutic sensitivities. They are ideal models for investigations of the importance of environment on the behavior and response to therapy of a tumor. They have the obvious disadvantage of diverse, non-neural origins, but the importance of the cellular origin of an experimental neoplasm will vary with the questions being asked.

L1210 Leukemia. This tumor is discussed in this review because of the abundance of studies of its behavior in the nervous system and because of the importance of therapeutic studies using this model for subsequent clinical and experimental approaches to brain tumors. The L1210 leukemia was developed in the late 1940s by the application of 0.2% methylcholanthrene to the skin of DBA mice [1]. The disease produced was typical of human leukemia in that circulating tumor cells were accompanied by diffuse widespread visceral tumor. The systemic tumor was found to be highly sensitive to anti-folate agents, and this was the basis for the development of L1210 as an animal screen for the testing of potential anticancer agents. Extensive characterization of the growth kinetics and chemotherapeutic sensitivities of the tumor followed, and it is now probably the most well characterized experimental tumor.

Systemic Growth and Transplantation. L1210 can be successfully transplanted by intravenous (iv), intraperitoneal (ip), subcutaneous (sc), intramuscular (im), or intracerebral (ic) routes. In early studies dilutions of whole blood from affected animals were used for tumor passage, but subsequent isolation of the tumor cells permitted precise determination of dosage. There is rapid dissemination of tumor cells after implantation at any site, and both the rate of dissemination and the survival of the animals are directly related to the number of cells implanted. After implantation of 10^6 cells ip, for example, median survival is 7.0 days, while after 10^4 cells it is 10.3 days [2]. It was shown in elegant studies by Skipper *et al.* that a single tumor cell was sufficient to kill a recipient animal and did so within 15 days in 50% of animals [2]. Intravenous administration reduces median survival to 5.3 days after 10^4 cells and to 13.1 days after a single cell.

The calculated doubling time of the ip tumor based on survival studies is 0.48 days [2]. Labelled thymidine uptake shows a doubling time of 11.3 – 13.7 hours with a growth fraction (GF) approaching 100%. The calculated cell cycle time is 13 hours and the duration of S-phase is 7 hours [3]. These data were obtained on day 6 after ip implantation of 10^5 cells, i.e., at a time when the tumor burden was approaching a lethal level. This tumor is thus a rapidly proliferating, high GF tumor which has a propensity for widespread dissemination.

Intracerebral Growth. The time at which tumor cells appear in the brain after systemic implantation is a function of the total number of systemic tumor cells and of the route of tumor implantation. After ip implantation of 10^6 L1210 cells, brain involvement is detectable in 2 – 4 days [4]. This can be delayed by the early administration of methotrexate, which reduces systemic tumor burden without having any significant effect on established CNS tumor [5].

After ic implantation of 10^4 cells the median survival of animals is 8.4 days, and after implantation of 10 cells it is 12.8 days [2]. The calculated doubling time of the ic tumor is 0.42 days so that the tumor growing in the brain is kinetically indistinguishable from the systemic tumor. The tumor in the brain grows preferentially in the leptomeninges in a pattern similar to that seen in human CNS leukemia.

Transplantable Brain Tumors produced by Polycyclic Hydrocarbons

Background. A wide range of ic tumors has been produced by the local application of a variety of polycyclic aromatic hydrocarbons (PAH). The rate of successful induction and the type of tumor produced vary with the animal species, the specific carcinogen, and the method of implantation [6]. However, even under identical experimental conditions the incidence of tumors, their histological characteristics, and latency to their development have varied widely, so that therapeutic studies are impractical. Methylcholanthrene, benzpyrene, and dibenzanthracene have been the most frequently used chemicals.

With rare exception direct ic implantation of these agents is required for the production of brain tumors. In most instances cylindrical pellets of the carcinogen have been implanted, and phagocytosed carcinogen crystals have been identified in the pre-neoplastic lesions [7]. The latency period from implantation to the development of neurological symptoms has averaged between 200 and 400 days in most studies. The carcinogenic and mutagenic effects of PAH appear to be related to interaction of the agent or a metabolite with DNA [8].

The 'Ependymoblastoma' Group. Seligman and Shear first induced brain tumors by ic PAH implantation in 1939 [9]. In their series 11 of 20 five-month-old C_3H mice implanted with cylindrical pellets of 20-methylcholanthrene developed gliomas, two developed fibrosarcomas, and no tumors developed in seven animals. There was considerable histologic variation among the gliomas, and latency ranged from 227 to over 500 days. One of these serially transplanted gliomas is now designated glioma 261 (GL 261).

Zimmerman and Arnold used the same techniques to induce several intracranial tumors which varied with the site of carcinogen implantation [10]. Gliomas of various types developed in the cerebrum, three of the 13 tumors in the cerebellum were medulloblastomas, and a variety of sarcomas developed in the meninges. Individual tumors were often of a mixed histological picture, and in a later paper Zimmerman described the isolation and maintenance of histologically distinct portions of such a tumor as separate, stable tumor lines [11]. Subsequent lineage of these tumors is obscure, but an ependymoblastoma (Ep) from this original group was maintained and later used for chemotherapeutic studies. Ependymoblastoma A (EpA) is a mutant subline derived from the original Ep as

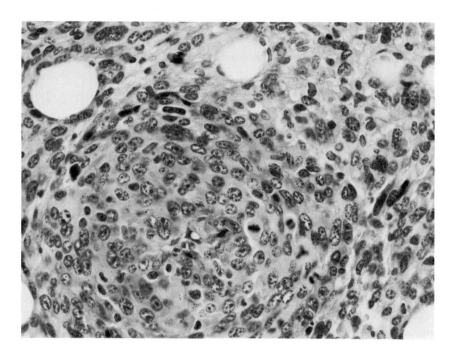

Figure 1. Sheets of undifferentiated cells with no differentiation features of ependyma. EpA model, hematoxylin and eosin × 400 (Courtesy of Dr. William Shapiro).

described by Ausman *et al.* [12]. It is said to be histologically identical to the parent line, but it grows faster. Glioma 26 (GL 26) was also produced by Zimmerman *et al.* in 1948 by ic implantation of methylcholanthrene in C57 black mice, as reported by Sugiura [13].

These four lines, GL 261, Ep, EpA, and GL 26, are collectively referred to as the ependymoblastoma model [14].

Morphology. Since the original induction of these tumor lines, morphologic descriptions have been few. They were said to be histologically identical to one another and to consist of 'small uniform polygonal cells with oval, darkly staining nuclear chromatin and scanty cytoplasm' [12]. Further, 'the cells were arranged in sheets' and 'pseudorosettes were present'. This appearance was thought to be 'typical of an ependymoblastoma'. It does not possess in its present state of passage distinctive cellular or cytoarchitectural features of ependymal differentiation, but rather consists of sheets of undifferentiated cells (Figure 1) which could have originally been derived from ependymal cells, oligodendrocytes, neuroblasts, or even non-neuroectodermal epithelium. We would agree with Rubinstein who has stated that it is 'poorly differentiated' and 'does not demonstrate ependymal differentiation' [15].

One distinctive morphological feature of the ependymoblastoma whose biological significance remains unclear is the unequivocal presence of B-type retrovirus particles with all the characteristics of mammary tumor virus [16, 17]. In addition to mammary tumor virus particles, antigens of C-type retroviruses cross-reactive with those of Rauscher leukemia virus have been reported in ependymoblastoma tumor homogenates [7].

Transplantation and Growth. These tumors are readily maintained by subcutaneous transplantation. For ic implantation, sc or ic tumors are dissected and 1 mm^3 fragments are placed by freehand injection through a 19-gauge needle into the subcortical region of the mouse brain. Complications from this procedure are few, and large numbers of animals can be implanted in a relatively short time. The tumors grow as a mass in the brain and eventually over the brain's surface and into the subcutaneous tissue at the injection site.

In spite of the erratic location of these tumors, a powerful advantage for therapeutic studies has been the reproducible mortality distribution they produce. In the original study, the median day of death of over 600 animals implanted with the Ep tumor was 29 days with a standard deviation of 3.0 days, and fewer than 1% of the animals lived for more than 60 days [12]. The corresponding figures for the other three tumors were: Gl 26 – 24 ± 3.0 days, 0% 60-day survivors; GL 261 – 24 ± 1.9 days, 0.9% 60-day survivors; and EpA – 19 ± 1.2 days, 2% 60-day survivors. The stability of this aspect of the lines is indicated by a recent chemotherapy experiment in which the median survivals were: EpA – 19.5 and 20.0 days; GL 26 – 20.5 and 22.0 days; and GL 261 – 20.5 and 23.0 days [18]. Shapiro has estimated the generation times of three of the tumors to be: Ep – 69 hrs.; GL 261 – 57 hrs.; EpA – 45 hrs. [14].

Rubin *et al.*, reported growth of the Ep tumor in tissue culture after explantation [17]. Continuous culture was maintained for up to 13 months, though details of the growth characteristics were not given. Tumors produced by reinjection of these cultured cells into C57B1 mice were 'histologically identical to the original'. Albright *et al.*, estimated the *in vitro* doubling time of GL 26 to be 36 hrs. [19].

Special Studies. The disposition of drugs in these tumors has been investigated as a measure of permeability. The anti-metabolite 5-FU was shown to reach levels in the Ep tumor 11 times greater than in normal brain [20]. Likewise, methotrexate, an agent which does not cross the intact blood-brain barrier to any significant extent, was found to concentrate in ic Ep in comparison to normal brain [21] as well as to inhibit the uptake of ^{14}C-labelled deoxyuridine in ic GL 26 [22]. Tator has also shown that tritiated methotrexate concentrates in an ic ependymoblastoma to a greater extent than in normal brain. However, his data indicate that the drug accumulates to a lesser extent in ic than in sc tumors and that its distribution in the ic tumor is not uniform [23]. Finally, as indicated below,

Table 1. Characteristics of 5 animal brain tumor models which have been refined for therapeutic studies

Tumor	Animal host	Method of induction and transmission	Morphology	In vivo kinetics
L1210	DBA mice	Methylcholanthrene, transplantable	lymphoid leukemia	TD = 0.5 days
Ependymoblastoma group*	C57BL mice	Intracerebral methylcholanthrene, transplantable	'Ependymoblastoma'	TD = 2 – 4 days
9L	Fischer 344 rats	Methyl-nitrosourea, transplantable	'Gliosarcoma'	TD = 2 days GF = 0.45 – 0.50
ASV	Fischer 344 rats	B – 77 strain ASV, autochthonous	Anaplastic astrocytoma	variable
Xenografts	Athymic BALB/c or NIH Swiss mice	Heterotransplantation	Variable	variable

*Includes Ep, EpA, GL 26, and GL 261.

the water-soluble nitrosourea, ACNU, has been shown to prolong survival of mice bearing ic GL 26, GL 261, or EpA [18].

Scheinberg *et al.* demonstrated in a series of studies that the PAH-induced ependymoblastomas were immunogenic [24, 25]. The growth of both sc and ic tumor was inhibited by preimmunization with tumor and adjuvant.

Tumors produced by the N-Nitroso Compounds

Background. The neuro-oncogenic effects of systematically administered nitrosoureas have been studied in detail over the past 15 years. Since the demonstration by Druckrey of the high incidence of nervous system tumors in rats after the administration of some nitroso compounds, information on the occurrence and nature of these tumors under different conditions has accumulated [6, 26]. Both N-methylnitrosourea (MNU) and N-ethyl-nitrosourea (ENU) are neuro-oncogenic, although the effects of ENU are more age-related. Repeated iv or ip administration of (MNU) has reliably produced neuroectodermal tumors in adult rats and rabbits [26]. ENU is most effective when given in the prenatal period at doses significantly below adult toxicity levels [27]. The single iv administration of ENU to a pregnant female rat in doses from 1 – 50 mg/kg has produced central and peripheral nervous system tumors in the offspring with the incidence of tumors proportional to the dose of the carcinogen [28].

The full range of histologic types of nervous system tumors has been produced by the nitroso compounds, with the exception that neoplastic transformation of the neuronal cells has been unusual [29]. Tumors have often been multi-focal, although certain brain regions such as the periventricular areas appear to be more susceptible [30]. Both astrocytic and oligodendrocytic tumors have been described, and in many cases a mixture of cell types is present. Ependymal tumors and mixed glial and mesodermal tumors ('gliosarcomas') have also been seen. The degree of anaplasia has varied widely [6, 28].

The thrust of research using nervous system tumors induced with nitrosoureas has shifted into general areas of cancer biology, particularly into mechanisms of tumor initiation, tumor progression, and organotropy. Specific DNA adducts that have delayed repair in brain compared to other organs may be important in initiating and in explaining the preferential neuro-oncogenicity of the compounds (for a review see [8]). While mechanisms of progression are not well defined, Laerum *et al.* [31, 32] have elegantly described the progression of transformed cells and the appearance of tumorigenic cell populations *in vitro* following *in vivo* exposure to ENU.

Both autochthonous (primary) and transplanted ENU and MNU-induced tumors have been used for therapeutic studies [33, 34, 6, 35, 36; *vide infra*]. Primary nitrosourea tumors, while possessing attractive features of autochthonous brain tumors, have many drawbacks as therapeutic models.

Many different histologic types of tumors are induced, multiple tumors are more frequent than solitary ones, latency periods are long and variable, standardization of tumor size at time of initiation of treatment is difficult, and very large treatment group sizes are necessary to control for the broad mortality distribution of therapy control groups.

Transplanted and *in vitro* maintained nitrosourea tumors have become widely used tools in neurobiology and neuro-oncology. The C_6 rat glioma, originally induced with MNU, is the most widely studied glial cell line in neurobiology [37]. It was induced, unfortunately, in non-inbred Wistar-Furth rats. It can therefore only be transplanted in irradiated rats [38] or in neonatal rats [39]. Since such animals remain or become partially immunologically intact, host resistance and tumor rejection mediated against strong histocompatibility antigens complicates evaluation of anti-tumor effects of therapeutic agents. It is possible that the use of athymic mice might allow more easily interpretable *in vivo* studies with the well-characterized C_6 line.

There are an enormous number of *in vitro* cell lines derived from nitrosourea-induced brain tumors, many of which are transplantable and which could be used for therapeutic studies. Investigators who have developed such lines include Schubert, Herschmann, Wechsler, Stavrou, Lantos, Mennel, and Thust.

The 9L Gliosarcoma. The 9L gliosarcoma was derived from a line of brain tumors produced in Fischer 344 rats by weekly injections of 5 mg/kg MNU [40]. The group of tumors so produced were propagated both in tissue culture and in the subcutaneous tissue of newborn rats, and the cells have retained viability after prolonged freezing. The most commonly used line, designated '9L', differs from the original '#9' tumor [41] and from tumors derived from the 9L stock which have been harvested and transplanted in other laboratories [42], reflecting the instability of the line under varying conditions. Most of the work on the 9L line has been done either at the Brain Tumor Research Center in San Francisco [43] or at the University of Rochester, and these lines are designated 9L/SF and 9L/Ro, respectively. 9L was said to be 'the only brain tumor model that has been developed for both *in vivo* and *in vitro* study' [44], although this claim can now be made for many of the tumor lines growing in athymic mice discussed below.

Morphology. The tumors produced by the ic injection of a 9L cell suspension have been considered anaplastic astrocytomas [43]. The tumors are densely cellular with individual cells varying from oval to spindle-shaped. Mitoses are prominent, but endothelial proliferation and necrosis are not conspicuous. These latter features distinguish the 9L from the original #9 tumor which contained both necrosis with palisading and marked endothelial proliferation [41]. The lack of significant necrosis in the 9L tumor may be related to its lack of hypoxic cells and is of some experimental advantage, as will be discussed later.

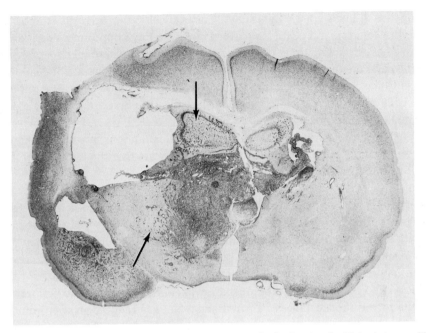

Figure 2. Whole mount of a coronal section of an F344 rat brain showing the 9L brain tumor. Note the cells at the margin (arrows) growing in a sarcomatous pattern, hematoxylin and eosin.

The morphology of the ic tumor changed somewhat after continuous culture. Spindle cells became more prominent, and both reticulin and collagen were produced. This portion of the tumor was distinct from the more typical glial portion, and the tumor was reclassified as a gliosarcoma [15]. Present passage levels (Figure 2) derived from cells obtained from Dr. Takao Hoshino have cellular and growth properties of sarcomas. The tendency to grow along blood vessels around the tumor margin is a highly characteristic growth feature of sarcomas transplanted intracerebrally. Such growth patterns are not seen commonly in autochthonous animal or human gliomas.

Kinetics and Growth Characteristics. In monolayer cell culture the 9L tumor has a doubling time of between 18 and 20 hours with a GF of 0.95 [43]. The cell-cycle time is 19.5 – 20.0 hours, and the colony forming efficiency (CFE) is 60 – 80% [42]. These cells have also been grown in a multicellular spheroid culture system, and many of the kinetic and therapeutic data in this system approximate those of the *in vivo* tumor. In 200 – 500 μm diameter spheroids the cell cycle time is 20.5 hrs., the GF is 0.50, and the CFE is 40 – 50% [45]. Subcutaneous tumors formed after implantation of 10^6 cells reach a diameter of 2 – 3 cm in 3 – 4 weeks [44].

After ic injection of 4×10^4 cells there is virtually a 100% tumor incidence, killing the animals in 23 ± 2 days (survival with the early 9L tumor was somewhat longer), and death is preceded by a 5 – 6 day symptomatic period [43]. Tumors have doubling times of approximately 48 hrs as determined by successive weights, and the tumors weigh between 200 and 250 mg at the time of death [42]. The labelling index of the ic tumors is between 15 and 20% [43]. The cell-cycle time is 20 hours, as determined by labelled mitoses curves, and the calculated potential doubling time is 42 hours with a calculated GF of 0.35 – 0.46 [43]. The CFE of the ic tumors is 10 – 30% under optimal conditions including 30% fetal calf serum and irradiated feeder cells [42]. It appears that the hypoxic cell fraction of tumors smaller than 250 mg is < 1% [46].

Permeability. The permeability characteristics of the 9L tumor have been investigated using standard markers such as inulin, urea, horseradish peroxidase (HRP), and common chemotherapeutic agents [47, 48]. While capillaries in the ic tumor mass are clearly abnormally permeable, the exchange of substances between blood and the surrounding edematous brain including infiltrating tumor ('brain-adjacent-to-tumor') is complex. Levin *et al.*, found that the permeability for the diffusible substances ^{14}C-urea and ^{22}Na was reduced in the brain surrounding the tumor in comparison to normal brain [47]. They suggested that delivery of water soluble chemotherapeutic agents to this region of actively proliferating tumor cells might be dependent on diffusion from the main tumor mass. These data were confirmed in a sense by Groothuis *et al.*, in which the distribution of HRP, a 44,000 molecular weight substance which is normally excluded from the brain, was analyzed in the 9L tumor [48, 49]. They found HRP in an irregular pattern, and in no instance was the brain adjacent to the tumor "completely permeated." This differed from the sc 9L tumor in which there was 100% HRP permeability.

It is clear that the permeability of the ic 9L tumor is complex and that as a consequence appropriate exposure of all tumor cells to chemotherapeutic agents might be the critical factor in therapeutic success. It remains to be seen to what extent this pattern of permeability can be generalized to other brain tumors.

Virus-induced Brain Tumors

Background. A number of viruses are capable of producing ic tumors in animals after direct inoculation. The first were the sarcomas induced by Vasquez-Lopez using Rous sarcoma virus in adult chickens [50]. Subsequently considerable experience with avian sarcoma virus (ASV)-induced gliomas in mammals has accumulated [6]. Hamsters, rabbits, dogs, mice, monkeys, marmosets, rats, cats, guinea pigs, and gerbils have all been susceptible to the neuro-oncogenic effects of this agent. Two other RNA viruses, murine sarcoma virus and simian sarcoma virus, have also induced gliomas in mammals after ic inoculation [51, 52]. A number of DNA viruses have also had neuro-oncogenic effects, including human adenovirus 12, which produced neuroblastic tumors, and JC virus, a human papovavirus derived from the brain of a patient with progressive multifocal leukoencephalopathy [53, 54]. Details of the brain tumors produced by these and other viruses have been reviewed elsewhere [55, 56], but only the rat ASV-induced astrocytoma has been refined to the point of reproducible tumor induction and survival necessary for therapeutic studies.

The ASV Astrocytoma. Rabotti in the early 1960's first produced gliomas in mammals by ic ASV inoculation [57, 58]. In a series of studies in the 1970s, Bigner *et al.*, refined this model in rats and demonstrated conclusively that most tumors induced by ic inoculation of ASV were astrocytic in origin [59], essentially 100% tumor incidence could be produced by the appropriate virus concentration and inoculation site [60], and survival of inoculated animals could be standardized sufficiently to allow controlled therapeutic studies [61]. These therapeutic studies, as well as detailed morphological and immunological characterization of the tumors, have made the ASV-induced astrocytoma a major model for the investigation of human brain tumors (Figure 3).

Tumor Induction. Brain tumors are produced by ASV only after direct ic inoculation. The use of a highly concentrated virus suspension and standardized inoculation near the subependymal region in neonatal Fischer 344 (CDF) rats resulted in a 96% incidence of intracranial tumors with a latency of less than three months in one study. The majority of tumors so produced were anaplastic astrocytomas, though a variable percentage of gliosarcomas, sarcomas, and other less common tumors such as oligodendrogliomas was seen [62]. No neuronal tumors have been produced [63]. Alteration of any of the inoculation parameters changes the timing and the morphology of tumors produced. Inoculation at 100 days of age, for example, decreases the incidence of tumors within 150 days to 50% [60]. That adult rats were still highly susceptible to the neuro-oncogenic effects of the virus was demonstrated in a study in which 528 day-old rats were inoculated and followed until death 26-415 days after inoculation. Twenty of 30 animals had brain tumors at death [64]. Tumors produced in mature animals have been less often multiple, histologically more uniform, and have shown greater differentiation. Changing the inoculum size has also influenced the latency. Animals inoculated at nine days with 8.7×10^4 focus-forming units of ASV survived an average of 174 days after inoculation, while nine day-old animals receiving 8.7×10^3 units survived 308 days. Interestingly, though the survival in young animals receiving less virus was similar to that of older animals receiving 8.7×10^4 units, the tumors produced were age rather than dosage specific [64]. In the beagle, the type of tumor produced was largely a function of the inoculation site [65].

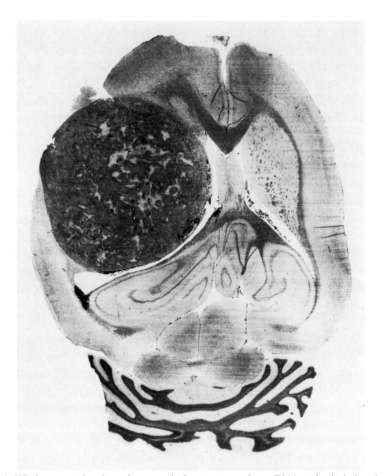

Figure 3. Whole mount showing a large cortical astrocytoma in an F344 rat brain induced with intracerebral inoculation of ASV, hematoxylin and eosin. (From Figure 1, Copeland *et al.*, Amer J Path 83:149–176, 1976, with permission of author and publisher).

ASV in its infectious form is no longer detectable 48 hours after ic inoculation of mammals [66]. Moreover, no retrovirus particles are present in the brain tumors induced in mammals with ASV [59, 67, 68]. Nevertheless, ASV structural antigens are present in tumor cells and the genome can be demonstrated by cocultivation of the brain tumor cells or with susceptible chick cells or by injection of brain tumor cells into susceptible avian hosts [9].

Biology. The ASV-induced anaplastic astrocytoma model is the most widely used autochthonous experimental brain tumor that has been refined for therapeutic studies. The biological variability of an autochthonous model probably mimics the heterogeneous human gliomas more closely than do the transplantable models. For example, ASV tumors were abnormally permeable in one study, but the permeability to HRP was highly variable, both within and among tumors [48]. This permeability did not appear to correlate with either histologic type or tumor size, but it was related to the degree of vascularity of the tumors. Ultrastructurally areas in ASV-induced brain tumors contain both endothelial fenestrations similar to those present in normal choroid plexus and gaps in the endothelial walls similar to those seen in anaplastic human gliomas [70].

Recent autoradiographic studies of ASV-induced gliomas have also shown that in general blood flow was depressed and 'blood-to-tissue transport' was increased in comparison to normal brain [71] and that the 'proliferation' of such tumors, based on thymidine uptake, was highly variable indicating that the marginal zone was not the only area of active tumor growth [72].

Finally, animals bearing ASV-induced gliomas had depressed cell-mediated immunocompetence in a pattern similar to that seen in human glioma patients. Specifically, the mitogenic responsiveness to phyto-hemaglutin of spleen cells from animals with these tumors was depressed, and the degree of depression was related to the size of the tumor [73]. However, this laboratory measure of immunosuppression may not be biologically significant. ASV tumor bearing rats maintain ASV-specific tumor associated immunity to rat and hamster ASV-induced tumors [74]. Moreover, ASV tumor bearing F344 rats are very efficient hosts in which to induce experimental allergic encephalitis with guinea pig spinal cord, indicating that their T-lymphocytes and other reticuloendothelial effector cells are capable of mediating significant immunopathologic lesions [75].

Human Tumor Xenografts

Background. There have been many attempts to transplant human brain tumors into animals. In general, successful growth has required either transplantation to an immunologically protected site, such as the anterior chamber of the eye, or immunosuppression of the recipient animals. Greene and Arnold reported the successful transplantation of 3 human glioblastomas into the anterior chamber of the guinea pig eye [76], and Greene later reported successful growth of 8 human glioblastomas in the brains of mice and guinea pigs [77]. However, numerous other similar attempts at heterotransplantation have met with only limited success [6]. Jänisch and Schreiber found that successful ic transplantation of human brain tumors into albino rats could only be achieved in newborn animals. Wilson and Barker transplanted 20 human neural tumors intrathecally into guinea pigs, but only a single medulloblastoma grew [78]. It is clear that the brain is partially protected from immunological rejection mechanisms and that some heterografts will grow after ic implantation. However, takes have been irregular and serially transplantable lines have been difficult to achieve, limiting the usefulness of this model for therapeutic purposes.

The Athymic Mouse System. A breakthrough in the successful growth of heterologous tissues occurred with the discovery and characterization of the athymic 'nude' mouse mutant [79]. These animals were found to be congenitally athymic permitting growth of a variety of foreign tissues including neoplasms [80]. Many human tumors have now been grown in these animals with preservation of morphologic and biochemical features [81, 82].

Human Gliomas in Athymic Mice. Bigner *et al.* [83] reported successful subcutaneous growth of six of 15 established human glioma cell lines in athymic mice. In a subsequent study of three of these cell lines, Bullard *et al.*, documented serial subcutaneous transplantability, retention of distinctive morphological characteristics, a human glucose-6-phosphate dehydrogenase isoenzyme pattern, and the achievement of reproducible growth curves [84]. They also commented that these cell lines grew intracerebrally, although mortality distributions were not determined. One of the serially transplantable cell lines, U-251 MG, expressed glial fibrillary acidic protein (GFAP) both in tissue culture and in the mouse. The achievement of reproducible growth curves and the accessibility of the subcutaneous tumors for sequential measurements established the suitability of this model for therapeutic studies (Figure 4).

Wara *et al.* [85], Rana *et al.* [86], and Reid *et al.* [87] have described successful sc transplantation of individual human glioblastomas (GBM) into athymic mice. In general, these tumors retained their morphological features, although in one instance serial sc transplantation of a GBM was associated with a progressive increase in cell density and dominance by anaplastic spindle cells [87]. Shapiro *et al.* [88] successfully transplanted seven of seven malignant gliomas subcutaneously and six of seven intracerebrally. Two of these tumors, both gliosarcomas, were serially transplanted and grew progressively faster until stabilization beyond the third animal passage. The volume doubling time of one of these tumors in a fifth and sixth passage was estimated to be 5.5 days. Schold *et al.*, reported the successful sc transplantation of 16 of 17 anaplastic human gliomas [89]. The tumors often required a prolonged period before growing, over 300 days in one instance, but 11 of the tumors were successfully transplanted into at least a second animal generation. The tumors growing in athymic mice retained morphological resemblance to their parent human tumors, although there was a tendency for the mouse-borne tumors to be more homogeneous. Fifteen of the 17 human gliomas expressed GFAP when grown in mice (Figure 4). In serial transplantation, the doubling times of these tumors ranged from 4.0 to 19.1 days. Most tumors retained their morphological characteristics in serial transplantation, although one became progressively more 'sarcomatous' [90]. The pattern of GFAP expression also changed with serial transplantation. Often a smaller proportion of the tumor expressed GFAP as the tumors were passed. That this was not invariably the case, however, indicated the heterogeneous behavior of these serially transplantable tumors.

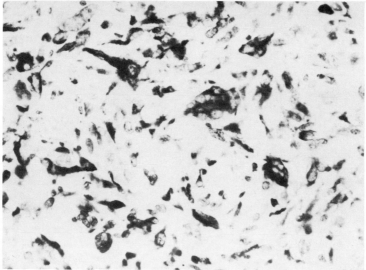

Figure 4. Top: An athymic mouse with a subcutaneous human glioma growing in its flank.
  Bottom: Immunoperoxidase stain with rabbit anti-glial fibrillary acidic protein (GFAP) antiserum of a human glioma growing in an athymic mouse. GFAP positive and negative cells reminiscent of surgical biopsies of human glioblastoma multiforme are present. PAP × 400.

THERAPEUTIC STUDIES OF EXPERIMENTAL BRAIN TUMORS

L1210

The successful chemotherapy of ic L1210 with the nitrosourea compounds was the foundation for the use of these agents against tumors of the brain in humans. Two ic systems have been used with this tumor. The most common is the direct ic implantation of leukemic cells. Ten thousand cells in the brain produce a median survival of 8–9 days, and although the tumor has disseminated at the time of death, the nervous system tumor is thought to be the cause of death. In order to avoid the trauma of ic implantation, another system has occasionally been used. Based on the observation of a greater effect of MTX against systemic L1210 than against the tumor in the brain, animals are treated with an effective but non-curative dose of the drug after receiving tumor cells ip. This produces sufficient cell kill to allow the development of leptomeningeal infiltration by tumor. This

Table 2. Selected results of the treatment of intracerebral L1210 leukemia

Treatment*	Number of cells implanted	% T/C	Reference
BCNU[a]	10^4	208,211	91
BCNU	10^5	250	125
CCNU	10^5	>400	125
meCCNU	10^5	200	125
ACNU	10^6	200	93
PCNU	10^6	208	93
PCB	10^5	114	125
AZQ[b]	10^6	179,184	92
CPA	10^5	157	125
MTX	10^5	129	125
5-FU	10^5	186	125
5-FU	10^6	150	93
6-TG	10^6	133	93
Ara-C[c]	10^5	>400	125
Ara-C	10^6	175	93
Ara-C[a]	10^4	211	2
Ara-C[d]	10^4	329	2
ACNU + 6-TG	10^6	300	93
PCNU + 6-TG	10^6	curative[e]	93
ACNU + Ara-C	10^6	367	93
ACNU + 5-FU	10^5	193	93
DAG	10^5	100–200[f]	93

*Treatment was at the maximum tolerated dose on day 1 or day 2 after ic implantation unless otherwise stated.
[a] Daily × 15 days
[b] Daily × 9 days
[c] q3 hr × 8 every 4 days × 3
[d] q3 hr × 8 every 4 days ×
[e] 9 of 9 animals were alive on day 60
[f] estimated from log cell kill

system most closely mimics the clinical situation, but it is not known whether therapeutic results differ between the two systems. In either case, for therapeutic studies median survival of the treated animals is compared to that of control animals and is expressed as a ratio (% T/C = (median survival of treated animals/median survival of control animals) × 100).

Schabel *et al.* showed that while CPA, 6-MP, 5-FU, and other common chemotherapeutic agents had marginal effects on survival, l-methyl-nitrosourea and BCNU produced % T/C of 160 and 208, respectively, after ic implantation of 10^4 cells [91]. Subsequently, other nitrosoureas were shown to have similar effects (Table 2). The effect on survival produced by these lipid-soluble, non-ionized agents was essentially identical to that produced in similarly staged systemic disease. Other agents of interest in brain tumor therapy have had some effect on ic L1210, but none has been as effective as the nitrosoureas. AZQ has been effective [92], and clinical trials with this agent are just beginning. Procarbazine and dianhydrogalactitol, on the other hand, have had only marginal success against ic L1210, in spite of their apparent activity against human brain tumors [93, 94, 95].

Nitrosourea-based combinations against ic L1210 have recently been reported by Schabel *et al.* [93]. They observed survival prolongation in excess of that produced by single agents with ACNU and either 6-TG, Ara-C, or 5-FU. The combination of PCNU and 6-TG was "curative" (i.e., survival > 60 days) in 9 of 9 treated animals. These results suggest that marginal activity of an agent when used alone should not exclude its use in combinations.

The 'Ependymoblastoma' Group

More therapeutic information is available on this group of tumors than on any other brain tumor model. Shapiro and Ausman tested a series of agents against these four tumors in the early 1970s. They used freehand ic implants of tumor fragments and were able to obtain reproducible mortality distributions for each of the tumors. They reported the effects of CCNU, BCNU, CPA, mithramycin, MTX, vincristine, and 5-FU [96, 97]. Drugs were administered ip with the schedules varying with the agents, but the first treatment was generally on day two after tumor implantation. BCNU, CCNU, and occasionally CPA were effective, but the others were not. These results reinforced the impression of the potential value of the nitrosoureas suggested by the L1210 experience, adding evidence to the notion of the importance of the blood-brain barrier in brain tumor chemotherapy. However, Shapiro [22] and Levin [20] showed that both MTX and 5-FU accumulated in these tumors, suggesting that mechanisms other than poor drug delivery might also account for the failure of these agents. One of the ependymoblastomas, EpA, was subjected to extensive chemotherapeutic testing by Geran *et al.* [98]. They tested 177 agents against this tumor, confirming the earlier data and documenting significant therapeutic effects of a number of

Table 3. Selected results of the treatment of transplantable murine ependymoblastomas

Tumor	BCNU	CCNU	meCCNU	PCNU	ACNU	PCB[3]	DAG	CPA	Mithra	5-FU	MTX	VM-26[4]	VCR	DTIC	AZQ[3]	ARA-C	References
Ep	+	NT[2]	NT	NT	NT	NT	NT	−, +	−	NT	NT	NT	NT	NT	NT	NT	96
EpA	+, ++	++	++	++	++	++	++	−	−	−	−	NT	−	++	++	−[3]	18, 92, 96, 97, 98
GL 26	+	++	++	++	−, +	NT	+	+	−	−	−	+	−	+	NT	NT	14, 18, 96, 97, 126, 127
GL 261	+	++	+	++	+	NT	NT	+	−	−	−	−	−	NT	NT	+[5]	18, 22, 96, 97, 126, 128

[1] Treatment was on day 2 after tumor implantation, unless otherwise stated. Results were tabulated as follows: % T/C > 250 = ++, % T/C 125 − 250 = +, % T/C < 125 = −.
[2] NT = not tested.
[3] Daily days 1 − 5.
[4] Twice weekly for 2 weeks beginning on day 2.
[5] Every other day × 10 doses.

other agents, including procarbazine, 8-azaguanine, dibromodulcitol, DAG, and meCCNU (Table 3).

A partial listing of other single-agent chemotherapeutic studies of these tumors is provided in Table 3. These experiments were conducted under diverse conditions so the results are often not directly comparable, but an assessment of the activity of a number of drugs was made for comparison. In general, three other nitrosoureas, meCCNU, PCNU, and ACNU, were active against these tumors, although the activity of ACNU against GL 26 was marginal. There were also other instances of different responses among these tumors. DAG, for example, was highly active against EpA, but it was much less effective against GL 26.

Levin and Wilson have reported the activity of a number of nitrosourea-based combinations against GL 26 [99]. Therapeutic results were expressed in terms of the number of long-term survivors (> 120 days) rather than % T/C, so they cannot be compared directly to the results in Table 3, but it was clear that combinations of BCNU-PCB, CCNU-PCB, and BCNU-DAG were more effective than treatment with any single agent. A number of combinations produced 100% long-term survival. These studies indicated that an agent of only marginal activity when used alone, DAG for example, could significantly improve therapeutic results when combined with a nitrosourea.

In a series of studies against an unspecified ependymoblastoma from this group, Laurent et al. [100] and Muller and Tator [101] have demonstrated the benefit of adding amphotericin B to CCNU. The mechanism may be related to enhanced uptake of the cyclohexyl moiety of the drug, although the data are inconclusive. This regimen has not yet been tested against any other experimental brain tumor. Fractionated radiotherapy has also been shown to improve survival of mice bearing ic Ep [102].

The 9L Gliosarcoma

Chemotherapy. Selected results of the chemotherapy of the ic 9L gliosarcoma are listed in Table 4. Four nitrosoureas, BCNU, CCNU, PCNU, and meCCNU, have been shown to significantly prolong survival of rats bearing this tumor [43, 103]. The effectiveness of BCNU has been shown to depend on the timing of its administration. Treatment on day 16 is superior to identical treatment on day 10, a finding the authors attributed to incomplete vascularization of the smaller tumors [104]. DAG was ineffective against 9L [105], in contrast to the results with EpA.

Clonogenic Assay. Rosenblum et al., have refined an *in vivo – in vitro* clonogenic assay of the chemosensitivity of the 9L tumor [106]. Briefly, the CFE of treated ic 9L tumors is determined and compared to that of untreated tumors. Under optimal conditions including irradiated feeder cells and a high concentration of fetal calf serum, the CFE of the untreated tumor is 10–30% [107]. The reduction in CFE is presumed to be a measure of log cell kill and it correlates with animal survival [106].

The surviving fraction of 9L cells after a single dose of BCNU at the LD_{10} is between 10^{-3} and 10^{-4} in this system. The effect is dose-related, and a single large dose of the drug is more effective than an equitoxic series of smaller doses. A calculated cell kill of at least 90% is required to produce a detectable increase in survival [106]. However, an agent which produces less than 90% cell kill, such as 5-FU, produces additional survival prolongation when used in combination with BCNU [108].

Radiotherapy. The radiosensitivity of the 9L tumor has been investigated in some detail. The radiation dose-response curves of both the *in vitro* and *in vivo* tumors have indicated that 9L is a relatively radioresistant tumor with an insignificant hypoxic cell fraction [109]. Nevertheless, single or fractionated doses of irradiation have significantly prolonged the survival of rats bearing the 9L tumor. Barker *et al.*, reported that treatment with a single dose of 2000r on day 16 produced % T/C of 148 and 187 in successive experiments [110]. Wheeler *et al.*, showed that three different fractionation schedules of radiation improved survival with no significant differences among the schedules [111]. Henderson *et al.*, have recently demonstrated a significant dose-response relationship for single radiation doses between 1200r and 1800r [112]. The radiosensitizer misonidazole

Table 4. Selected results of the treatment of intracerebral 9L gliosarcoma

Treatment*	% T/C	Reference
BCNU	160	103
BCNU (day 9)	184	43
BCNU (day 10)	162	104
BCNU	222	104
BCNU	157, 195	110
CCNU	152	103
meCCNU	139	103
PCNU	189	103
DAG	107	105
1800r	182	112
2000r	148, 187	110
BCNU, 2000r	>415, 335, >530 300, 357, 398**	110

* Drugs were administered at the maximum tolerated dose on day 14 or 16 after tumor implantation, unless otherwise noted. Radiation was administered as a single fraction to the whole head.
** Results were obtained by varying the timing of drug administration relative to radiation. There was no detectable schedule dependency.

did not significantly improve the response to radiation, perhaps because of the lack of a hypoxic cell fraction in these tumors [113].

Both Wheeler *et al.* [46] and Barker *et al.* [110] have investigated combinations of BCNU and radiation against 9L. Wheeler found that BCNU significantly improved the mortality distribution compared to either drug or radiation alone and that this effect was highly schedule-dependent. The greatest potentiation was seen when BCNU preceded radiation by 16 hours, a figure which was essentially identical to that observed by the same investigators using 9L *in vitro* [114]. Barker *et al.*, on the other hand, reported that while the combination of BCNU and radiation significantly improved median survival compared to either modality alone, this effect was independent of scheduling. In their series, 17 of 60 animals receiving combined treatment were free of tumor at day 122 or 125 and were considered 'cures' [110]. These authors suggested that schedule dependency of combination therapy may depend on the growth state of the tumor at the time of treatment and that this may account for different results with biologically different 9L tumors. Similarities and differences in the therapeutic responses of these 9L sublines have been discussed recently by Wheeler and Wallen [115].

The ASV Astrocytoma

Chemotherapy and Radiotherapy. For therapeutic studies rats are inoculated at either two or five days of life and randomly assigned to treatment groups one month later, when the average tumor diameter is 2 mm. Results from a number of studies are listed in Table 5. The nitrosoureas, BCNU, PCNU, and CCNU, significantly prolonged survival, while 5-azacytidine and meCCNU had significant but less pronounced effects [6]. Procarbazine was ineffective at the doses used. A transplantable virus-induced dog brain tumor has also been used for chemotherapy experiments, and although it was necessary to use smaller experimental groups, a number of agents produce significant survival prolongation [116].

Irradiation with fractionated doses totalling from 2000r to 5750r also produces significant survival prolongation. The combination of irradiation and BCNU is better than either treatment alone, producing % T/C of from 163 to 188 [117, 118].

Immunotherapy. Observations of the immunological deficiencies of animals bearing ASV-induced gliomas led Mahaley *et al.* to investigate the efficacy of immunotherapy in this model [119, 120]. In a series of studies they demonstrated that both BCG and histoincompatible sarcoma cells which contained ASV-determined cell surface transplantation antigens were ineffective in prolonging survival. The addition of the immunostimulant levamisole to a combination of BCG and sarcoma cells produced a statistically significant increase in survival in

Table 5. Selected results of the treatment of the intracerebral rat avian sarcoma virus astrocytoma

Treatment*	% T/C	Reference
BCNU	131	119
BCNU	117,126	118
BCNU	161	6
CCNU	136	6
PCNU	143	6
meCCNU	119	6
5-azacytidine	124	6
procarbazine	<100	6
2000r	123	118
BCNU, 2000r	164	118
2300r	116	118
BCNU, 2300r	168	118
BCNU, 2300r	163	117
4600r	124,154	117
BCNU, 4600r	188	117
5750r	166	117
BCG, sarcoma cells, levamisole	137	120

* Treatment was initiated 1 month after virus inoculation into neonatal rats. Radiation was administered in equal fractions over 2 – 5 weeks.

1 group of animals (% T/C = 137), but levamisole did not improve the survival of animals treated with a radiotherapy-BCNU combination. These results paralleled the clinical ineffectiveness of a levamisole regimen.

Xenografts

There have been few reports of the treatment of human glioma xenografts in nude mice. Shapiro *et al.*, treated two human gliosarcomas growing intracerebrally in nude mice with a number of agents, including BCNU, meCCNU, PCNU, and PCB [88]. Intravenous administration of BCNU on day 23 to animals bearing tumors derived from patient 'S.R.' increased median survival from 33 to 85 days (% T/C = 258), while meCCNU given under identical conditions produced a % T/C of over 300. PCB, at the doses used, was without effect. In a second tumor (derived from patient 'L.M.') BCNU and PCNU produced only modest results (%T/C = 123 and 137, respectively), while PCB doubled median survival. The authors concluded that they had demonstrated that drugs effective clinically against brain tumors were also effective against human brain tumors growing in nude mide and that different tumors of similar histology responded differently to the same agents. In a subsequent report these investigators illustrated similar variability in the response of sc tumors growing in nude mice and reported an insignificant effect of cis-platinum against two tumors [121].

Bullard et al., reported the results of the treatment of tumors derived from three human glioma cell lines growing subcutaneously in nude mice [84]. The tumors were treated after serial sc passage when reproducible growth curves had been achieved. The three tumor lines responded differently to a single dose of BCNU. All mouse-borne sc tumors derived from U-251 MG regressed after treatment, and the median growth delay to 1 cm^3 tumor volume was 15.0 days. Two of seven tumors derived from U-118 MG regressed, while none of ten from D-54 MG did so. The median growth delays for these two lines were 14.5 days and 3.5 days, respectively. In subsequent studies, these three tumor lines have been treated with two other agents, PCB and mithramycin, and a similar variability of response has been seen. PCB produced tumor regression in 10 of 10 U-251 tumors, eight of 10 D-54 tumors, and two of 10 U-118 tumors. Mithramycin had no significant effect on any of the tumor lines. Similarly, treatment of two transplanted gliomas in this laboratory with the same three agents has produced variable responses. Procarbazine produced tumor regressions in 9 of 10 animals bearing the transplanted human gliosarcoma N-519 and in 2 of 10 animals bearing the human GBM N-456. BCNU produced no tumor regressions in either N-519 or N-456, while mithramycin produced only 1 regression against N-519.

Houchens et al., have treated an intracranially implanted human glioma cell line in nude mice with a series of agents [122]. They found significant survival prolongation only with the nitrosoureas BCNU, CCNU, and PCNU. They have also reported therapeutic synergism between radiation and BCNU in this model (% T/C = 205 and 218) [123].

These preliminary results of the chemotherapy of human gliomas in nude mice indicate that each tumor may have an individual profile of chemotherapeutic sensitivity. The biochemical basis for the sensitivity profile would have important implications for the choice of appropriate chemotherapy, if one assumes that the tumor growing in the animal has retained important biological characteristics. This system also has obvious promise as a pre-clinical screen for potentially active chemotherapeutic agents, although the ability to maintain a large, healthy colony of mice would be essential. Furthermore, the relevance of this system for the treatment of human brain tumors should eventually be demonstrated in correlative studies if the expense and effort required to maintain these animals are to be justified.

Overall Evaluation of Therapeutic Results

A number of factors govern the outcome of the chemotherapy of any tumor. The drug must be effectively delivered on an appropriate schedule to a tumor which has intrinsic sensitivity to its action. Success of any drug against any tumor is limited by these considerations. Unsuccessful results in an individual model therefore require elaboration. The brain tumor models described here differ from one another not only in species, cell of origin, and location, but they have

Table 6. Comparative chemotherapeutic results of selected animal brain tumor models*

%ILS Tumor	BCNU	CCNU	PCNU	ACNU	PCB	MTX	DAG	ARA-C
EpA	224	>476	>400	>500	280	100	>440	133
Gl 26	135	>333	273	132	ND	102	153	ND
Gl 261	235	>400	400	146	ND	104	ND	136
9L	222	ND	189	ND	ND	ND	107	ND
ASV	161	ND	143	ND	<100	ND	ND	ND
L1210	250	234	208	200	114	105	100–200	175

ND = No published data.
*In general, these data represent the optimal results from individual studies. The figures are not strictly comparable in that drugs were administered at different times and on different schedules. For details, consult the individual studies.

different permeability and kinetic characteristics, and they very likely differ in their sensitivity and resistance to the biochemical effects of chemotherapeutic agents.

Each of these models has features which make it useful as a model for brain tumor therapy. Successful treatment of ic L1210 leukemia clearly requires pharmacological characteristics permitting drug entry into the brain, and large numbers of animals can be evaluated in a short time using a well-characterized tumor. The ependymoblastoma group has been extensively characterized with regard to its chemotherapeutic sensitivity, and a number of the drugs which were found effective by Geran et al. [98] have not yet been thoroughly evaluated clinically. Its growth characteristics probably more closely resemble those of human brain tumors than do those of L1210, and the histologic differences between the experimental ependymoblastoma and the most common human brain tumors are not important if the therapeutic information from the model can be applied clinically. The EpA is currently the closest approximation to an active large scale screen for compounds potentially active against brain tumors. The 9L gliosarcoma has the advantages of being in a larger animal and of being extensively characterized in vivo, in monolayer cell culture, and in a multicellular spheroid system. Detailed knowledge of its radiosensitivity, both alone and in combination with nitrosoureas, has already been applied to clinical situations. Its biological instability is only a problem if individual experiments are not well controlled, and its current sarcomatous character may not detract from clinical applications of therapeutic results in it. The therapeutic sensitivities of the ASV astrocytoma have not been as extensively characterized as those of the L1210, Ep, or 9L tumors, but the autochthonous nature of the ASV tumor, the refinement of its production, and its histologic similarity to anaplastic human gliomas make it an extremely valuable experimental tumor. Its responses to nitrosoureas, though significant, have been modest, like the results in the chemotherapy of human gliomas [124]. It is also the only common brain tumor model in which immunotherapy has been extensively investigated. The potential hazards of the virus require expert handling and facilities, but the expense involved is justified considering the advantages. The nude mouse system is in the infancy of its evaluation as a brain tumor model. Its value as a kind of *in vivo* test tube for the cultivation of neoplastic glial cells of human origin is unquestioned, and its potential for the determination of the chemosensitivities of these cells is great. This is also an expensive model, however, and proper containment facilities are required for the maintenance of this fragile species. If fruitful human tumor lines can be established in this system the expense would be justified in institutions with the proper environment.

Proper evaluation of any therapeutic studies of brain tumor models must emphasize comparison with clinical results of therapeutic modalities. Data are still limited on human brain tumor sensitivies, however, and with the exception of the value of the nitrosoureas, none of the therapeutic results in animal tumor systems has yet been translated into effective clinical regimens.

SUMMARY AND CONCLUSIONS

Of the great variety of experimental brain tumors that have been produced, only a handful have been developed to the point of suitability for therapeutic studies. To be useful a tumor should be stable and its biological behavior should be predictable. Reproducible mortality distribution of untreated animals is a prerequisite for controlled studies of treatment modalities. There should also be theoretical justification for relating the experimental tumor to human brain tumors. A number of other characteristics are of uncertain, although potentially great, importance in the evaluation of a brain tumor model. Should the tumor be of neuro-epithelial origin, or is the environment of the tumor of sufficient importance that any of a number of tumors transplanted into the brain are equally valuable as models? How closely should the histology of an experimental tumor mimic that of the most common and malignant primary human brain tumor, the glioblastoma multiforme? What are the most appropriate kinetic parameters in an experimental brain tumor? Can useful therapeutic information be obtained from tumors of brain origin growing in more accessible compartments, such as the subcutaneous space? Is there an advantage to treating human neoplastic glial cells, such as those grown in nude mice, over tumor cells of animal origin? To what extent does the transplantation procedure affect therapeutic results, and are autochthonous tumors more appropriate? What is the importance of the animal species used in relating pharmacologic and therapeutic data to the human situation?

These are unsettled questions, and although a great deal of data are available on the various models which have been used for therapeutic investigations, it has been difficult to relate these data to the human tumor because of the lack of detailed information on the *in situ* human gliomas. It may turn out that each of the models reviewed here as well as others yet to be developed relate to particular aspects of the overall clinical situation in humans.

APPENDIX

Abbreviations of Chemotherapeutic agents used in the Text and Tables

ACNU	= 1-(4-amino-2-methyl-5-pyrimidinyl)methyl-3-(2-chloroethyl)-3-nitrosourea
Ara-C	= cytosine arabinoside, NSC-63878
AZQ	= aziridinylbenzoquinone, NSC-182986
BCNU	= 1,3-bis(2-chloroethyl)-1-nitrosourea, NSC-409962
CCNU	= 1-(2-chloroethyl)-3-cyclohexyl-1-nitrosourea, NSC-79037
CPA	= cyclophosphamide, NSC-26271
DAG	= dianhydrogalactitol, NSC-132313
DTIC	= 5-(3,3-dimethyl-1-triazeno)imidazole-4-carboxamide, NSC-45388
5-FU	= 5-fluorouracil, NSC-19893
meCCNU	= 1-(2-chloroethyl)-3-(4-methyl-cyclohexyl)-1-nitrosourea, NSC-95441
6-MP	= 6-mercaptopurine, NSC-755
MTX	= methotrexate NSC-740
PCB	= procarbazine, NSC-77213
PCNU	= 1-(2-chloroethyl)-3-(2,6-dioxo-3-piperidyl)-1-nitrosourea, NSC-95466
6-TG	= 6-thioguanine, NSC-752
VCR	= vincristine, NSC-49842
VM-26	= teniposide, NSC-122819

ACKNOWLEDGEMENTS

The authors express their appreciation to Dr. James A. Swenberg for reviewing the manuscript.

REFERENCES

1. Law LW, Dunn TB, Boyle PJ, Miller JH: Observations on the effect of a folic-acid antagonist on transplantable lymphoid leukemias in mice. J Natl Cancer Inst 10:179–192, 1949.
2. Skipper HE, Schabel FM Jr, Wilcox WS: Experimental evaluation of potential anticancer agents. XXI. Scheduling of arabinosylcytosine to take advantage of its S-phase specificity against leukemia cells. Cancer Chemother Rep 51:125–165, 1967.
3. Wodinsky I, Kensler CJ, Venditti JM: Comparative kinetics and chemotherapy of the slow growing B16 melanoma and the fast growing L1210 leukemia. Proc Am Assoc Cancer Res 13:8, 1972.
4. Skipper HE, Schabel FM Jr, Trader MW, Thomson JR: Experimental evaluation of potential anticancer agents. VI. Anatomical distribution of leukemic cells and failure of chemotherapy. Cancer Res 21:1154–1164, 1961.

5. Chirigos MA, Humphreys SR: Effect of alkyating agents on meningeal leukemia L1210 arising in methotrexate-treated mice. Cancer Res 26:1673–1977, 1966.
6. Bigner DD, Swenberg JA (eds): Jänish and Schreiber's Experimental Tumors of the Central Nervous System. Upjohn, Kalmazoo, 1977.
7. Zimmerman HM: Experimental models of neoplasia in the central nervous system. In: Primary Intracranial Neoplasms. Sher JH, Ford DH (eds), Spectrum Publications, New York, 1979, pp 1–32.
8. Kleihues P, Bigner DD: Tumors of the nervous system. In: The Molecular Basis of Neuropathology. Thompson RHS, Davison AN (eds), Edward Arnold, London, 1981, pp 81–103.
9. Seligman AM, Shear MJ: Studies in Carcinogenesis. VIII. Experimental production of brain tumors in mice with methylcholanthrene. Am J Cancer 37:364–395, 1939.
10. Zimmerman HM, Arnold H: Experimental brain tumors. I. Tumors produced with methylcholanthrene. Cancer Res 1:919–938, 1941.
11. Zimmerman HM: The nature of glimas as revealed by animal experimentation. Am J Pathol 31:1–29, 1955.
12. Ausman JI, Shapiro WR, Rall DP: Studies on the chemotherapy of experimental brain tumors: development of an experimental model. Cancer Res 30:2394–2400, 1970.
13. Sugiura K: Tumor transplantation. In: Methods of Animal Experimentation, vol. 2, Gay Wi (ed), Academic Press, New York, 1965, pp 171–222.
14. Shapiro WR: Chemotherapy of brain tumors: results in an experimental murine glioma. In: Models of Human Neurological Diseases, Klawans HL (ed), Excerpta Medica, Amsterdam, 1974, pp 121–143.
15. Rubinstein LJ: Correlation of animal brain tumor models with human neuro-oncology. Nat Cancer Inst Monogr 46:43–49, 1977.
16. Rubin R, Sutton CH, Zimmerman HM: Experimental ependymoblastoma (fine structure). J Neuropathol Exp Neurol 27:421–438, 1968.
17. Rubin RC, Ames RP: Mammary tumor virus in experimental ependymoblastoma. Prog neuropathol 2:335–349, 1973.
18. Hasegawa H, Shapiro WR, Posner JB, Basler G: Effect of 1-(4-amino-2-methyl-5-pyrimidinyl) methyl-3-(2-chloroethyl)-3-nitrosourea hydrochloride on experimental brain tumors. Cancer Res 39:2687–2690, 1979.
19. Albright L, Madigan JC, Gaston MR, Houchens DP: Therapy in an intracerebral murine glioma model, using Bacillus Calmette-Guerin, neuraminidase-treated tumor cells, and 1-(2-chloroethyl)-3-cyclohexyl-1-nitrosourea. Cancer Res 35:658–665, 1975.
20. Levin VA, Chadwick M: Distribution of 5-fluorouracil-2-^{14}C and its metabolites in a murine glioma. J Natl Cancer Inst 49:1577–1584, 1972.
21. Levin VA, Clancy TP, Ausman JI, Rall DP: Uptake and distribution of ^{3}H-methotrexate by the murine ependymoblastoma. J Natl Cancer Inst 48:875–883, 1972.
22. Shapiro WR: The effect of chemotherapeutic agents on the incorporation of DNA precursors by experimental brain tumors. Cancer Res 32:2178–2185, 1972.
23. Tator CH: Retention of tritiated methotrexate in a transplantable mouse glioma. Cancer Res 36:3058–3066, 1976.
24. Scheinberg LC, Suzuki K, Edelman F, Davidoff LM: Studies in immunization against a transplantable cerebral mouse glioma. J Neurosurg 20:312–317, 1963.
25. Scheinberg LC, Levine MC, Suzuki K, Terry RD: Induced host resistance to a transplantable mouse glioma. Cancer Res 22:67–72, 1962.
26. Druckrey H, Ivankovic S, Preussmann R, Zulch KJ, Mennel HD: Selective induction of malignant tumors of the nervous system by resorptive carcinogens. In: The Experimental Biology of

Brain Tumors, Kirsch WM, Grossi-Paoletti E, Paoletti P (eds), Charles C. Thomas, Springfield, 1972, pp 85–147.
27. Druckrey H, Ivankovic S, Preussmann R: Teratogenic and carcinogenic effects in the offspring after a single injection of ethylnitrosourea to pregnant rats. Nature 210:1378–1379, 1966.
28. Swenberg JA, Koestner A, Wechsler W, Denlinger RH: Quantitative aspects of transplacental tumor induction with ethylnitrosourea in rats. Cancer Res 32:2656–2660, 1972.
29. Rubinstein LJ: Current concepts in neuro-oncology. Adv Neurol 15:1–25, 1976.
30. Jones EL, Searle CE, Smith WT: Tumours of the nervous system induced in rats by the neonatal administration of n-ethyl-n-nitrosourea. J Path 1109:123–139, 1973.
31. Laerum OD, Rajewsky MF: Sequential phenotypic changes in neural target cell populations. In: Biology of Brain Tumors, Laerum OD, Bigner DD, Rajewsky MF (eds), UICC, Geneva, 1978, pp 129–141.
32. Haugen A: Chemical carcinogenesis and ultrastructure of fetal brain cells in culture. Thesis, University of Bergen Press, Bergen, 1978.
33. Allen N: Experimental therapy of brain tumors. In: Neurochemistry and Clinical Neurology, Battistin L, Hashim G, Lajtha A (eds), A.R. Liss, New York, 1980, pp 441–454.
34. Spence AM, Geraci JP: Combined cyclotron fast-neutron and BCNU therapy in a rat brain-tumor model. J Neurosurg 54:461–467, 1981.
35. Swenberg JA: Treatment of primary brain tumors with CCNU. Proc Am Assoc Cancer Res 14:25, 1973.
36. Denlinger RH, Nichol CA, Cavallito JC, Sigel CW: Chemotherapy of primary brain tumors in rats with two lipid soluble diamino pyrimidine folate antagonists. Proc Am Assoc Cancer Res 17:95, 1976.
37. Herschmann HR: Characteristics of cell culture lines derived from chemically induced neural tumors. In: Biology of Brain Tumors, Laerum OD, Bigner DD, Rajewsky MF (eds), UICC, Geneva, 1978, pp 172–183.
38. Cravioto H, Ransohoff J: Nitrosourea induced gliomas: tissue culture and transplantation studies. In: Experimentelle Neuro-Onkologie, Schreiber D, Janisch W (eds), VEB Broschurenverarbeitung, Leipzig, East Germany, 1975.
39. Chelmicka-Schorr E, Arnason BGW, Holshouser SJ: C-6 glioma growth in rats: suppression with a β-adrenergic agonist and a phosphodiesterase inhibitor. Ann Neurol 8:447–449, 1980.
40. Benda P, Someda K, Messer J, Sweet NH: Morphological and immunochemical studies of rat glial tumors and clonal strains propagated in culture. J Neurosurg 34:310–323, 1971.
41. Schmidek HH, Nielsen SL, Schiller AL, Messer J: Morphological studies of rat brain tumors induced by N-nitrosomethylurea. J Neurosurg 34:335–340, 1971.
42. Wheeler KT, Barker M, Wallen CA, Kimler BF, Henderson SD: Evaluation of 9L as a brain tumor model. In: Methods in Tumor Biology: Tissue Culture and Animal Tumor Models, Sridner R (ed), Marcel Dekker, New York, in press.
43. Barker M, Hoshino T, Gurcay O, Wilson CB, Nielsen SL, Downie R, Eliason J: Development of an animal brain tumor model and its response to therapy with 1,3-bis(2-chloroethyl)-1-nitrosourea. Cancer Res 33:976–986, 1973.
44. Weizaecker M, Deen DF, Rosenblum ML, Hoshino T, Gutin PH, Barker M: The 9L rat brain tumor: description and application of an animal model. J Neurol 224:183–192, 1981.
45. Deen DF, Hoshino T, Williams ME, Muraoka I, Kuebel KD, Barker M: Development of a 9L rat brain tumor cell multicellular spheroid system and its response to 1,3-bis(2-chloroethyl)-1-nitrosourea and radiation. J Natl Cancer Inst 64:1373–1382, 1980.
46. Wheeler KT Jr, Kaufman K, Feldstein M: Response of a rat brain tumor to fractionated therapy with low doses of BCNU and irradiation. Int J Radiat Oncol Biol Phys 5:1553–1557, 1979.
47. Levin VA, Freeman-Dove M, Landahl HD: Permeability characteristics of brain adjacent to tumors in rats. Arch Neurol 32:785–791, 1975.

48. Groothuis DR, Fischer JM, Vick NA, Bigner DD: Comparative permeability of different glioma models to horseradish peroxidase. Cancer Treat Rep, 65 (suppl. 2):13–18, 1981.
49. Groothuis DR, Fischer JM, Lapin G, Bigner DD, Vick NA: Permeability of different experimental brain tumor models to horseradish peroxidase. J Neuropath Exp Neurol 41:164–185, 1982.
50. Vasquez-Lopez E: On the growth of Rous sarcoma inoculated into the brain. Am J Cancer 26:29–55, 1936.
51. Yung WK, Blank WK, Vick NA, Schwartz NA: "Glioblastoma." Induction of a reproducible autochthonous tumor in rat brains with murine sarcoma virus. Neurol 36:76–83, 1976.
52. Johnson L Jr, Wolfe LG, Whisler WW, Norton T, Thakkar B, Deinhardt F: Induction of gliomas in marmosets by Simian sarcoma virus, type 1 (SSV-1). Proc Am Assoc Cancer Res 16:119, 1975.
53. Walker DL, Padgett BL, ZuRhein GM, Albert AE: Human papovavirus (JC): induction of brain tumors in hamsters. Science 181:674–676, 1973.
54. Rieth KG, KiChiro G, London WT, Sever JL, Houff SA, Kornblith PL, McKeever PE, Buonomo C, Padgett BL, Walker DL: Experimental gliomas in primates: a computed tomography model. J Comput Assist Tomog 4:285–290, 1980.
55. Bigner DD: Role of viruses in the causation of neural neoplasia. In: Biology of Brain Tumors, Laerum OD, Bigner DD, Rajewsky MF (eds), UICC, Geneva, 1978, pp 85–111.
56. Bigner DD, Pegram CN: Virus-induced experimental brain tumors and putative associations of viruses with human brain tumors: a review. Adv Neurol 15:57–83, 1976.
57. Rabotti GF, Raine WA: Brain tumors induced in hamsters inoculated intracerebrally at birth with Rous sarcoma virus. Nature 204:898–899, 1964.
58. Rabotti GF, Grove AS Jr, Sellers RL, Anderson WR: Induction of multiple brain tumors (gliomas and leptomeningeal sarcomata) in dogs by Rous sarcoma virus. Nature 209:884–886, 1966.
59. Vick NA, Bigner DD, Kvedar JP: The fine structure of canine gliomas and intracranial sarcomas induced by the Schmidt-Ruppin strain of the Rous sarcoma virus. J Neuropath Exp Neurol 30:354–367, 1971.
60. Copeland DD, Vogel FS, Bigner DD: The induction of intracranial neoplasms by the inoculation of avian sarcoma virus in perinatal and adult rats. J Neuropath Exp Neurol 34:340–358, 1975.
61. Bigner DD, Self DJ, Frey J, Ishizaki R, Langlois AJ, Swenberg JA: Refinement of the avian oncornavirus-induced primary rat brain tumor model for therapeutic screening. Rec Res Cancer Res 51:20–34, 1975.
62. Wilfong RF, Bigner DD, Self DJ: Brain tumor types induced by the Schmidt-Ruppin strain of Rous sarcoma virus in inbred Fischer rats. Acta Neuropath 25:196–206, 1973.
63. Copeland DD, Bigner DD: Glial-mesenchymal tropism of in vivo avian sarcoma virus neuro-oncogenesis in rats. Acta Neuropath 41:23–25, 1978.
64. Copeland DD, Bigner DD: Influence of age at inoculation on avian oncornavirus-induced brain tumor incidence, tumor morphology, and postinoculation survival in F344 rats. Cancer Res 37:1657–1661, 1977.
65. Bigner DD, Kvedar JP, Shaffer TC, Vick NA, Engel WK, Day ED: Factors influencing the cell type of brain tumors induced in dogs by Schmidt-Ruppin Rous sarcoma virus. J Neuropath Exp Neurol 31:583–595, 1972.
66. Bigner DD, Pegram CN, Vick NA, Copeland DD, Swenberg JA: Characterization of the avian sarcoma virus induced mammalian brain tumors model for immunologic and chemotherapeutic studies. VIIth International Congress of Neuropathology. Excerpta Medica, Amsterdam, 1975, pp 445–451.
67. Burger PC, Bigner DD, Self DJ: Morphologic observations of brain tumors in PD4 hamsters induced by four strains of avian sarcoma virus. Acta Neuropathol 26:1–21, 1973.
68. Copeland DD, Talley FA, and Bigner DD: The fine structure of intracranial neoplasms induced

by the inoculation of avian sarcoma virus in neonatal and adult rats. Am J Pathol 83:149–176, 1976.
69. Bigner DD, Vick NA, Kvedar JP, Mahaley MS Jr, and Day ED: Virus-cell relationships in dog brain tumors induced with Schmidt-Ruppin Rous Sarcoma virus. Progr Exp Tumor Res 17:40–58, 1972.
70. Vick NA, Bigner DD: Microvascular abnormalities in virally-induced canine brain tumors: structural bases for altered blood-brain barrier function. J Neurol Sci 17:29–39, 1972.
71. Groothuis DR, Molnar P, Blasberg R, Fenstermacher J, Patlak C, Bigner DD, Fischer J, Vick N: Quantitative autoradiographic measurements of blood flow and permeability in virally induced gliomas. Neurol 31, Part 2:44, 1981 (abstract).
72. Pasternak JF, Groothuis DR, Fischer JM, Bigner DD, Vick NA: Topologic variability of cellular proliferation in experimental rat gliomas. Neurol 31, Part 2:44, 1981 (abstract).
73. Roszman TL, Brooks WH, Markesbery WR, Bigner DD: General immunocompetence of rats bearing avian sarcoma virus-induced intracranial tumors. Cancer Res 38:74–77, 1978.
74. Adams DO, Gilbert RW, Bigner DD: Cellular immunity in rats with primary brain tumors: inhibition of macrophage migration by soluble extracts of avian sarcoma virus-induced tumors. J Natl Cancer Inst 56:1119–1123, 1976.
75. Meyers ME, Roszman TL, Brooks WH, Bigner DD: Experimental allergic encephalomyelitis in rats with primary gliomas induced with avian sarcoma virus. Fed Proc 37:413, 1978.
76. Greene HSN, Arnold H: The homologous and heterologous transplantation of brain and brain tumors. J Neurosurg 2:315–331, 1945.
77. Greene HSN: The transplantation of human brain tumors to the brains of laboratory animals. Cancer Res 13:422–426, 1953.
78. Wilson CB, Barker M: Intrathecal transplantation of human neural tumors to the guinea pig. J Natl Cancer Inst 41:1229–1240, 1968.
79. Pantelouris EM: Absence of thymus in a mouse mutant. Nature 217:370–371, 1968.
80. Giovanella BC, Stehlin JS Jr, Williams LJ Jr, Lee S-S, Shepard RC: Heterotransplantation of human cancers into nude mice: a model system for human cancer chemotherapy. Cancer 42:2269–2281, 1978.
81. Povlsen CO, Visfeldt J, Rygaard J, Jensen G: Growth patterns and chromosome constitutions of human malignant tumours after long-term serial transplantation in nude mice. Acta Path Microbiol Scand Sect A 83:709–716, 1975.
82. Fogh J, Tiso J, Orfeo T, Sharkey FE, Daniels WP, Fogh JM: Thirty-four lines of six human tumor categories established in nude mice. JNCI 64:745–751, 1980.
83. Bigner DD, Markesbery WR, Pegram CN, Westermark B, Ponten J: Progressive neoplastic growth in nude mice of cultured cell lines derived from anaplastic human gliomas. J Neuropathol Exp Neurol 36:593, 1977.
84. Bullard DE, Schold SC Jr, Bigner SH, Bigner DD: Growth and chemotherapeutic response in athymic mice of tumors arising from human glioma-derived cell lines. J Neuropath Exp Neurol 40:410–427, 1981.
85. Wara WM, Begg A, Phillips TL, Rosenblum ML, Vasquez D, Wilson CB: Growth and treatment of human brain tumors in nude mice – preliminary communication. In: The Use of Athymic (Nude) Mice in Cancer Research, Houchens DP, Ovejera AA (eds), Gustav Fischer, New York, 1978, pp 251–256.
86. Rana MW, Pinkerton H, Thornton H, Nagy D: Heterotransplantation of human glioblastoma multiforme and meningioma to nude mice. Proc Soc Exp Biol Med 155:85–88, 1977.
87. Reid LM, Holland J, Jones C, Wolf B, Niwayama G, Williams R, Kaplan NO, Sato G: Some of the variables affecting the success of transplantation of human tumors into the athymic nude mouse. In: The Use of Athymic (Nude) Mice in Cancer Research, op. cit., pp 107–121.
88. Shapiro WR, Basler GA, Chernik NL, Posner JB: Human brain tumor transplantation into nude mice. J Natl Cancer Inst 62:447–453, 1979.

89. Schold SC Jr, Bullard DE, Bigner SH, Jones TR, Bigner DD: Growth, morphology, and serial transplantation of anaplastic human gliomas in athymic mice. J Neuro-Oncology (in press).
90. Jones TR, Bigner SH, Schold SC Jr, Eng LF, Bigner DD: Anaplastic human gliomas grown in athymic mice: morphology and glial fibrillary acidic protein expression. Am J Pathol 105:316–327, 1981.
91. Schabel FM Jr, Johnston TP, McCaleb GS, Montgomery JA, Laster WR, Skipper HE: Experimental evaluation of potential anticancer agents. VIII. Effects of certain nitrosoureas on intracerebral L1210 leukemia. Cancer 23:725–733, 1963.
92. Khan AH, Driscoll JS: Potential central nervous system antitumor agents. Aziridinylbenzoquinones 1. J Med Chem 19:313–317, 1976.
93. Schabel FM Jr, Laster WR Jr, Trader MW, Corbett TH, Griswold DP Jr: Combination chemotherapy with nitrosoureas plus other anticancer drugs against animal tumors. In: Nitrosoureas – Current Status and New Developments, Prestayko AW, Crooke ST, Baker LH, Carter SK, Schein PS, (eds), New York, Academic Press, 1981, pp 9–26.
94. Eagan RT, Childs DS Jr, Layton DD Jr, Laws ER Jr, Bisel HF, Holbrook MA, Fleming TR: Dianhydrogalactitol and radiation therapy: treatment of supratentorial glioma. J Am Med Assoc 241:2046–2050, 1979.
95. Green SB, Byar DP, Walker MD, Pistenma DA, Alexander E Jr, Batzdorf U et al.: Comparisons of carmustine, procarbazine, and high-dose methylprednisolone as additions to surgery and radiotherapy for the treatment of malignant glioma. Cancer Clin Trials, (in press).
96. Shapiro WR, Ausman JI, Rall DP: Studies on the chemotherapy of experimental brain tumors: evaluation of 1,3-bis(2-chloroethyl)-1-nitrosourea, cyclophosphamide, mithramycin, and methotrexate. Cancer Res 30:2401–2413, 1970.
97. Shapiro WR: Studies on the chemotherapy of experimental brain tumors: evaluation of 1-(2-chloroethyl)-3-cyclohexyl-1-nitrosourea, vincristine, and 5-fluorouracil. J Natl Cancer Inst 46:359–368, 1971.
98. Geran RI, Congleton GF, Dudeck LE, Abbott BJ, Gargus JL: A mouse ependymoblastoma as an experimental model for screening potential antineoplastic drugs. Cancer Chemother Rep Part 2, 4:53–87, 1974.
99. Levin VA, Wilson CB: Correlations between experimental chemotherapy in the murine glioma and efectiveness of clinical therapy regimens. Cancer Chemother Pharmacol 1:41–48, 1978.
100. Laurent G, Dewerie-Vanhouche J, Machin D, Hildebrand J: Inhibition of RNA synthesis in murine ependymoblastoma by the combination of amphotericin B and 1-(2-chloroethyl)-3-cyclohexyl-1-nitrosourea. Cancer Res 40:939–942, 1980.
101. Muller PJ, Tator CH: The effect of amphotericin B on the survival of brain tumor-bearing mice treated with CCNU. J Neurosurg 49:579–588, 1978.
102. Edelman FL, Aleu F, Scheinberg LC, Evans JC, Davidoff LM: Effect of irradiation on the growth of intracerebral gliomas in mice. J Neuropath Exp Neurol 23:1–17, 1964.
103. Levin VA, Kabra P: Effectiveness of the nitrosoureas as a function of their lipid solubility in the chemotherapy of experimental rat brain tumors. Cancer Chemother Rep 58:787–792, 1974.
104. Tel E, Barker M, Levin VA, Wilson CB: Effect of combined surgery and BCNU (NSC-409962) on an animal brain tumor model. Cancer Chemother Rep 58:627–631, 1974.
105. Levin VA, Freeman-Dove MA, Maroten CE: Dianhydrogalactitol (NSC-132313): pharmacokinetics in normal and tumor-bearing rat brain and antitumor activity against three intracerebral rodent tumors. J Natl Cancer Inst 56:535–539, 1976.
106. Rosenblum ML, Wheeler KT, Wilson CB, Barker M, Knebel KD: In vitro evaluation of in vivo brain tumor chemotherapy with 1,3-bis(2-chloroethyl)-1-nitrosourea. Cancer Res 35:1387–1391, 1975.
107. Rosenblum ML, Deen DF, Hoshino T, Dougherty DA, Williams ME, Wilson CB: Comparison of clonogenic cell assays after in vivo and in vitro treatment of 9L gliosarcoma. Br J Cancer 41, Suppl IV:307–308, 1980.

108. Rosenblum ML, Hoshino T, Levin VA, Wilson CB: Planning of brain tumor therapy based on laboratory investigations: sequential BCNU 5-FU treatment of malignant gliomas. Presented at the 50th meeting of the American Association of Neurological Surgeons, Boston, April 5–9, 1981.
109. Leith JT, Schilling WA, Wheeler KT: Cellular radiosensitivity of a rat brain tumor. Cancer 35:1545–1550, 1975.
110. Barker M, Deen DF, Baker DG: BCNU and x-ray therapy of intracerebral 9L rat tumors. Int J Radiat Oncol Biol Phys 5:1581–1583, 1979.
111. Wheeler KT, Kaufman K: Influence of fractionation schedules on the response of a rat brain tumor to therapy with BCNU and radiation. Int J Radiat Oncol Biol Phys 6:845–849, 1980.
112. Henderson SD, Kimler BF, Morantz RA: Radiation therapy of 9L rat brain tumors. Int J Rad Oncol Biol Phys 7:497–502, 1981.
113. Reddy EK, Kimler BF, Henderson SD, Morantz RA: Combinations of misonidazole and radiation therapy in a rat brain tumor (9L) system. In: Radiation Sensitizers, Brady LW (ed), Masson Publishing, USA, New York, 1980, pp 457–460.
114. Wheeler KT, Deen DF, Wilson CB, Williams ME, Sheppard S: BCNU-modification of the in vitro radiation response in 9L brain tumor cells of rats. Int J Radiat Oncol Biol Phys 2:79–88, 1977.
115. Wheeler KT, Wallen CA: Is cell survival a determinant of the in situ response of 9L tumors? Br J Cancer 41, Suppl IV: 299–303, 1980.
116. Merker PC, Wodinsky I, Geran RI: Review of selected experimental brain tumor models used in chemotherapy experiments. Cancer Chemother Rep Part 1, 59:729–736, 1975.
117. Steinbok P, Mahaley MS Jr, U R, Varia MA, Lipper S, Mahaley J, Dalzell JG, Bigner DD: Treatment of autochthonous rat brain tumors with fractionated radiotherapy: the effects of graded radiation doses and of combined therapy with BCNU or steroids. J Neurosurg 53:68–72, 1980.
118. Steinbok P, Mahaley MS Jr, U R, Zinn DC, Lipper S, Mahaley JL, Bigner DD: Synergism between BCNU and irradiation in the treatment of anaplastic gliomas: an in vivo study using the avian sarcoma virus-induced glioma model. J Neurosurg 51:581–586, 1979.
119. Mahaley MS Jr, Gentry RE, Bigner DD: Immunobiology of primary intracranial tumors. Part 2: The evaluation of chemotherapy and immunotherapy protocols using the avian sarcoma virus glioma model. J Neurosurg 47:35–43, 1977.
120. Mahaley MS Jr, Steinbok P, Aronin P, Dudka L, Zinn D: Immunobiology of primary intracranial tumors. Part 4: Levamisole as an immune stimulant in patients and in the ASV glioma model. J Neurosurg 54:220–227, 1981.
121. Shapiro WR, Basler GA: Chemotherapy of human brain tumors transplanted into nude mice. In: Multidisciplinary Aspects of Brain Tumor Therapy, Paoletti P, Walker MD, Butti G, Knerich R (eds), Elsevier/North Holland, Amsterdam, 1979, pp 309–316.
122. Houchens DP, Ovejera AA, Riblet SM: Human brain tumor xenografts is nude mice as a chemotherapy and/or immunotherapy model. Eur J Cancer, in press.
123. Slagel DE, Feola J, Houchens DP, Ovejera AA: Combined modality treatment using radiation and/or chemotherapy in an athymic (nude) mouse-human medulloblastoma and glioblastoma xenograft model. Cancer Research 42:812–816, 1982.
124. Walker MD, Green SB, Byar DP, Alexander E Jr, Batzdorf U, Brooks WH et al.: Randomized comparisons of radiotherapy and nitrosoureas for the treatment of malignant glioma after surgery. N Engl J Med 303:1323–1329, 1980.
125. Wodinsky I, Merker PC, Venditti JM: Responsiveness to chemotherapy of mice with L1210 lymphoid leukemia implanted in various anatomic sites. J Natl Cancer Inst 59:405–408, 1977.
126. Shapiro WR: The chemotherapy of intracerebral vs. subcutaneous murine gliomas: a comparative study of the effect of VM-26. Arch Neurol 30:222–226, 1974.
127. Levin VA, Crafts D, Wilson CB, Kabra P, Hansch C, Boldrey E, Enot J, Neely M: Imidazole

carboxamides: relationship of lipophilicity to activity against intracerebral murine glioma 26 and preliminary phase II clinical trial of 5-(3,3-bis(2-chloroethyl)-1-triazeno)imidazole-4-carboxamide (NSC-82196) in primary and secondary brain tumors. Cancer Chemother Rep, Part 1, 59:327 – 331, 1975.
128. Basler GA, Shapiro WR: Comparison of 2,2'-anhydro-1-^{b}-D-arabinofuranosyl-5-fluorocytosine and cytosine arabinoside in the treatment of murine brain tumor. Cancer Treat Rep 60:875 – 879, 1979.

Note: This chapter includes material available as of January 29, 1982.

3. Chemotherapy of Brain Tumors – Basic Concepts

WILLIAM R. SHAPIRO and THOMAS N. BYRNE

1. INTRODUCTION

The chemotherapy of cancer differs from the treatment of bacterial infections in one major way: In cancer, the target of therapy is the living tissues of the host; the toxicity accruing as a side effect of therapy comes about from the same mechanisms that make the treatment efficacious. In the therapy of bacterial infections the treatment is directed against the bacterium and the side effects are usually incidental chemical reactions with host tissue. The therapeutic index, i.e. the ratio between the efficacy and the toxicity, is much higher for treatments directed against bacteria than it is for treatments directed against cancer. Thus, any consideration of anticancer therapy requires that the therapist pay even stricter attention to the therapeutic index than he might in treating bacterial infections.

To improve the therapeutic index for cancer treatment, one may either reduce the toxicity or make the treatment more specific to the target. Unfortunately, those aspects-chemical, kinetic, genetic-that might make chemotherapy more specific have been so elusive that we are forced to use toxic chemicals whose effects on the target tumor tissue are qualitatively identical to those on normal tissues and quantitatively so little different that the therapeutic index often approaches 1.

While the above considerations apply to all cancer, the therapy of brain tumors is more complicated. The nature of the tumor and the problems of drug delivery impose additional difficulties on the chemotherapist. It is to these aspects that we will direct most of our attention.

The terms, benign and malignant, must be defined differently for primary brain tumors than for systemic cancer. An intracranial tumor such as a meningioma is benign in the classical sense in that it is a slow-growing, histologically benign neoplasm with few mitoses, little necrosis or vascular hyperplasia and does not metastasize. However, a meningioma is often silent for many years, and when it is clinically recognized it may be so large as ultimately to prove fatal. It is difficult to call such a tumor 'benign'. Conversely, a glioblastoma multiforme is histologically one of the most malignant tumors. It is pleomorphic, induces considerable vascular hyperplasia (occasionally to the point of sarcomatous change), undergoes necrosis, induces surrounding cerebral edema, infiltrates into the surrounding brain and is relatively rapidly growing. Although these tumors rarely metastasize, they are almost always impossible to cure by surgical resection. Furthermore, when they prove fatal, they are much smaller than systemic cancers.

Another distinction between systemic cancer and brain tumors is the lethal tumor size. When a patient develops systemic cancer his body must deal with a 'cancer burden' consisting of a population of cancer cells in the blood or bone marrow (leukemia) or in a solid single mass or multiple metastatic masses (carcinoma or sarcoma). For the average adult it is thought that a 1 kg burden of systemic tumor is lethal. In contrast, an intracranial neoplasm shares its space with the brain which already occupies 1200 – 1400 cc in the skull. In this confined space 100 g of tumor and associated edema is almost always lethal. By the time a patient develops neurological symptoms the tumor is usually 30 – 60 g in size. The lethality of a brain tumor is thus more dependent on its size than on its histologic grade of malignancy, and the distinction between benign and malignant tumors is less important for intracranial tumors than it is for systemic cancer.

Two aspects of intracranial tumors which make them different from systemic cancer in ways that influence the potential for chemotherapy are the blood-brain barrier and cerebral edema. The blood-brain barrier is a physiologically and anatomically defined entity that prevents certain chemical species from crossing from the vascular space into the normal brain. Such barriers are not limited to the brain, however, as a similar barrier occurs in the testis [1]. We are concerned with the changes of the blood-brain barrier as a tumor grows and the influence of the intact and partially-intact barrier on the entry of chemotherapeutic agents. Cerebral edema occurs because the brain does not contain a lymphatic system. The normal transudate that is formed across cerebral capillary membrane enters the extracellular space of the brain and is returned to the vascular tree by the less efficient cerebrospinal fluid system. As a brain tumor develops, it converts normal brain blood vessels with their tight junctions into leaky tumor vessels that permit a marked increase in the amount of transudate. This extra fluid, so called 'cerebral edema', contributes to the mass of the tumor and, therefore, to the clinical signs and symptoms the tumor produces. There are several modes of treatment available to reduce cerebral edema, e.g. corticosteroid hormones and osmotic diuretics, that do not affect the tumor burden itself, and it is necessary to differentiate their effects from those following directed antitumor treatment. Cerebral edema also affects chemotherapy more directly. We shall deal with these aspects below.

Finally, in considering the efficacy of chemotherapy, we must consider the measurement of the response itself. Such measurements must be indicative of changes in tumor burden if they are to tell us anything about the relationship of the therapy to the tumor. In the therapy of malignant brain tumors, the two evaluable variables are the clinical state of the patient and the size of the tumor as seen by CT scan. Unfortunately, a number of factors affect the clinical state other than the size of the tumor, such as edema, seizures, fixed neurological deficits and damage from surgery or radiation therapy. Furthermore, the CT scan defines the minimal size of the tumor itself, failing to visualize tumor that has not disturbed the barrier. Thus, it is often difficult to define the results of chemotherapy in brain tumors by the clinical state of the patient, by the CT scan or even by both.

Most investigations of brain tumor chemotherapy utilize survival as that endpoint which is most easily measured and most likely to be related to the patient's tumor burden. Recent results from the Brain Tumor Study Group (BTSG) demonstrate a number of confounding variables related to survival. For example, in one such study [2], several variables significantly affected survival. Patients younger than 45 years of age lived longer than patients over 65 years of age irrespective of therapy. The death rate (number of deaths/10 patient-months) for patients under 45 years of age was 0.42, while for patients older than 65 years of age the death rate was 1.41. This was a highly significant difference. Other important prognostic indicators included personality change, duration of symptoms, performance status postoperatively, and the pathologic diagnosis, i.e. whether the tumor was a glioblastoma or a grade 3 astrocytoma. A subsequent BTSG study [3] demonstrated several more variables that affected survival. Any consideration of chemotherapeutic trials requires an awareness of such variables.

In this review we shall first examine aspects of the tumor that affect the way chemotherapeutic agents interact with the tumor cells. We shall next examine the influence of the blood-brain barrier and the presence of cerebral edema as they interact with chemotherapeutic agents. We shall then review the drugs that have been used in brain tumor chemotherapy and conclude with a review of the systems used to test chemotherapeutic agents for use in malignant glioma patients.

2. CELLULAR ASPECTS OF BRAIN TUMOR CHEMOTHERAPY

Since most of the experience with brain tumor chemotherapy has been with the malignant gliomas, we shall confine our remarks to these tumors. The future may well permit comments about chemotherapy of more benign neuroectodermal tumors, but the malignant astrocytoma or glioblastoma multiforme is an invariably fatal disease and constitutes today's major challenge for brain tumor chemotherapy.

Characterization of human intracranial tumors has improved in recent years by the use of increasingly sophisticated techniques. Such studies indicate that intracranial neoplasms are similar in a number of ways to tumors found elsewhere in the body. Among these characteristics are the kinetics of tumor growth, biochemical properties including a number of specific or associated protein and lipid constituents, and immunological phenomena associated with the tumors. In such studies there has been a tendency to regard the neoplasm as a single, homogeneous pathological entity. However, recent observations from the laboratory indicate that malignant gliomas are not homogeneous in cellular composition, but rather are composed of heterogeneous cell populations. Tumor heterogeneity has increasingly been demonstrated in animal tumors and this heterogeneity impacts on studies of other characteristics of the tumor cells [4]. Any understanding of the principles of chemotherapy requires a consideration of differences among tumor cells within a patient's tumor population and of differences among the tumors of different patients.

Differences among cell populations within tumors have been noted in a variety of studies in both animals and humans. One of the most obvious characteristics by which tumor cells may differ is morphology. An extreme example of intratumor structural variability is teratocarcinoma, a neoplasm in which representatives of any type of adult somatic tissue as well as embryonal malignant cells may be found [5]. In other investigations directed at the question of heterogeneity, several distinct cellular populations have been isolated from a single mouse-mammary-tumor-virus (MMTV)-associated mammary tumor [6], and similar populations have been isolated from the ethylnitrosourea-induced rat neuro tumor RT4 [7]. Most of the subpopulations of cells maintain their distinctive cellular morphology in multiple passages *in vitro*, and demonstrate karyotypic diversity with karyotypes of isolated mammary tumor populations ranging from diploid to hyperploid. This difference in karyotypic deviations suggests a genetic basis for the heterogeneity. As noted below, such karyotypic diversity is also present in human malignant gliomas. In addition, heterogeneity may be demonstrated in the differences in melanin content of hamster melanomas, and in hormone-responsive mouse mammary tumors containing mixed populations of high estrogen and progesterone receptor positive, hormone-dependent cells and low or negative receptor cells [4]. The MMTV-associated mammary tumor can be divided into subpopulations of cells which express different MMTV-coated cell surface antigens [6]. One of the most profound examples of tumor heterogeneity was shown by Fidler and Kripke [8]. These investigators demonstrated that animal tumors can be composed of subpopulations of cells with diverse metastatic potential. The demonstration of tumor subpopulations differing in such fundamental neoplastic properties as invasiveness and metastasis provides a framework for the observed ability of cancer to change during the course of disease. For our purposes, perhaps the most important characteristic of animal tumor heterogeneity is the heterogeneous chemotherapeutic responsiveness of subpopulations of tumor cells. Such heterogeneity has been demonstrated in the MMTV-associated tumor lines by Heppner et al. [9].

Tumor heterogeneity in human cancer has been demonstrated recently in a number of instances. For example, tumors removed from metastatic sites may bear little morphological resemblance to the primary tumor. Metastatic tumors may grow at different rates and be more or less malignant than the primary tumor. Baylin et al. [10] demonstrated differences in enzyme-containing activity in metastatic tumors from different sites examined at autopsy in patients. Siracky [11] noted differences in cell sensitivity in vitro to a variety of commonly-used drugs between cell suspensions of primary and metastatic ovarian carcinoma. Estrogen receptor activity has been shown to be different in primary breast cancers, recurrences and metastases in the same patient [12]. Even within the same tumor, one can find antigenic differences among human cell populations taken from different locations at operation [13]. Other differences have been found among protein constituents using electrophoresis [14], membrane-associated antigens of human malignant melanomas [15] and chemosensitivity differences among cell populations derived from a single melanoma nodule [16]. Such differences have also been found in limited studies in the chromosomal population of tumors derived from ovarian adenocarcinoma [17].

Heterogeneity of human brain tumors is suggested by their pleomorphic histologic appearance. Rubinstein pointed out that malignant gliomas not infrequently demonstrate fields that suggest a proliferation of separate clones of anaplastic cells [18]. The term 'glioblastoma multiforme' emphasizes the remarkable diversity of cells that morphologically comprise that tumor. Many gliomas contain fields of benign tumor reflecting the frequent clinical progression from a slowly to a rapidly growing neoplasm. One common mixed glioma is the oligodendroglioma-astrocytoma, which contains fields of neoplastic oligodendroglial cells alternating or interdigitating with areas of astrocytoma cells. The most extreme example of such an admixture of cell types is found in the gliosarcoma with its complex of malignant glial and mesenchymal cells of disparate pattern and rate of growth.

Our discussion has involved heterogeneity of a tumor at a given time, but heterogeneity also implies progressive change over time and raises the question of the origin and mechanism of such heterogeneity. Does tumor heterogeneity imply heteroclonal origin [19]? The answer for leukemias is no; they are monoclonal in origin. However, this question is difficult to address for solid tumors and has not been examined in brain tumors. Our discussion is limited to those instances in which different populations of cells are found at surgery or autopsy.

As noted below, we performed experiments in which human gliomas were implanted intracerebrally and subcutaneously into nude mice. Some of our studies demonstrated different chemosensitivities for tumors derived from individual patients [20]. One possible explanation for this differential chemosensitivity was that contained within each patient's tumor were cells that grew readily in the nude mouse and were incidentally sensitive to different drugs. With the background of the literature described above and our own observations in nude mice, we began to examine the possibility that human glioblastoma is composed of heterogeneous populations of cells. Recently, Bigner *et al.* [21] demonstrated heterogeneity among cell lines derived from human gliomas; however, such lines usually contain variants or mutant cells newly generated in culture and may bear little resemblance to the stem or substem population of cells present at the time of resection. Therefore, they do not adequately provide information on the heterogeneity of the brain tumors from which they were derived.

In order to study early-passage tumor cells and their clones, we developed methods to determine the extent of variability actually existing among cell types within the same glioma and among cell types from different gliomas [22]. Tumor tissue obtained from previously untreated patients undergoing resection were immediately dissociated mechanically into single cells which were then karyotyped and dilution plated *in vitro*. The immediate karyotyping of dissociated tumor cells permitted the determination of the chromosomal complement for each tumor. Further, clones established from the single dissociated cells could be identified by their karyotype as cellular representatives of the brain tumor. Karyotyping thus characterized the clonal populations and permitted the determination of the extent of cellular heterogeneity, the evolution of each cell type and the behavior of individual clones versus mixed population of clones in culture. In the initial series of investigations 8 tumors were analyzed [22]. Chromosomal complements of each cell type ranged from near diploid (2n) to hypo- or hypertetraploid (4n) in chromosome number. The frequency distribution of these tumor lines is illustrated in Figure 1. While cellular heterogeneity has been noted previously among astrocytomas [23], the new studies used banding techniques that permitted an analysis of individual cells differing in specific chromosomal complement but not necessarily in chromosomal number. This analysis demonstrated that cells with the same modal number often contained two or more subpopulations, implying extensive variability in the cell chromosomal complements. Furthermore, the distribution of karyotypes varied within each tumor. Some tumors were predominantly aneuploid, others were predominantly hypodiploid, and the remaining were some combination of both. These observations indicated that human gliomas are not only heterogeneous in their cell populations but are different from each other.

Cells with so many combinations of chromosomes will almost certainly show altered gene expression and should demonstrate phenotypic variability. In subsequent examinations, clones grown from the initial dissociated cells and identified by karyotype as having been present in the patient's original tumor, were characterized phenotypically by examining certain protein constituents, cell growth and chemosensitivity. Initial studies [24] demonstrated clonal variations

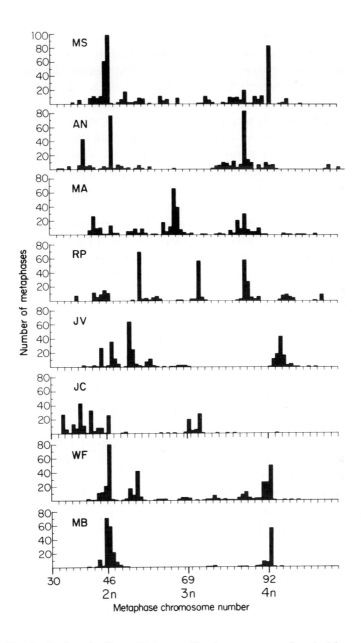

Figure 1. The distribution of cells containing specific chromosome numbers in 8 human gliomas. Each glioma represents 250 or more metaphases. Three to 21 karyotypically distinct clonal populations were identified from each glioma. Thus, each tumor is heterogeneous and differs from all of the others. (Reproduced with permission [22]).

in cell kinetics, glial fibrillary acidic protein (GFAP) and chemosensitivity. The last was very important because it implied that heterogeneity impinges on therapy. By microcytotoxicity assay, clones from individual tumors were subjected to different concentrations of the drugs BCNU or cisplatin. Heterogeneity of dose-response curves was demonstrated with ED_{50}'s ranging from 45 to 94 µg/ml for BCNU and 60 to 120 µg/ml for cisplatin. As noted elsewhere in this volume, such concentrations of drugs are some ten-fold higher than are likely to be achieved *in vivo* in patients. In an effort to examine the problems of heterogeneity of chemosensitivity response at doses achievable in patients, we utilized a colony forming assay [25]. Clones from other gliomas were subjected in this assay to doses of BCNU or cisplatin, and again, heterogeneity of chemosensitivity was seen. The doses were considerably lower; the differences in ED_{50}'s among the clones varied from 4.5 to 7.0 µg/ml for BCNU and from 0.35 to 1.4 µg/ml for cisplatin. Because the drug doses spanned a narrow range it was crucial to estimate the differences among the clonal populations at a statistically significant level and this was achieved. The differences were small, however, probably reflecting the fact that these tumors were derived at surgery from patients who had not had therapy with either of the drugs. The experiments, however, demonstrated a mechanism for the development of resistant cell populations. If drug levels achievable within a tumor fail to be high enough to kill all cell populations, relatively resistant populations would be spared. Preliminary work by Rankin et al. [26] demonstrated the rapid development of highly resistant clonal populations when human glioma cells were subjected to BCNU exposure at 1µg/ml. The rapidity with which the resistance developed could be increased by stepwise increase in drug concentrations. Thus, the inherent heterogeneous chemosensitivities within gliomas may account for the later development of a highly resistant tumor.

Other studies from our laboratory have demonstrated that the cell populations may change as they are maintained *in vitro*. However, the rate of change is similarly heterogeneous [27]. It is possible to demonstrate relatively stable cell populations that maintain their original karyotypes for up to 12 passages and very unstable cell populations whose karyotypes shift with each passage. Such studies indicate that if the biology and therapy of gliomas is to be examined *in vitro* in a way that relates to the tumor's presence in a human, they must be studied in early passage. This concept bears directly on the examination of human gliomas for chemosensitivity (see below).

Human gliomas have been characterized in the laboratory in a number of ways, although usually without considering the problems of heterogeneity. It is worth reviewing these studies with the assumption that in part they represent first approximations of what may be occurring within the patient.

The rate of growth of human brain tumors has been studied by a number of investigators [28]. Highly malignant glioblastomas tend to have higher labeling indices than do more differentiated anaplastic astrocytomas [29]. The increased labeling index of the more malignant tumors correlates with a graver prognosis. Flow cytometry studies performed on malignant tumors have tended to confirm the presence of heterogeneous populations varying from 2n to 4n DNA content [30].

Four growth parameters characterize the behavior of any neoplastic population [31]: The cell cycle time, the interval from one mitosis to the next; growth fraction, the proportion of the tumor cells actually dividing; tumor doubling time, the time to double tumor size; and cell loss. While glioblastomas appear to proliferate at a rapid rate, a high rate of cell loss prevents the tumor from doubling its volume in less than a week. The kinetics of tumor cell growth impinge directly on the value of the efficacy of chemotherapeutic agents.

Chemotherapeutic agents may be cell cycle specific, i.e. are effective only during certain phases of the cycle, usually when DNA is synthesized, or cell cycle nonspecific, i.e. the drug may be effective both during DNA synthesis and at other times during the cycle period. Some drugs such as BCNU that are cell cycle nonspecific are known to be more effective during DNA synthesis, however. Modifying the growth fraction of a tumor with cell cycle nonspecific agents, radiation or surgery may improve the efficacy of cell cycle specific agents because the number of cells in active division may be increased. This may be especially valuable in combination chemotherapy where cell cycle specific agents can be combined with cell cycle nonspecific agents in order to reduce the total cell population and increase the percentage of cells in active division. Levin *et al.* [32] took advantage of such cell cycle kinetics in the rat 9L tumor by treating it with BCNU and 5-FU and applying the results to humans.

Other characteristics of individual tumor cells that the chemotherapist might use to advantage in order to improve chemotherapy relate to the biochemical differences among tumor cells and the differences between tumor cells and normal cells. Unfortunately, while a number of such differences have been demonstrated most are quantitative rather than qualitative and so far are not usable in brain tumor chemotherapy. Two exceptions which are worth noting are cyclic AMP and polyamines.

Cyclic AMP levels have been found to be lower in brain tumors than in normal brain [33]. The reduced cyclic AMP levels may be associated with the dedifferentiated state of the tumors, since a direct relationship can be found between cyclic AMP concentrations and the level of cell differentiation, both in nonneural and neural tissues [34].

Administration of β-adrenergic agonists, which raise cyclic AMP levels by direct stimulation of adenyl cyclase, increases the differentiation of glial tumor cells [35]. That such findings may be applicable *in vivo* was demonstrated by Chelmicka-Schorr *et al.* [36], who treated rats harboring implanted C6 gliomas with isoproterenol, a beta-agonist. This agonist suppressed tumor growth. The effect was accentuated by papavarine, a phosphodiesterase inhibitor. Whether such a mechanism can be applied to the treatment of human tumors remains to be determined.

The other biochemical difference relates to the polyamines, putrescine and spermidine. Both polyamines can be demonstrated to be increased in the cerebrospinal fluid (and in the serum) of patients harboring malignant gliomas [37] and medulloblastomas [38]. Since polyamines are biochemically related to cell division, increased polyamine levels may be an indication of either the presence or recurrence of tumors in the nervous system. Research directed toward modifying polyamine levels in tumors by drugs may yield a new source of chemotherapeutic agents.

3. THE BLOOD-BRAIN BARRIER AND THE PHARMACOKINETICS OF BRAIN TUMOR CHEMOTHERAPY

We noted above two differences that distinguish brain tumors from systemic cancer: the blood-brain barrier and the absence of a lymphatic system. The blood-brain barrier impinges on chemotherapy because it affects drug delivery to the tumor cells. The absence of a lymphatic system impinges on chemotherapy because brain tumors produce cerebral edema, which alters the pharmacokinetics of drug entry. We shall deal with these two aspects in the next section.

3.1. The Blood-Brain Barrier in Malignant Gliomas

In 1962, Rall and Zubrod [39] described the pharmacologic characteristics of drugs required for therapy of central nervous system neoplasms including small molecular size, less than 200 daltons, high lipid solubility and minimal ionization. Their observations were based on treating meningeal leukemia in which the blood-brain barrier is, for the most part, preserved. The situation is different for brain tumors. In patients with brain tumor, radioisotopes and radio-contrast materials readily cross into the tumor while failing to enter the surrounding brain tissue. Such penetration represents the mechanism by which the brain tumor is perceived separately from the brain.

Evidence from the laboratory similarly supports the absence of a significant barrier in the bulk of a brain tumor. Vick *et al.* [40] found endothelial abnormalities in the blood vessels of an autochthonous viral-induced experimental astrocytoma in the rat that allowed the passage of high molecular weight horseradish peroxidase into the extracellular space of the tumor. Ausman *et al.* [41], using an implanted choriocarcinoma growing in the brains of monkeys, similarly found that the blood-brain barrier was broken down within the bulk of brain tumors permitting the penetration of albumin and inulin. Levin *et al.* [42] suggested that the brain adjacent to tumor also had a diminished blood-brain barrier, but Vick *et al.* [40] were unable to find such an area in their experimental system.

The question remains as to the quantitative permeability of water-soluble chemotherapeutic agents into brain tumor, the brain adjacent to tumor and the brain distant from tumor. These questions must be resolved if the pharmacokinetics of both lipid-soluble drugs that normally cross an intact barrier and water-soluble drugs that do not cross are to be modeled. Quantitative definition of blood-brain barrier breakdown has been attempted by a number of investigators, most notably Levin et al. [42]. Recently, new techniques utilizing quantitative autoradiography (QAR) have permitted even finer-detailed measurements of regions of interest in and around brain tumors. In this technique, after animals bearing experimental brain tumors are given timed infusions of radiolabeled compounds, they are sacrificed, and their brains are frozen. Autoradiograms are performed on 20 micron cross sections of the brain. The autoradiograms are then "read" in a computerized analyzer that permits the quantitative examination of entry of the radiolabeled materials into regions of interest as small as 50 μm in diameter.

Utilizing a number of animal brain tumor models, Blasberg et al. [43] have performed preliminary experiments on changes of blood-brain barrier associated with the tumors. One model of interest is the metastatic brain tumor model [44]. This model is based on the introduction of tumor cells into the carotid artery of rats. The tumor, Walker 256 carcinoma, is highly sensitive to small doses of cyclophosphamide. The inoculation of tumor cells into the carotid artery ordinarily produces tumor growth in the jaw and base of the skull of the animals, causing death. By administering small doses of cyclophosphamide the extracranial tumor growth is minimized and brain tumors develop. The experimental advantage of this tumor model is that the brain is not specifically manipulated by direct mechanical introduction of tumor cells; instead the tumors develop in the brain from intravascular tumor emboli. An autochthonous brain tumor can also be studied using the direct intracerebral inoculation of avian sarcoma virus [45]. The two models permit a comparison of the effects of primary vs. metastatic tumor on blood-brain barrier and blood flow characteristics.

The QAR observations initially utilized three tracer compounds. 2-Deoxyglucose radiolabeled with ^{14}C (^{14}C-2-DG) was used to define the local utilization of glucose by tumors and surrounding brain. ^{14}C-alpha-aminoisobutyric acid (^{14}C-AIB) was used to measure blood-brain barrier breakdown. AIB shares with other amino acids an active transcellular uptake system for entry into cells. AIB, however, does not cross an intact blood-brain barrier. When the barrier is broken down the AIB enters the CNS and is taken up by living cells in the area of the barrier breakdown, where it can be measured by autoradiography. The site of localization of the ^{14}C-AIB thus corresponds to the site of blood-brain barrier breakdown. Blood flow in the tumor and brain is measured by the rapid infusion of ^{14}C-iodoantipyrine (^{14}C-IAP). IAP rapidly crosses all membranes and its rate of entry into tissues is directly proportional to the blood flow and inversely proportional to the partition coefficient between the tissue and the blood. By performing the experiments in 20 – 30 seconds, blood flow may be determined directly.

Using these three measures, preliminary observations have been made on the metastatic model [43, 45]. They indicate the following: Utilization of glucose as measured by 2-DG is markedly increased in Walker 256 as it is in many tumors. Presumably, this is due to its active glycolytic pathways. There is often depression of glucose metabolism in the brain immediately surrounding the tumor. Local tissue blood flow in the rat brain tumors of large size, 4 – 5 mm, demonstrates variability within the tumor. In the necrotic regions of the tumor, blood flow was reduced to the range of 7 – 18 ml/100g-min. In the regions of viable tumor adjacent to and surrounding the necrotic regions, the blood flow was 18 – 33 ml/100g-min. Along the edge of the tumor near the brain, the blood flow was higher, ranging from 33 – 46 ml/100g-min. In the brain tissue remote from the tumor, blood flow in the rat was normal. The cortical blood flow was between 100 and 200 ml/100g-min, and that in white matter was between 33 and 55. In the brain tissue adjacent to the tumor, at a distance $\leq$ 1 mm, blood flow ranged between 45 and 90 ml/100g-min. In contrast to the general reduction of blood flood in and around large brain tumors, blood flow in very small brain tumors, less than 1 mm, was unaffected; measurements were in the 90 – 200 ml/100g-min range. The blood-brain barrier breakdown as measured by blood-to-brain transfer constants (k_i) was as follows. For normal brain, the k_i was equal to or less than 5 ml/100g-min. For the central regions of large tumors, 4-5 mm, the k_i was between 25 and 90 ml/100g-min. Along the edge of the tumor mass the breakdown was highly heterogeneous with values ranging from 15 to 44 ml/100g-min. Of major importance was that in very small tumors and in the tissue immediately surrounding small tumors, <1 mm, the blood-brain barrier breakdown was minimal, ranging from 5 to 12 ml/100g-min.

These results indicate the following: In the central bulk of brain tumor the blood flow is reduced and the blood-brain barrier is disrupted. Along the edge of the tumors, blood flow begins to approach normal and blood-brain barrier breakdown is relatively minimal. In very small tumors blood flow is normal and blood-brain barrier breakdown is either nonexistent or minimal. While these preliminary data require confirmation, they indicate the quantitative nature of the breakdown that occurs in brain tumors. They also imply that the blood-brain barrier is a problem in the delivery of drugs which cannot cross an intact barrier. The problem, however, is limited to the regions along the edge of large tumors, and to very small tumors whose blood vessels presumably retain the intact nature of normal brain capillaries.

This discussion above on the blood-brain barrier introduces the whole problem of drug delivery. This problem is not unique to brain tumors but is an issue in cancer chemotherapy in general. A number of factors, such as lipid solubility, protein binding, molecular size and metabolism, influence the delivery of any chemotherapeutic compound into a cancer cell where it exerts its effect. Quantitatively, drug delivery must be measured in terms of an integrated tumor-cell drug exposure (concentration × time, C × T) and an intracellular half-life ($T_{1/2}$) of the compound. An immediate problem in determining either the C × T or the $T_{1/2}$ is the nature of the active principle. Thus, a number of compounds are rapidly broken down into metabolites, some or all of which may not be active. This is especially true for the nitrosoureas, whose parent compounds rapidly disappear in the bloodstream. Since it is unknown whether the metabolites themselves are active, it is difficult to measure or estimate the intracellular activity of such compounds. On the other hand, drugs like methotrexate are not substantially metabolized *in vivo* at usual doses and may be actively measured, both extra- and intracellularly [46]. For this reason, most studies of drug entry into experimental tumor systems have been based on measurements of parent compound. Human studies have been limited to blood and cerebrospinal fluid concentrations. The few measurements of drug entry into human tissues, and especially brain tumors, have not taken into account the need to know C × T rather than individual measurements at single points in time.

Despite these limitations, two important considerations apply to drug delivery: The effect of blood flow and blood-brain barrier breakdown. For drugs to be delivered to tumor cells, blood flow must be adequate. In normal tissues blood flow is clearly adequate or the tissue would die of lack of nutrients and oxygen. This is not necessarily true for tumor cells, which are capable of surviving periods of relative hypoxia. Drug delivery to such cells may be retarded and these cells may 'see' concentrations of drug that do not kill the cell but perhaps alter its metabolism in such a way that the cell becomes resistant (see above). As noted, blood flow within intracerebral tumors appears to be quite variable. Whether cell viability relates to the blood flow variability is not known but it is presumed that within regions of relatively low flow, some cell death occurs. Necrosis is apparent in these circumstances, but some cells may survive only to become resistant after drug exposure. For small lipid-soluble drugs, like the nitrosoureas, blood-brain barrier breakdown is not crucial and drug delivery is directly dependent on blood flow. Levin *et al.* [47] have calculated that for BCNU, flow is probably adequate to permit cell kill for all but a small fraction of the tumor cells. For the latter, the cells are far enough from the capillaries that the drug breaks down before it can achieve lethal concentrations. The cell, therefore, survives, and may be more resistant to the drug.

For drugs that cannot cross an intact blood-brain barrier, the problem is one of entry from regions of barrier disruption. In regions in which the tumor is very small and where the blood flow and the blood-brain barrier are normal, or in the heterogeneous regions along the edge of the tumor, the entry of water-soluble agents like methotrexate is markedly limited. Such agents must make their way to the brain tumor cells in these regions by diffusion from regions of the tumor where the barrier has been disrupted. The effectiveness of such agents is directly dependent, therefore, on the diffusion capability of such agents. Levin et al. [47] have studied such drugs and suggest that, in their systems, water-soluble agents are unlikely to penetrate into regions in which brain tumor cells infiltrate. One problem with these models is that they are based on the 9L rat gliosarcoma inoculated subcutaneously. It is not clear that the measurements made in tumors in this location can be applied directly to intracerebral tumors. Nevertheless, this kind of modeling is crucial for an understanding of drug pharmacokinetics.

Before leaving the problems of blood-brain barrier it is worthwhile to review several new techniques to circumvent the problem. The first is that one might attempt purposely to disrupt the barrier. This has been done by Neuwelt et al. [48] in normal dogs and by Hasegawa et al. [49], who studied CNS penetration of methotrexate into rats with meningeal carcinomatosis with and without hyperosmolar intracarotid mannitol. The general principle is the same in both kinds of studies. Hyperosmolar agents are delivered rapidly into the carotid artery of an animal or man. The hyperosmolality leaches water out of the capillary endothelial cells, resulting in the disruption of some of the tight junctions between the endothelial cells. This permits the escape of material from the capillaries into the surrounding brain. The capillaries appear to be affected for about 20 minutes, but this time is dependent on the nature of the agent and its osmolality. It can readily be demonstrated that vital dyes that normally do not cross the blood-brain barrier enter the brains of animals so inoculated. Neuwelt has confirmed that such is also the case with CT contrast-enhancing agents in humans [50]. Once the barrier is open, drugs that normally do not cross can be administered and may enter regions of the brain hitherto excluded by an intact barrier. In studies by Hasegawa et al. [49], methotrexate concentration increased markedly on the side of the brain ipsilateral to the carotid artery mannitol infusion and there was also an increase in methotrexate concentration in the spinal fluid of the animals. A small increase in survival was associated with the additional effectiveness of the methotrexate afforded by the better entry. However, when the blood-brain barrier was opened in solid-brain-tumor-bearing animals given methotrexate, no improvement in survival could be demonstrated. It was not possible to show increased methotrexate concentration in the tumor itself, but, rather, only in the normal brain around the tumor [51]. As noted above, this method would be useful if it could increase the drug concentration in the regions of relatively normal brain into which tumor cells were infiltrating, or in regions of brain where tumor was just beginning as, for example, in metastatic tumors. Balanced against the merits of increasing drug concentration in the brain must be the observation that entrance of chemotherapeutic agents into surrounding normal brain may induce brain injury, as has been observed with methotrexate [52].

Another technique to circumvent the problems of blood flow and permeability is to increase the concentration of drug in the blood delivered to the brain tumor.

This may be accomplished by direct intracarotid or high-dose intravenous drug administration. Both methods have been tried using BCNU in patients with brain tumors. BCNU is lipid-soluble and thus rapidly clears the blood. When administered via the carotid artery, its high extraction fraction permits almost 4 times as much BCNU to cross into the tumor as occurs after the same dose is administered intravenously. One may choose to give the same or higher doses in order to increase local $C \times T$, or one may choose to use a lower intracarotid dose, thereby reducing systemic toxicity. Recently Greenberg *et al.* [53] reported on the intracarotid administration of BCNU at 200 – 400 mg/m^2 in patients with brain tumors. Although some response was seen, the technique induced severe eye damage. Madajewicz *et al.* [54] reported the use of intracarotid BCNU at doses of 100 mg/m^2. They observed responses in patients with metastatic brain tumors without eye toxicity. At this dose the major advantage is to reduce systemic toxicity rather than increase brain tumor $C \times T$ above that which would be expected with intravenous BCNU at standard doses (240 mg/m^2 over 3 days).

High-dose intravenous BCNU uses doses up to 1400 mg/m^2 with autologous marrow rescue. The technique has been reported by Takvorian *et al.* [55] in patients with brain tumor. Myelosuppression remained severe, but lung and liver toxicity was dose-limiting.

3.2. Cerebral Edema

As noted above, fluid that accumulates from leaking capillaries within the brain tumor cannot be removed except by slow diffusion through the brain toward the cerebrospinal fluid pathways. This fluid, or cerebral edema, itself produces symptoms by adding to the mass effect of tumors. Cerebral edema appears to be of three types as initially defined by Klatzo [56] and by Fishman [57]. Vasogenic edema is the type seen in the vicinity of brain tumors. For the most part it is confined to the cerebral white matter and comes about from an increase in permeability of the tumor's capillary endothelial cells. Normally, cerebral capillaries have tight junctions between the endothelial cells. When vasogenic edema is present the junctions appear to be open [or perhaps, as suggested by Vick *et al.* [40], there are new fenestrae] which permit fluid to leak from the capillaries into the surrounding brain. Cytotoxic edema is exemplified by hypoxic injuries in which the cellular elements themselves appear to swell. Hypoxic edema appears not to occur in brain tumors. Finally, interstitial edema, as defined by Fishman [57], is an increase in fluid content associated with hydrocephalus.

Cerebral edema is important with respect to brain tumors in two ways. The first is that cerebral edema almost certainly affects blood-brain barrier function in the region of the brain around the tumor. Levin *et al.* [42], in early studies, demonstrated that the brain adjacent to rat 9L tumors is much different in its physiological characteristics from brain remote from tumors. In part, this difference must relate to the presence of extra fluid in the extracellular space of such tumors. The problem of drug entry into such regions is, therefore, compounded by the occurrence of cerebral edema. The second reason that cerebral edema is important is the fact that the edema produces symptoms. Corticosteroid hormones can readily improve symptoms, presumably by reducing cerebral edema, and, in patients with brain tumors treated with chemotherapeutic agents, it is necessary to differentiate between the effects of the chemotherapeutic agent and the effects of the corticosteroid hormones.

4. CHEMOTHERAPEUTIC AGENTS IN USE

The progress in the past few years in oncology has not been in the discovery of many new drugs but rather in a better understanding of such chemotherapeutic concepts as mechanisms of drug action, drug synergism and antagonism, adjuvant chemotherapy and drug toxicity [58]. These concepts are crucial to the treatment of almost all cancer since the only tumors that can predictably be cured by single-agent chemotherapy are choriocarcinoma in women with methotrexate, and Burkitt's lymphoma with cyclophosphamide [59]. The great advances made in the treatment of acute lymphocytic leukemia in children, Hodgkin's disease, non-Hodgkin's lymphomas, testicular cancer and many childhood solid tumors have been due to combination chemotherapy.

Most of the studies cited in this section on chemotherapeutic agents have evaluated the efficacy of single drugs against malignant brain tumors. While the results have not been overly impressive, it must be remembered that many of these agents showed only minimal activity when used alone against many of the tumors cited above. When they were used in combination with schedules recognizing cell kinetics, cell cycle phase specificity, therapeutic synergism, cellular heterogeneity and toxicity, there were remarkable improvements in survival.

There are several agents which are being tested for the first time against malignant gliomas. However, if the experience in managing these tumors is similar to that in other areas of oncology, we are likely to be successful much sooner if we study ways of improving the efficacy of our current drugs through combination chemotherapy and new chemo-radiotherapy strategies.

4.1. Alkylating Agents

An alkylating agent, nitrogen mustard, was the first modern chemotherapeutic agent used in the treatment of malignant gliomas [60]. Carmustine (BCNU), also an alkylating agent, is currently recognized as the most effective single chemotherapeutic agent against this tumor [2]. Other drugs which have shown promise, such as dianhydrogalactitol, cisplatin, and spirohydantoin, are also believed to act via alkylation of genetic material.

While alkylating agents are generally divided into two subgroups, classical alkylating agents and nitrosoureas, they share a number of characteristics. First, they are electrophilic or spontaneously decompose into electrophilic species *in vitro* and *in vivo*. For example, nitrogen mustard rapidly undergoes cyclization in aqueous solution to form carbonium ions which are highly reactive and electrophilic. Although there are many nucleophilic sites with which the carbonium ion may react, several lines of evidence indicate that DNA is the primary target [61, 62]. Nitrosoureas undergo a similar series of reactions. Alkylation of DNA is presumed to be the mechanism whereby these agents are teratogenic, carcinogenic and cytotoxic.

The classical alkylating agents, including nitrogen mustard and chlorambucil, most frequently form covalent bonds with the 7-position of guanine [58]. These drugs are bifunctional, i.e. they contain two alkylating moieties, and can link two guanine bases in the form of intrastrand or interstrand cross-linkages. When guanine is alkylated in the 7-position, miscoding with thymidine occurs. When two strands of DNA are cross-linked, DNA cannot replicate. Other consequences of alkylation include inhibition of glycolysis, respiration, and protein synthesis, but it is the cross-linking of DNA which appears most cytotoxic [63].

Most alkylating agents are cell cycle nonspecific. The classical alkylating agents have been found to be quite hydrophilic [62]. This spurred the development of more lipophilic agents which might be more effective in solid tumors and more readily cross into the central nervous system (CNS). Spirohydantoin mustard is an example of a more lipophilic alkylating agent that might be effective against malignant gliomas. The clinical experience with the classical alkylating agents against malignant gliomas has been reviewed recently [64].

The nitrosoureas constitute the second group of alkylating agents. They include BCNU, CCNU, methyl-CCNU, PCNU, streptozotocin and chlorozotocin (see Table 1). Their mode of action appears to be the same as that for the classical alkylating agents, i.e. alkylation of chromosomal material. These drugs are more lipid-soluble, however, and readily cross into the CNS. BCNU, CCNU and methyl-CCNU have been extensively studied clinically against brain neoplasms [64]. BCNU will be discussed as the prototype for this group as it has been found to be the most successful drug in the group against gliomas.

BCNU (Figure 2) has a molecular weight of 214 daltons and is a bifunctional alkylating agent. It can, therefore, cross-link DNA as do the classic bifunctional alkylating agents. The drug's antitumor activity is thought to be related to its alkylating activity [65]. The isocyanate portion of the drug may carbamoylate lysine residues of cellular proteins, which has been related to the drug's toxicity.

BCNU is generally regarded as a cell cycle nonspecific agent, but Barranco *et al.* [66] reported that Chinese hamster cells were most sensitive during the S-phase. The drug has been reported to cause a G_2-phase arrest [67]. Cellular uptake of BCNU in L5178Y lymphoblasts *in vitro* has been shown to be by passive diffusion [68].

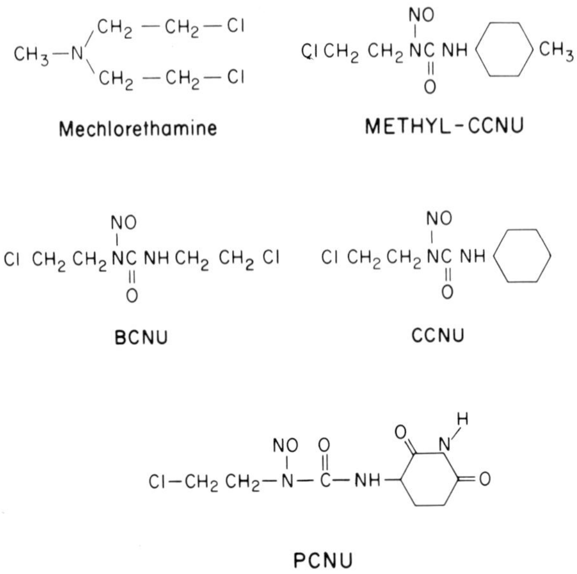

Figure 2. Structural formulas of mechlorethamine, methyl-CCNU, BCNU, CCNU and PCNU.

Inherent or acquired resistance to nitrosoureas is a common clinical occurrence. Decreased uptake of drugs by cells has been found to be a mechanism for resistance to nitrogen mustard and methotrexate [69]. In order to determine if this mechanism was operative in gliomas resistant to nitrosoureas, Rosenblum *et al.* [70] compared the influx and efflux of nitrosoureas in sensitive and resistant rat 9L brain tumor cells. They found no difference between sensitive and resis-

Table 1. Common and chemical names of chemotherapeutic agents used in text.

Common name	Chemical name
Carmustine (BCNU)	1,3-Bis(2-chloroethyl)-1-nitrosourea
Lomustine (CCNU)	1-(2-Chloroethyl)-3-cyclohexyl-1-nitrosourea
Semustine (methyl-CCNU)	1-(2-Chloroethyl)-3-(4-methylcyclohexyl)-1-nitrosourea
PCNU	1-(2-Chloroethyl)-2-(2,6-dioxo-3-piperidyl)-1-nitrosourea
Procarbazine	N-isopropyl-α-(2-methylhydrazino)-p-toluamide hydrochloride
Cisplatin (DDP)	Cis-diamminedichloroplatinum II
Dianhydrogalactitol (DAG)	1,2:5,6-Dianhydrogalactitol
VM-26	4′-Demethyl-epipodophyllotoxin-9-(4,6-0-thenylidine-β-D glucopyranoside)
Vincristine	Vincristine sulfate
5-FU	5-Fluorouracil
HU	Hydroxyurea
Methotrexate (MTX)	4-Amino-4-deoxy-10-methyl pteroylglutamic acid

tant cells in these parameters. They did find that the steady-state drug concentration in the sensitive cells was more than 1.5 times the intracellular drug concentration in the resistant cells. They concluded that the mechanism of resistance was altered intracellular biodistribution and biotransformation rather than membrane transport. Their studies also showed that cross-resistance existed between BCNU, CCNU, methyl-CCNU and PCNU, a new nitrosourea. Partial cross-resistance to spirohydantoin was also present, and there was no cross-resistance to radiation therapy. Altered DNA repair processes have also been cited as a mechanism of resistance to alkylating agents [69].

Biotransformation and spontaneous decomposition of the drug are rapid, with a plasma half-life of less than 15 minutes in mice, which is probably similar to that in humans [71]. Using ^{14}C-labelled BCNU, DeVita *et al.* [71] found the isotope present in the CSF at concentrations approaching those seen in plasma at one hour in humans.

The dose-limiting side effect is delayed myelosuppression, which occurs at 3 – 5 weeks and lasts 1 – 3 weeks. Pulmonary fibrosis is the other major frequent side effect occurring in up to 20% of patients [72]. Risk factors for the development of BCNU-induced pulmonary fibrosis have been reported [72].

Nitrosoureas have been associated with renal toxicity. However, this is uncommon [73] and the toxicity appears to be dose-related. In a Mayo Clinic series [74], only 4 of 857 patients were believed to have nitrosourea-induced nephrotoxicity. However, three of the four affected patients had received cumulative doses of over 1000 mg/m^2 of methyl-CCNU. The most important parameter available for predicting renal toxicity is decrease in kidney size [73]. It is possible that as patients survive longer a delayed renal toxicity may become evident.

BCNU has been found to be effective in the treatment of multiple myeloma, Hodgkin's disease and non-Hodgkin's lymphoma. BCNU has been used in a number of phase II and III studies which are detailed elsewhere [64]. When used in combination with irradiation, it was shown by the BTSG to improve the 18-month survivorship to 19%, in comparison to 4% for irradiation alone [75].

Wheeler *et al.* [76] studied the effect of BCNU in combination with irradiation in a rat gliosarcoma, 9L, *in vitro*. They found a synergistic effect between BCNU and irradiation, especially when the cells were treated with BCNU 15 hours prior to irradiation. The mechanism of this synergism is not known, but plausible explanations include interference of repair processes which can proceed when either agent is used alone or synchronization of cells by BCNU to a phase when they are more radiosensitive. As mentioned above, BCNU has been found to cause a G_2-phase arrest [67] and cells are known to be most sensitive to irradiation during G_2- and M-phase [77].

Another investigational approach to improve results with existing drugs has been to combine nitrosoureas with hyperthermia. In a murine ependymoblastoma model, Thuning *et al.* [78] found a synergistic effect in the combination of hyperthermia and nitrosourea.

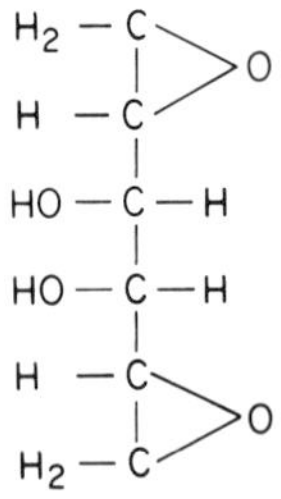

Dianhydrogalactitol

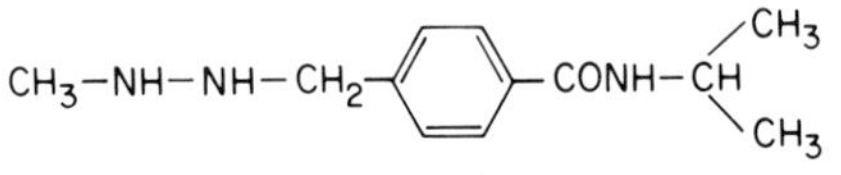

Procarbazine

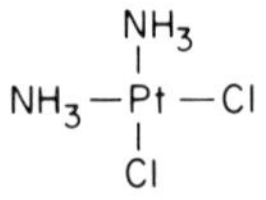

Cisplatin

Figure 3. Structural formulas of dianhydrogal actitol, procarbazine and cisplatin.

CCNU (Figure 2) is similar to BCNU in mechanism of action and metabolism. Pulmonary toxicity has not been a common toxicity. Methyl-CCNU (Figure 2) is similar to BCNU and CCNU in most pharmacological and biochemical respects. It was shown by the BTSG to be no more effective than BCNU [2]. PCNU (Figure 2) is a new nitrosourea soon to undergo phase II trials in the BTSG.

4.2. Procarbazine

Procarbazine (Figure 3), an N-methylhydrazine, was originally developed as a potential monoamine oxidase inhibitor (see Table 1). The agent is water-soluble and has a molecular weight of 258 daltons.

Its mode of action as an antineoplastic agent is unclear, but DNA, RNA and protein synthesis are inhibited. Nucleic acid synthesis appears to be inhibited before protein synthesis [58]. Chromosomal breakage has been seen after administration of the drug. The alterations of DNA are thought to be characteristic of those seen after ionizing irradiation and bifunctional alkylating agents. As might be anticipated from its action on DNA, procarbazine is carcinogenic and teratogenic.

Procarbazine is absorbed readily through the gastrointestinal tract and quickly crosses the blood-brain barrier with peak CSF levels occurring 30 – 90 minutes after administration [79]. Metabolism of the drug occurs in the liver and through renal excretion.

The most significant toxicity is myelosuppression. CNS toxicity includes ataxia, nightmares and hallucinations. These side effects may be related to monoamine oxidase inhibition. A number of interactions occur between procarbazine and other neuropharmacologically active drugs. Procarbazine may have disulfuram-like activity and, therefore, may cause gastrointestinal side-effects if ingested with alcohol. It may also cause a rash.

Procarbazine has been the subject of phase II and III evaluations against malignant gliomas [64]. The BTSG found that in combination with RT, it was as effective as BCNU plus RT [3].

4.3. Dianhydrogalactitol

Dianhydrogalactitol (DAG), a highly water-soluble drug (Figure 3) with a molecular weight of 146 daltons, has been found to be active against the murine intracranial ependymoblastoma [80]. The drug is believed to act briefly as an alkylating agent, causing interstrand and intrastrand cross-linking of DNA [81]. DAG causes inhibition of DNA, RNA and protein synthesis [82]. It is a cell cycle nonspecific drug. Eckhardt *et al.* [83] found that DAG rapidly crosses the blood-brain barrier and enters brain tumors. The major limiting toxicity is bone marrow suppression.

In a phase II trial Espana *et al.* [84] demonstrated activity of DAG against malignant gliomas. Eagen *et al.* [85] performed a randomized phase III study of the drug in combination with split or continuous irradiation against malignant gliomas. The group receiving DAG and either form of RT had a median survival of 67 weeks compared to 35 weeks for those undergoing RT alone.

4.4. Cisplatin

Cisplatin (see Table 1) is a water-soluble planar inorganic compound with a molecular weight of 300 daltons. It has the empiric formula $N_2Cl_2PtH_6$ (Figure 3). It is the first heavy metal to achieve the status of an anticancer agent.

Cisplatin appears to act through interstrand and intrastrand cross-linking of DNA in a manner similar to the bifunctional alkylating agents [58]. The electrically neutral compound diffuses into cells where the chloride ions are lost by hydrolysis. Two reactive ligand sites are then formed at which covalent linkages occur with nucleic acids or proteins. As with other alkylating agents, cisplatin primarily binds the guanine base. Cisplatin is a cell cycle nonspecific agent.

Cisplatin is administered intravenously. Originally it was given as an intravenous push; however, the major dose-limiting toxicity, nephrotoxicity, was found to be greatly reduced if a high urine flow was maintained [86]. Currently the drug is diluted with saline and mannitol and given slowly. Renal tubular dysfunction is the major toxicity, which is dose-related and cumulative. Ultimately the renal damage may be irreversible. Ototoxicity manifested by high tone hearing loss and tinnitus may also occur. Severe myelosuppression is rarely seen, which could make cisplatin a good drug for administration with marrow toxic drugs.

Cisplatin has been found to have activity against germinal neoplasms of the testis, lymphomas, ovarian carcinoma and squamous cell carcinoma of the head and neck. In a phase II study of cisplatin with malignant glioma, it was found to show objective activity in 4 of 6 patients [87]. Cisplatin is currently undergoing phase II trials in the BTSG.

4.5. Hydroxyurea

Hydroxyurea (Figure 4) is a cell cycle specific drug which interferes with DNA synthesis through inhibition of ribonucleoside diphosphate reductase [88]. This enzyme catalyzes the reductive conversion of ribonucleotides to deoxyribonucleotides which are necessary for DNA synthesis. It readily crosses the blood-brain barrier and is excreted in the urine [89]. It has a molecular weight of 76 daltons.

HU has been shown to potentiate radiation damage *in vitro* [90]. This observation led to a study of the efficacy of HU in combination with radiation therapy in the treatment of patients with glioblastoma multiforme (GM) and nonglioblastoma multiforme malignant glioma (NGM) [91]. Postoperatively, all patients received BCNU and RT but they were randomized to receive RT with or without concurrent HU. The endpoint of the study was median time to tumor progression (MTP). The results show that for patients with NGM no improved MTP was found in the group receiving HU. For patients with GM, however, there was a statistically significant prolongation in MTP in patients receiving BCNU, RT and HU (42 wks) in comparison to those receiving only BCNU and RT (31 wks). Patients who underwent a subtotal or total resection of tumor and received HU experienced an MTP of 49 weeks. The authors speculated that the GM tumors had a higher growth fraction which would make them more vulnerable to a cell cycle specific drug.

4.6. VM-26

VM-26 (Figure 4) is a semisynthetic podophyllotoxin derived from the *Podophyllum peltatum* root (see Table 1). It has a molecular weight of 656 daltons and is highly lipophilic.

VM-26 is a mitotic spindle poison like the vinca alkaloids. It has been found to be taken up by cells by passive diffusion and binds to high- and low-affinity binding sites [92]. The drug, however, has been found to concentrate in cells at levels 16 to 20 times that seen extracellularly [93]. Podophyllotoxin binds to tubulin, a microtubule subunit. *In vitro* VM-26 has been shown to cause a premitotic delay with cells irreversibly accumulating in the G_2-phase of the cell cycle [94]. It is a cell cycle specific drug.

Despite its high lipophilicity, only minimal amounts have been found to cross into the CNS. Creaven and Allen [95] found CSF drug levels < 1% of simultaneous peak plasma levels. However, in another patient who had undergone a craniotomy and irradiation, the CSF levels were approximately 27%

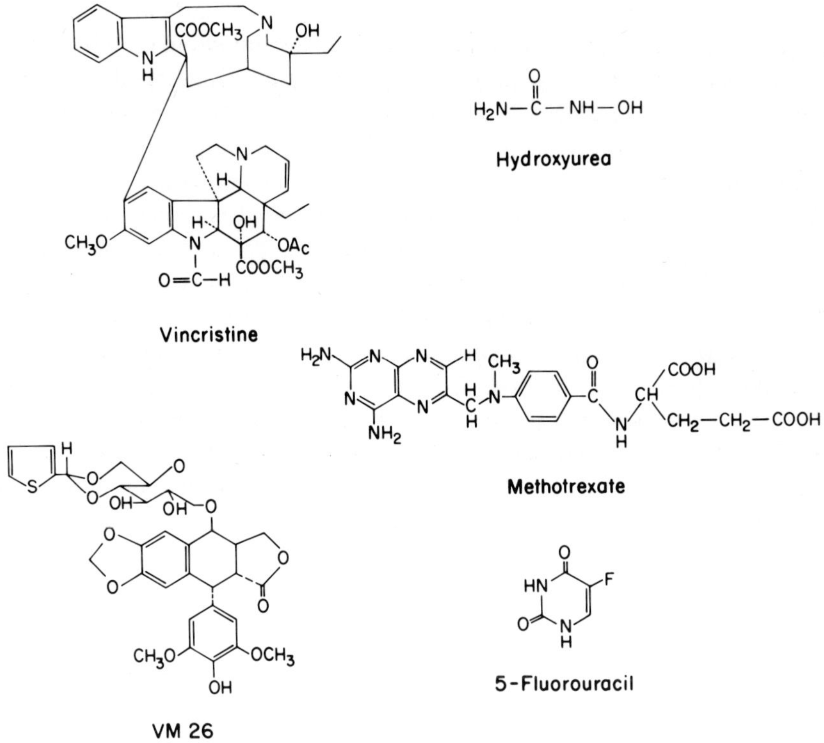

Figure 4. Structural formulas of vincristine, hydroxyurea, methotrexate, VM-26 and 5-fluorouracil.

of that in plasma, indicating that the drug could have a role in the management of gliomas.

The major dose-limiting toxicity of VM-26 is hematologic, though this is not cumulative. Hypotension may occur during and after rapid intravenous infusions, but usually not with infusions over 30 minutes.

Skylansky *et al.* [96] reported objective improvement in 5 of 13 patients receiving VM-26 for malignant intracranial tumors. Improvement generally occurred during the middle of the first course of therapy. There was relatively mild hematological toxicity except in patients who had previously received nitrosourea. Kessinger *et al.* [97] also reported that 48% of patients with malignant gliomas, who had failed RT and BCNU, responded to VM-26.

4.7. Vincristine

Vincristine (Figure 4) is a water-soluble vinca alkaloid with a molecular weight of 923 daltons. Its antineoplastic mechanism is thought to be that of a mitotic spindle poison [58]. It is a cell cycle specific drug. It binds to tubulin during the S-phase and causes metaphase arrest. Penetration of the drug into the CSF under normal conditions is poor. Peripheral neuropathy is the dose-limiting toxicity and appears to be mediated through damage to microtubules in peripheral nerves [98]. Vincristine is a marrow-sparing drug.

No controlled study of vincristine as a single agent against malignant glioma has been reported; however, a number of studies have found it to have some activity [64]. It is frequently used in combination chemotherapy. The concepts outlined above regarding scheduling of cell cycle specific agents after cell cycle nonspecific drugs are also operative for vincristine. Thus, any form of therapy that reduces the number of viable cells in a tumor will recruit cells in G_0 to divide. Cell cycle specific agents would be expected to be more effective soon after treatment with drugs most effective against cells in G_0, e.g. nitrosoureas or procarbazine. With this in mind, vincristine has been used in conjunction with these latter two agents [99]. Shapiro and Young found that this "triple therapy" yielded a median survival of 63 weeks and 44% of patients were alive at 18 months. Vincristine is a marrow-sparing drug and it may be used with highly myelotoxic drugs.

4.8. 5-Fluorouracil

5-Fluorouracil (5-FU) is a pyrimidine analog (Figure 4) differing from uracil by a fluorinated number-5 carbon. It has a molecular weight of 130 daltons. By acting as a false pyrimidine 5-FU competes with uracil for the enzyme thymidylate synthetase and thereby inhibits production of thymidine. Thymidylate synthetase catalyzes the transfer of a methyl group from N^{5-10}-methylene tetrahydrofolate to deoxyuridylic acid (dUMP) to form deoxythymidylic acid (dTMP). Prior to binding with thymidylate synthetase, 5-FU must be converted to the corresponding ribonucleoside and ribonucleotide. RNA synthesis is also disturbed since 5-FU competes with uracil in the production of RNA. 5-FU is a cell cycle specific drug since it blocks DNA synthesis.

5-FU has been shown to cross the blood-brain barrier after intravenous administration despite the fact that it is not lipid-soluble and approximately 24% ionized at physiological pH (100). 5-FU is degraded in the liver. Toxic manifestations of 5-FU include stomatitis, diarrhea, myelosuppression, alopecia, and, rarely, an acute cerebellar syndrome [58]. The drug has been used in cases of hepatoma, carcinoma of the colon, ovary, cervix, urinary bladder, pancreas, prostate and head and neck.

In their studies involving rat 9L gliosarcoma, Levin *et al*. [32] found synergism between BCNU and 5-FU against the 9L tumor. In their human studies involving recurrent gliomas they gave 5-FU two weeks after a single dose of BCNU and found greater effect with this combination than with a prior group of patients treated with BCNU and procarbazine. The mechanism of this synergism was unclear but may relate to the concept of an enhanced effect of a cell cycle specific agent after an alkylating agent as discussed above. This pilot study has led to the inclusion of BCNU followed by 5-FU in a BTSG phase II trial.

Finally, McDonnell *et al.* [101] found that 5-FU administered intravenously caused a localized opening of the blood-brain barrier in cats which lasted 7 to 8.5 hours. This finding has not yet been tested clinically.

4.9. Folate Analogs

Methotrexate, the 4-amino N-10 methyl analog of folic acid (Figure 4), was the first antimetabolite found to be effective against neoplastic disease [102]. It has a molecular weight of 455 daltons. Its efficacy is due to its ability to tightly bind to and inhibit dihydrofolate reductase, a property largely due to the 4-amino moiety. Activity of dihydrofolate reductase is necessary for the production of tetrahydrofolate (FH_4), the active form of the coenzyme, which functions in the transfer of one-carbon fragments. Synthesis of thymidylate requires the donation of a methyl group from N^{5-10}-methylene tetrahydrofolate to the uracil moiety of dUMP. The methyl group is transferred as formaldehyde which requires reduction by the pteridine ring of FH_4. The result is oxidation of FH_4 to form dihydrofolate. It is apparent, therefore, that FH_4 must be resynthesized from dihydrofolate if DNA synthesis is to continue. One-carbon transfers requiring FH_4 are also necessary for purine synthesis. Methotrexate is a cell cycle specific drug and functions only during S-phase.

Methotrexate enters cells through an active transport carrier system [103]. One mechanism of methotrexate resistance is reduced uptake of the folate analog by these systems [58]. A second mechanism of resistance was shown by Alt *et al.* [104] to involve increased levels of dihydrofolate reductase. These investigators found, through nucleic acid hybridization, a significant increase in the number of gene copies for dihydrofolate reductase which correlated with the elevated levels of dihydrofolate reductase.

Methotrexate, at conventional doses, enters the central nervous system very poorly. In high doses, the drug has been found to cross the blood-brain barrier in significant amounts [58]. Much of the drug is excreted by the kidneys unchanged. Nephrotoxicity can occur at high doses, when conversion to the less soluble 7-hydroxy form can be appreciable. Hematopoiesis is also suppressed. In the treatment of bone or soft tissue sarcomas, Allen *et al.* [105] found that high-dose methotrexate could cause leucoencephalopathy in the absence of CNS irradiation or metastases.

Methotrexate has been shown to have activity against acute lymphoblastic leukemia, choriocarcinoma in women, lymphoma, osteogenic sarcoma, carcinomas of the head and neck, lung, breast and testis [58]. High-dose methotrexate has been used in a number of malignancies with leucovorin (N^5-formyl tetrahydrofolate) rescue with considerable success. At high plasma concentrations, methotrexate has been found to enter cells which have become resistant through reduced carrier-mediated transport by means of passive diffusion [106].

Since leucovorin shares the same transport system as methotrexate, it can selectively enter normal cells and thereby compete with methotrexate for binding of dihydrofolate reductase, resulting in normal activity of this enzyme. Intrathecal methotrexate has been used in the CNS prophylaxis of acute lymphoblastic leukemia and the treatment of leukemic and carcinomatous meningitis [107].

High-dose methotrexate has been found to be ineffective against malignant gliomas [108]. However, a more lipophilic folic acid analog, 2,4,diamino-5-(3', 4'-dichlorophenyl)-6-methyl pyrimidine (DDMP), readily enters the CNS and has been found in brain tumors in concentrations ten times greater than those seen in plasma [58]. Another lipid-soluble antifolate, triazinate, is currently undergoing phase II BTSG trials.

4.10. Antibiotics

The antitumor antibiotics are microbial fermentation products which are cytotoxic. None has been shown to be active against brain tumors and they are only briefly reviewed.

Doxorubicin (Adriamycin) is an anthracycline which has not been studied as a treatment for primary brain tumors. Metastatic brain tumors have been found to progress while peripheral metastases respond to doxorubicin [64].

Bleomycin was shown to be ineffective against malignant glioma in combination with irradiation in a controlled, randomized study [109]. Mithramycin is a lipid-soluble antibiotic with a molecular weight of > 1000 which was the subject of the first BTSG protocol [110]. It was shown to be ineffective.

4.11. Corticosteroids

There have been a number of animal studies evaluating the efficacy of corticosteroids as oncolytic agents in the treatment of brain tumors. Brzustowicz *et al.* [111] found that cortisone could cause growth suppression of subcutaneously transplanted ependymoblastoma in mice. Using a mouse intracerebral melanoma model, Kotsilimbas *et al.* [112] found that treatment with dexamethasone resulted in a reduction of wet and dry tumor weight. Gurcay *et al.* [113] found that methylprednisolone-treated rats harboring transplanted intracerebral gliomas showed reduced tumor weights. In addition, Shapiro and Posner [114] found dexamethasone to suppress the growth of intracerebral ependymoblastomas in mice. Survival of the treated animals was increased over controls. They also found that dexamethasone caused a decrease in the uptake of tritiated thymidine into tumor DNA.

Chen and Mealey [115] studied the effects of methylprednisolone on human glioma cells in tissue culture. They found that cell division was inhibited within 24 hours of incubation with the drug and that the degree of inhibition was proportional to the dosage and duration of exposure. Tritiated thymidine incorporation into DNA was inhibited 60% whereas uridine and leucine incorporation were less significantly suppressed. The authors concluded that inhibition of DNA synthesis was the primary site of action of methylprednisolone.

Reports showing the efficacy of corticosteroids against experimental brain tumors have not been unanimous. Geran *et al.* [80] tested a number of corticosteroids against an intracranial mouse ependymoblastoma model and found no effect.

The BTSG has recently completed a phase III study of high-dose methylprednisolone against glioblastoma multiforme. They found no beneficial effect of the drug [3].

5. SYSTEMS FOR MEASURING CHEMOTHERAPEUTIC AGENTS FOR BRAIN TUMORS

5.1. In Vivo Systems

Brain tumor models in animals have been reviewed elsewhere in this volume. Several models have been used to test chemotherapeutic agents for use in clinical brain tumor chemotherapy. The whole issue of the value of animal models to determine efficacy of chemotherapy for human use is one that is undergoing extensive review. A major study by the National Cancer Institute, originally through the Cancer Chemotherapy National Service Center (CCNSC), produced a number of useful agents. However, many of these agents were available even prior to the extensive CCNSC searching system and others were found that ultimately failed the CCNSC search. The most notable example of the latter is vincristine, a drug not effective against L1210 leukemia. The effort to develop animal models that could be used in brain tumor chemotherapy have extended for a period over a decade. The ependymoblastoma model and its use for chemotherapy testing as described by Ausman *et al.* [116] was further developed by Geran *et al.* [80]. Despite the fact that a number of agents were studied, little attempt was made to correlate the experimental model results with those in humans. Levin and Wilson [117] showed that intracerebral murine glioma 26 predicted human tumor responses to CCNU, procarbazine and vincristine, and the combination of BCNU and procarbazine.

Skipper has pointed out that no one human cancer is a model for any other human cancer [118]. Thus, no animal model can be more than an approximate model for human cancer. As noted by Schold elsewhere (this volume), both autochthonous tumor models and the nude mouse system for growing human tumors have been developed in an effort to model human brain tumors. Shapiro *et al.* [119] described the nude mouse system to test chemotherapeutic agents. They found that different patients' tumors were responsive to different drugs [20]. It is apparent from our discussion of heterogeneity that the problem of differences between tumors extends even to differences among cells within tumors. As a screening system, such models are limited, although they often yield information which can then be applied in large animal studies to determine drug toxicity.

The value of animal models for chemotherapy thus depends on the utilization of these models to answer specific questions. Models to study blood-brain barrier and pharmacokinetics are crucial since such studies cannot be undertaken easily in patients. The potential for the recently-developed positron emission tomographic scanner (PET) and its use for examining pharmacokinetics of brain tumor chemotherapy has yet to be applied. However, such studies are theoretically possible with the PET scanner [120], and correlation between the results of the PET scanner and those of model systems, using e.g. QAR techniques, may permit more precise measurements of pharmacokinetics of chemotherapeutic agents.

5.2. In Vitro Assay

The existence of an *in vitro* assay system to evaluate the degree of chemosensitivity of a given patient's tumor to a panel of chemotherapeutic agents would be quite advantageous [121]. If an individual's tumor could be reliably shown to be sensitive to specific drugs *in vitro*, oncologists would have the capability of selecting drugs with the same specificity with which one selects antimicrobial agents. Alternatively, if one could only reliably predict agents to which the tumor was resistant, patients could avoid many of the unpleasant and dangerous side effects associated with drugs which would be ineffective. Along with Salmon's studies [122] correlating *in vitro* chemosensitivity with *in vivo* response of ovarian carcinoma and multiple myeloma, there have been several reports of the development of similar assay systems for other tumors including brain neoplasms [123 – 125].

Interest in defining such a reliable system for malignant gliomas has extended over the past few years. Rosenblum *et al.* [126] demonstrated in a rat brain tumor model a correlation between the fraction of surviving clonogenic cells after BCNU therapy and animal survival. In their assay single cells were analyzed for clonogenic capacity *in vitro* after the administration of varying doses of BCNU to the animals. With increasing doses of the drug they noted increasing survival of the animals which directly correlated with a decrease in numbers of surviving clonogenic cells.

The same group recently described their experience with 15 glioblastoma patients in which they correlated *in vitro* chemosensitivity by colony forming assay to observed clinical response [70]. At BCNU concentrations believed to be achievable *in vivo*, no cell kill was seen in 8 tumors tested *in vitro*. All of these patients were clinically found to be BCNU unresponsive. Seven tumors were thought to be sensitive to BCNU *in vitro* but only 3 of the patients improved clinically when given the drug. The observation that the colony forming assay more reliably predicts for drug resistance than it does for chemosensitivity was also made by Salmon *et al.* [122] in ovarian carcinoma and multiple myeloma.

Yung et al. [25] have recently found that clones of cells within a malignant glioma defined by different karyotype may have different chemosensitivities as noted above.

An *in vitro* microcytotoxicity assay for gliomas has also been developed and correlations made with clinical data [127]. In this assay system, cells grown in a monolayer are briefly exposed to a drug and then reincubated. Twenty-four hours later the cells which remain attached to the dish surface are considered viable and counted. The authors defined a sensitive cell line as one in which there was a greater than 75% reduction in remaining cells as compared to controls at 6.6 µg/ml BCNU concentration. Using this definition, they found that 5 of 5 patients not sensitive *in vitro* did not respond clinically to BCNU. Six of nine patients showing sensitivity *in vitro* showed a clinical response *in vivo* when given the drug.

A third assay system for malignant gliomas has been developed [128]. In this system tumor cells are plated into microcytotoxicity dishes and exposed to drug. The effect is then measured by determining the uptake of ^{3}H-leucine or ^{35}S-methionine, both measures of protein synthesis. Many tumors were tested using CCNU, methyl-CCNU, procarbazine and vincristine. The results demonstrated variable sensitivities of the different tumors to all of the agents. Finally, an organ culture system has also been devised to test chemotherapeutic agents against human tumors [129].

There are several problems that must be faced if *in vitro* assay systems are to be useful in defining appropriate chemotherapy for patients:

1. The nature of the assay: A colony forming (clonogenic) assay defines cells that ultimately divide and therefore represent cells in a patient that can grow and prove fatal. Drugs that kill such cells are clearly designated by such assays as potentially useful. On the other hand, assays that measure protein or DNA synthesis define drugs that modify metabolic behavior but not necessarily produce cell death. Such drugs must also be shown to prevent cell division in other assay systems. Microcytotoxicity assays clearly define drugs that kill cells. However, such assays may not be sensitive enough. Lower drug concentrations are usually required to prevent cell division than to kill the cell outright. Microcytotoxicity assays utilize higher drug concentrations than do colony forming assays and thus may miss effective agents. Organ cultures do permit assay systems that more closely approximate the *in vivo* situation. However, Rubinstein and Herman [130] have demonstrated that organ cultures tend to produce differentiation and, therefore, may not measure drug chemotherapy as such.

The colony forming assay would appear to be the most sensitive and the most specific for *in vitro* testing for human gliomas. However, many cells may not form colonies. In our own studies, we find that astrocytes are often highly motile and wander away from the colony before it has achieved the 30–50 cells needed to define it. Only cells that form colonies can be counted in such systems, thus introducing selection. In the final analysis it is likely that multiple assay systems will be needed to fulfill the requirements for drug screening.

2. Drug concentrations: Salmon *et al.* [122] have suggested that unless drugs are used in the assays at concentrations achievable *in vivo*, results are of limited value. Unfortunately, we often know achievable plasma concentrations, but almost never know achievable tumor concentrations. Levin *et al.* [131] found BCNU plasma concentrations of 3 μg/ml or less while Kornblith *et al.* [127] used 6.6 μg/ml in their *in vitro* system. Until more data are available it would appear prudent to perform dose-response experiments and then retrospectively analyze the results. One problem with this approach relates to the tumor's heterogeneity.

3. Heterogeneity: All drug assays have used mass cultures derived directly from the patient's tumor. In our studies on clones from such patients, we have found heterogeneous drug responses, with cells that are more or less sensitive to a given agent. When all the cells of a mass culture are killed by a drug, then the tumor is "sensitive". When all of the cells survive, the tumor is "resistant". When the results demonstrate cell kill for only some of the cells, it is assumed that the rest are resistant. Unfortunately, this is difficult to prove because the assays often do not permit subculturing, and the remaining cells cannot be tested. Thus, more work is required if we are to understand the phenomenon of resistance in mass cultures. Correlation between clonal assays and mass culture assays would appear to be a good starting point.

4. Response criteria: All *in vitro* assays must be compared to the patient's response. The response of patients with advanced systemic cancer is more easily defined than that of patients with newly operated brain tumors. "Time to progression" or other such end-points are less precise than survival, yet as noted in the introduction, survival may also have limitations. In terms of comparison, how are we to compare the *in vitro* response of the tumor from a patient who lives 8 months with one from a patient who lives 20 months?

5. New problems: The current techniques use short drug exposure times. While adequate for cell cycle nonspecific agents, this is too short an exposure for cell cycle specific agents which must be in contact with the cells for several generation times, often a period of several hours. An additional problem is the need for certain drugs to be metabolized to an active form before they can be tested *in vitro*; an example is cyclophosphamide. Finally, techniques to evaluate combination chemotherapy using *in vitro* systems have yet to be worked out.

REFERENCES

1. Dixon RL, Lee IP: Pharmacokinetic and adaptation factors involved in testicular toxicity. Fed Proc 39:66–72, 1980.
2. Walker MD, Green SB, Byar DP, et al: Randomized comparisons of radiotherapy and

nitrosoureas for the treatment of malignant glioma after surgery. New Engl J Med 303:1323–1329, 1980.
3. Green SB, Byar DP, Walker MD, et al: Comparisons of carmustine, procarbazine, and high dose methylprednisolone as additions to surgery and radiotherapy for the treatment of malignant glioma. New Engl J Med (in press).
4. Heppner GH, Shapiro WR, Rankin JK: Tumor heterogeneity. In: Pediatric Oncology, Humphrey GB et al (eds). The Hague, Martinus Nijhoff Medical Division Publishers, 1981, vol 1, pp 99–116.
5. Pierce GB: Cellular heterogeneity of cancers. In: World Symposium on Model Studies in Chemical Carcinogenesis, T'so POR, DiPaolo JA (eds). 1972, vol B, pp 463–472.
6. Dexter DL, Kowalski HM, Blazar BA, et al: Heterogeneity of tumor cells from a single mouse mammary tumor. Cancer Res 38:3174–3181, 1978.
7. Imada M, Sueoka N: Clonal sublines of rat neurotumor R14 and cell differentiation. I. Isolation and characterization of cell lines and cell type conversion. Devel Biol 66:97–108, 1979.
8. Fidler IJ, Kripke ML: Metastasis results from preexisting variant cells within a malignant tumor. Science 197:893–895, 1977.
9. Heppner GH, Dexter DL, DeNucci T, et al: Heterogeneity in drug sensitivity among tumor cell subpopulations of a single mammary tumor. Cancer Res 38:3758–3763, 1978.
10. Baylin SB, Weisburger WR, Eggleston JC, et al: Variable content of histaminase, L-DOPA decarboxylase and calcitonin in small-cell carcinoma of the lung. New Engl J Med 299:105–110, 1978.
11. Siracky J: An approach to the problem of heterogeneity of human tumour-cell populations. Br J Cancer 39:570–577, 1979.
12. Pertschuk LP, Tobin EH, Brigati DJ, et al: Immunofluorescent detection of estrogen receptors in breast cancer. Cancer 41:907–911, 1978.
13. Byers VS, Johnston JO: Antigenic differences among osteogenic sarcoma tumor cells taken from different locations in human tumors. Cancer Res 37:3173–3183, 1977.
14. Rogan KM, Faldetta TJ, Boto W, et al: Heterogeneity in the membrane proteins of human lymphoid cell lines as seen in sodium dodecyl sulfate polyacrylamide electrophoresis slab gels. Cancer Res 38:3604–3610, 1978.
15. Sorg C, Bruegger J, Seibert E, et al: Membrane-associated antigens of human malignant melanoma IV: changes in expression of antigens on cultured melanoma cells. Cancer Immunol Immunother 3:259–271, 1978.
16. Barranco SC, Ho DHW, Drewinko B, et al: Differential sensitivities of human melanoma cells grown in vitro to arabinosylcytosine. Cancer Res 32:2733–2736, 1972.
17. Woods LK, Morgan RT, Quinn LA, et al: Comparison of four new cell lines from patients with adenocarcinoma of the ovary. Cancer Res 39:4449–4459, 1979.
18. Rubinstein LJ: Current concepts in neuro-oncology. In: Advances in Neurology, Thompson RA, Green JR (eds). New York, Raven Press, 1976, vol 15, pp 1–25.
19. Fialkow PJ: Clonal origin and stem cell evolution of human tumors. In: Genetics of Human Cancer, Mulvihill JJ, Mieler RW, Fraumeni JF Jr (eds). New York, Raven Press, 1977, pp 439–453.
20. Basler GA, Shapiro WR: Brain tumor research with nude mice. In: The Nude Mouse in Experimental and Animal Research, Fogh J, Giovanella B (eds). New York, Academic Press, 1982, vol II, pp 475–490.
21. Bigner DD, Bigner SH, Ponten JN, et al: Heterogeneity of genotypic and phenotypic characteristics of fifteen permanent cell lines derived from human gliomas. J Neuropath Exp Neurol 40:201–229, 1981.
22. Shapiro JR, Yung W-KA, Shapiro WR: Isolation, karyotype and clonal growth of heterogeneous subpopulations of human malignant glioma. Cancer Res 41:2349–2359, 1981.

23. Mark J: Chromosomal characteristics of neurogenic tumors in adults. Hereditas 68:61–100, 1971.
24. Shapiro WR, Yung WA, Basler GA et al: Heterogeneous response to chemotherapy of human gliomas grown in nude mice and as clones in vitro. Cancer Treat Rep, 65(Suppl 2):55–59, 1981.
25. Yung W-KA, Shapiro JR, Shapiro WR: Heterogeneous chemosensitivities of subpopulations of human glioma cells in culture. Cancer Res 42:992–998, 1982.
26. Rankin JK, Yung W-KA, Shapiro WR, et al: Effects of low dose BCNU chemotherapy on human glioma cell lines in vitro. Proc Amer Assoc Cancer Res 22:43, 1981.
27. Rankin JK, Shapiro WR, Posner JB: Cellular stability and chromosomal evolution of early passage cells from human gliomas. Proc Amer Assoc Cancer Res 21:55, 1980.
28. Steel GG: Growth kinetics of brain tumors. In: Brain Tumors, Scientific Basis, Clinical Investigation and Current Therapy, Thomas DGT, Graham DI (eds). London, Butterworth, 1980, pp 10–20.
29. Hoshino T, Wilson CB: Cell kinetic analysis of human malignant brain tumors (gliomas). Cancer 44:956–962, 1979.
30. Hoshino T, Nomura K, Wilson CB, et al: The distribution of nuclear DNA from human brain tumor cells. J Neurosurg 49:13–21, 1978.
31. Hoshino T, Wilson CB: Review of basic concepts of cell kinetics as applied to brain tumors. J Neurosurg 42:123–131, 1975.
32. Levin VA, Hoffman WF, Pischer TL, et al: BCNU-5-fluorouracil combination therapy for recurrent malignant brain tumors. Cancer Treat Rep 62:2071–2076, 1978.
33. Furman MA, Shulman K: Cyclic AMP and adenyl cyclase in brain tumors. J Neurosurg 46:477–483, 1977.
34. Friedman DL: Role of cyclic nucleotides in cell growth and differentiation. Physiol Rev 56:652–708, 1976.
35. Bottenstein JE, deVellis J: Regulation of cyclic GMP, cyclic AMP and lactate dehydrogenase by putative neurotransmitters in the C-6 rat glioma cell line. Life Sci 23:821–834, 1978.
36. Chelmicka-Schorr E, Arnason BGW, Holshouser SJ: C-6 glioma growth in rats: Suppression with a β-adrenergic agonist and a phosphodiesterase inhibitor. Ann Neurol 8:447–449, 1980.
37. Marton LJ, Heby O, Levin VA, et al: The relationship of polyamines in cerebrospinal fluid to the presence of central nervous system tumors. Cancer Res 36:973–977, 1976.
38. Marton LJ, Edwards MS, Levin VA, et al: Predictive value of cerebrospinal fluid polyamines in medulloblastoma. Cancer Res 39:993–997, 1979.
39. Rall DP, Zubrod CG: Mechanism of drug absorption and excretion. Passage of drugs in and out of the central nervous system. Ann Rev Pharmacol 2:109–128, 1962.
40. Vick NA, Khandekar JD, Bigner DD: Chemotherapy of brain tumors: The "blood-brain barrier" is not a factor. Arch Neurol 34:523–526, 1977.
41. Ausman JI, Levin VA, Brown WE, et al: Brain-tumor chemotherapy: Pharmacological principles derived from a monkey brain-tumor model. J Neurosurg 46:155–164, 1977.
42. Levin VA, Freeman-Dove M, Landahl HD: Permeability characteristics of brain adjacent to tumors in rats. Arch Neurol 32:785–791, 1975.
43. Blasberg RG, Gazendam J, Patlak CS, et al: Changes in blood-brain transfer parameters induced by hyperosmolar intracarotid infusion and by metastatic tumor growth. In: The Cerebral Microvasculatures, Eisenberg HM, Suddith RL (eds). New York, Plenum Publ Co, 1980, pp 307–319.
44. Ushio Y, Chernik N, Shapiro W, et al: Metastatic tumor of the brain: Development of an experimental model. Ann Neurol 2:20–29, 1977.
45. Blasberg RG, Groothius D, Molnar P: The application of quantitative autoradiographic measurements in experimental brain tumor models. Semin. Neurol. 1:203–221, 1981.
46. Chabner BA and Johns DG: Folate antagonists. In: Cancer 5 a Comprehensive Treatise, Becker FF (ed). New York, Plenum Press, 1977, Vol 5, pp 363–377.

47. Levin VA, Patlak CS, Landahl HD: Heuristic modeling of drug delivery to malignant brain tumors. J Pharmacokinetics Biopharmaceutics 8:257–296, 1980.
48. Neuwelt EA, Maravilla KR, Frenkel EP, et al: Osmotic blood-brain barrier disruption. Computerized tomographic monitoring of chemotherapeutic agent delivery. J Clin Invest 64:684–688, 1979.
49. Hasegawa H, Allen JC, Mehta BM, et al: Enhancement of CNS penetration of methotrexate by hyperosmolar intracarotid mannitol or carcinomatous meningitis. Neurol 29:1280–1286, 1979.
50. Neuwelt EA, Frenkel EP, Diehl JT, et al: Initial clinical studies of chemotherapy delivery to brain tumor patients after osmotic blood-brain barrier modification. Proc Amer Soc Clin Oncol 22:531, 1981.
51. Allen JC, Hasegawa H, Mehta BM, et al: Influence of intracarotid mannitol on intracerebral methotrexate (MTX) concentrations surrounding experimental brain tumors. Proc Amer Assoc Cancer Res 20:286, 1979.
52. Shapiro WR, Allen JC, Horten BC: Chronic methotrexate toxicity to the central nervous system. Clin Bull 10:49–52, 1980.
53. Greenberg HS, Ensininger WD, Kindt GW, et al: Intra-arterial BCNU chemotherapy for malignant astrocytomas. Neurol 31 (Part 2):69, 1981.
54. Madajewicz S, West CR, Park HC, et al: Phase II study-intra-arterial BCNU therapy for metastatic brain tumors. Cancer 47:653–657, 1981.
55. Takvorian T, Hochberg F, Canellos G, et al: The toxicity of high-dose BCNU with autologous marrow support. In: Nitrosoureas, Current Status and New Development, Prestayko AW, Crooke ST, Baker LG, et al (eds). New York, Academic Press, 1981, pp 155–169.
56. Klatzo I: Neuropathological aspects of brain edema. J Neuropath Exp Neurol 26:1–14, 1967.
57. Fishman RA: Brain edema. New Engl J Med 293:706–711, 1975.
58. Calabrese P, Parks RE: Chemotherapy of neoplastic diseases. In: The Pharmacological Basis of Therapeutics, Gilman AG, Goodman LS, Gilman A (eds). New York, Macmillan Publishing Co, Inc, 1980, pp 1249–1313.
59. Keiser LW, Capizzi RL: Principles of combination chemotherapy. In: Cancer 5 a Comprehensive Treatise, Becker FF (ed). Plenum Press, 1977, pp 163–185.
60. French JD, West PM, von Amerongen PK, et al: Effects of intracarotid administration of nitrogen mustard on normal brain and brain tissues. J Neurosurg 9:378–389, 1952.
61. Wheeler GP, Alexander JA: Effects of nitrogen mustard and cyclophosphamide upon the synthesis of DNA in vivo and in cell free preparations. Cancer Res 29:98–109, 1969.
62. Connors TA: Alkylating drugs, nitrosourea and dimethyltriazenes. In: Cancer Chemotherapy 1980: The EORTC Cancer Chemotherapy Annual 2, Pinedo HM (ed). Amsterdam – Oxford, Excerpta Medica, 1980, pp 27–65.
63. Connors TA: Alkylating drugs, nitrosourea and dialkyltriazenes. In: Cancer Chemotherapy 1979: The EORTC Cancer Chemotherapy Annual 1, Pinedo HM (ed). Amsterdam – Oxford, Excerpta Medica, 1979, pp 25–55.
64. Edwards MS, Levin VA, Wilson CB: Brain tumor chemotherapy: An evaluation of agents in current use for phase II and III trials. Cancer Treat Rep 64:1179–1205, 1980.
65. Kann HE: Carbamoylating activity of nitrosoureas. In: Nitrosoureas Current Status and New Developments, Prestayko AW, Crooke ST, Baker LH, et al (eds). New York, Academic Press, 1981, pp 95–106.
66. Barranco SC, Humphrey RM: The effects of BCNU on survival and cell progression in Chinese hamster cells. Cancer Res 31:191–195, 1971.
67. Tobey RA, Crissman HA: Comparative effects of three nitrosourea derivatives on mammalian cell cycle progression. Cancer Res 35:460–470, 1975.
68. Begleiter A, Lam HYP, Goldenberg GJ: Mechanism of uptake of nitrosoureas by L5178Y lymphoblasts in vitro. Cancer Res 37:1022–1027, 1977.

69. Hall TC: Prediction of responses to therapy and mechanisms of resistance. Sem Oncol 4:193–202, 1977.
70. Rosenblum M, Deen D, Levin V, et al: Resistance of brain tumors to chemotherapy-stem cell studies of human and animal tumors. Proc Amer Assoc Cancer Res 22:243, 1981.
71. DeVita VT, Denhan C, Davidson J, et al: The physiological disposition of carcinostatic 1,3-bis(2-chloroethyl)-1-nitrosourea in man and animals. Clin Pharmacol Ther 8:566–577, 1967.
72. Aronin PA, Mahaley MS, Rudnick SA, et al: Prediction of BCNU pulmonary toxicity in patients with malignant gliomas. New Engl J Med 303:183–188, 1980.
73. Macdonald JS, Weiss RB, Poster D, et al: Subacute and chronic toxicities associated with nitrosourea therapy. In: Nitrosoureas Current Status and New Developments, Prestayko AW, Crooke ST, Baker LH, et al (eds). New York, Academic Press, 1981, pp 145–154.
74. Nichols WC, Moertel CG: Letter. New Engl J Med 301:1181, 1979.
75. Walker M, Alexander E, Hunt WF, et al: Evaluation of BCNU and/or radiotherapy in the treatment of anaplastic gliomas. J Neurosurg 49:333–343, 1978.
76. Wheeler KT, Deen DF, Wilson CB, et al: BCNU modification of the in vitro response in 9L brain tumor cells of rats. Int J Radiat Oncol Biol Phys 2:79–88, 1977.
77. Deen DF, Williams ME, Wheeler KT: Comparison of the CCNU and BCNU modification of the in vitro radiation response in 9L brain tumor cells of rats. Int J Radiat Oncol Biol Phys 5:1541–1544, 1979.
78. Thuning CA, Baker NA, Warren J: Synergistic effect of combined hyperthermia and a nitrosourea in treatment of a murine ependymoblastoma. Cancer Res 40:2726–2729, 1980.
79. Spivack SD: Drugs 5 years later: Procarbazine. Ann Intern Med 81:795–800, 1974.
80. Geran RI, Congleton GF, Dudeck LE, et al: A mouse ependymoblastoma as an experimental model for screening potential antineoplastic drugs. Cancer Chemother Rep (Part 2), 4:53–87, 1974.
81. DeJager R, Brugarolas A, Hansen H, et al: Dianhydrogalactitol (NSC-132313): Phase II study in solid tumors. Eur J Cancer 15:971–974, 1979.
82. Hoogstraten B, O'Bryan R, Jones S: 1,2:5,6-Dianhydrogalactitol in advanced breast cancer. Cancer Treat Rep 62:841–842, 1978.
83. Eckhardt S, Csetenyi J, Horvath IP, et al: Uptake of labeled dianhydrogalactitol into human gliomas and nervous tissue. Cancer Treat Rep 61:841–847, 1977.
84. Espana P, Wiernik PH, Walker MD: Phase II study of dianhydrogalactitol in malignant glioma. Cancer Treat Rep 62:1199–1200, 1978.
85. Eagen RT, Childs DS, Layton DD, et al: Dianhydrogalactitol and radiation therapy, treatment of supratentorial glioma. JAMA 241:2046–2050, 1979.
86. Hill JM, Loeb E, MacLellan A, et al: Clinical studies of platinum coordination compounds in the treatment of various malignant diseases. Cancer Chemother Rep 59:647–659, 1975.
87. Kham A, McCullough D: Use of cis-platinum in CNS malignancy. Proc Amer Assoc Cancer Res 20:326, 1976.
88. Krakoff IH, Brown NC, Reichard P: Inhibition of ribonucleoside diphosphate reductase by hydroxyurea. Cancer Res 28:1559–1565, 1968.
89. Beckloff GL, Lerner HJ, Frost D, et al: Hydroxyurea (NSC-32065) in biological fluids: dose-concentration relationships. Cancer Chemother Rep 48:57–58, 1965.
90. Piver MS, Howes AE, Suite HD, et al: Effect of hydroxyurea on the radiation response of C3H mouse mammary tumors. Cancer 29:407–412, 1972.
91. Levin VA, Wilson CB, Davis R, et al: A phase III comparison of BCNU, hydroxyurea and radiation therapy to BCNU and radiation therapy for treatment of primary malignant gliomas. J Neurosurg 51:526–532, 1979.
92. Allen L: The role of drug disposition kinetics on cellular transport of the antineoplastic agent VM-26. Drug Metab Rev 8:119–135, 1978.

93. Bender RA: Vinca alkaloids and epipodophyllotoxins. In: Cancer Chemotherapy 1979, The EORTC Cancer Chemotherapy Annual I, Pinedo HM (ed). Amsterdam-Oxford, Excerpta Medica, 1979, pp 100–106.
94. Krishan A, Paika K, Frei E: Cytofluorometric studies on the action of podophyllotoxin and epipodophyllotoxins (VM-26, VP-16-213) on the cell cycle traverse of human lymphoblasts. J Cell Biol 66:521–530, 1975.
95. Creaven PJ, Allen LM: PTG, a new antineoplastic epipodophyllotoxin. Clin Pharmacol Ther 18:227–233, 1975.
96. Skylansky BD, Mann-Kaplan RS, Reynolds AF, et al: 4'-Dimethyl-epipodophyllotoxin-α-D-thenylidene-glucoside (PTG) in the treatment of malignant intracranial neoplasms. Cancer 33:460–467, 1974.
97. Kessinger A, Lemon HM, Foley JF: VM-26 as a second drug in the treatment of brain gliomas. Proc Amer Assoc Cancer Res 20:295, 1979.
98. Donoso JA, Green LS, Heller-Bettinger IE, et al: Action of the vinca alkaloids vincristine, vinblastine desacetyl, vinblastine amide on axonal fibrillar organelles in vitro. Cancer Res 37:1401–1407, 1977.
99. Shapiro WR, Young DF: Treatment of malignant glioma with CCNU alone and CCNU combined with vincristine sulfate and procarbazine hydrochloride. Trans Amer Neurol Assoc 101:217–220, 1976.
100. Bourke RS, West CR, Chheda G, et al: Kinetics of entry and distribution of 5-fluorouracil in cerebrospinal fluid and brain following intravenous injection in a primate. Cancer Res 33:1735–1746, 1973.
101. McDonnell LA, Potter PE, Leslie RA: Localized changes in blood-brain barrier permeability following the administration of antineoplastic drugs. Cancer Res 38:2930–2934, 1978.
102. Chabner BA: Antimetabolites. In: Cancer Chemotherapy 1979, the EORTC Cancer Chemotherapy Annual I, Pinedo HM (ed). Amsterdam – Oxford, Excerpta Medica, 1979, pp 1–10.
103. Chello PL, Sirotnak FA, Dorick DM: Kinetics and growth phase dependence of methotrexate and folic acid transport by L1210 leukemia cells. Proc Amer Assoc Cancer Res 20:219, 1979.
104. Alt FW, Kellems RE, Bertino JR, Schimke RJ: Selective multiplication of dihydrofolate reductase genes in methotrexate-resistant variants of cultured murine cells. J Biol Chem 253:1357–1370, 1978.
105. Allen JC, Rosen G, Mehta BM, et al: Leucoencephalopathy following high-dose IV methotrexate chemotherapy with leucovorin rescue. Cancer Treat Rep 64:1261–1273, 1980.
106. Goldman ID: The membrane transport of methotrexate and other folate compounds: Relevance of rescue protocols. Cancer Chemother Rep (Part 3) 6:63–72, 1975.
107. Glass JP, Shapiro WR, Posner JB: Treatment of leptomeningeal metastases. Neurology 28:351, 1978.
108. Shapiro WR: High-dose methotrexate in malignant gliomas. Cancer Treat Rep 61:691–694, 1977.
109. Kristiansen K, Hagen S, Kollevold T, et al: Combined modality therapy of operated astrocytomas grade III and IV. Confirmation of the value of postoperative irradiation and lack of potentiation of bleomycin on survival time. Cancer 47:649–652, 1981.
110. Walker MD, Alexander E, Hunt WE, et al: Evaluation of mithramycin in the treatment of anaplastic gliomas. J Neurosurg 44:655–667, 1976.
111. Brzustowicz RJ, Svien HJ, Bennett WA, et al: The effect of cortisone on the growth of transplanted ependymomas in mice. Proc Staff Meet Mayo Clin 26:121–128, 1951.
112. Kotsilimbas DG, Meyer L, Berson M, et al: Corticosteroid effect on intracerebral melanomata and associated cerebral edema: Some unexpected findings. Neurology 17:223–226, 1967.
113. Gurcay O, Wilson C, Barker M, et al: Corticosteroid effect on transplantable rat glioma. Arch Neurol 24:266–269, 1971.

114. Shapiro WR, Posner JB: Corticosteroid hormones: Effects in an experimental brain tumor. Arch Neurol 30:217–221, 1974.
115. Chen TT, Mealey J: Effect of corticosteroid on protein and nucleic acid synthesis in human glial tumor cells. Cancer Res 33:1721–1723, 1973.
116. Ausman JI, Shapiro WR, Rall DP: Studies on chemotherapy of experimental brain tumors: Development of an experimental model. Cancer Res 30:2394–2400, 1970.
117. Levin VA, Wilson CB: Correlations between experimental chemotherapy in the murine glioma and effectiveness of clinical therapy regimens. Cancer Chemother Pharmacol 1:41–48, 1978.
118. Skipper HE: Cancer chemotherapy is many things: GHA Clowes Memorial Lecture. Cancer Res 31:1173–1180, 1971.
119. Shapiro WR, Basler GA, Chernik NL, et al: Human brain tumor transplantation into nude mice. J Natl Cancer Inst 62:447–453, 1979.
120. DiChiro G, DeLaPuz R, Smith B, et al: ^{18}F-2-fluoro-2-deoxyglucose positron emission tomography of human cerebral gliomas. J Cerebral Blood Flow Metab 1:511–512, 1981.
121. Frei E, Lazarus H: Predictive tests for cancer chemotherapy. New Engl J Med 298:1358–1359, 1978.
122. Salmon SE, Hamburger AW, Soehnlen BJ, et al: Quantititations of differential sensitivities of human tumor stem cells to anticancer drugs. New Engl J Med 298:1321–1327, 1978.
123. Pavelic ZP, Slocum HK, Rustum YM, et al: Colony growth in soft agar of human melanoma, sarcoma, and lung carcinoma cells disaggregated by mechanical and enzymatic methods. Cancer Res 40:2160–2164, 1980.
124. Rosenblum ML, Vasquez DA, Hoshino T, et al: Development of a clonogenic cell assay for human brain tumors. Cancer 41:2305–2314, 1978.
125. Kornblith PL, Szypko PE: Variations in response of human brain tumors to BCNU in vitro. J Neurosurg 48:580–586, 1978.
126. Rosenblum ML, Wheeler KT, Wilson CB, et al: In vitro evaluation of in vivo brain tumor chemotherapy with 1,3-bis(2-chloroethyl)-1 nitrosourea. Cancer Res 35:1387–1391, 1975.
127. Kornblith PL, Smith BH, Leonard LA: Response of cultured human brain tumors to nitrosoureas. Cancer 47:255–265, 1981.
128. Freshney RI: Tissue culture of glioma of the brain. In: Brain Tumors, Scientific Basis, Clinical Investigation and Current Therapy, Thomas DGT and Graham DI (eds). London, Butterworth, 1980, pp 21–50.
129. Saez RJ, Campbell RJ, Laws ER: Chemotherapeutic trials on human malignant astrocytomas in organ culture. J Neurosurg 46:320–327, 1977.
130. Rubinstein LJ, Herman MM: Studies on the differentiation of human and experimental gliomas in organ cultures. In: Recent Results in Cancer Research, Hekmatpanah J (ed). Berlin, Springer-Verlag, 51:35–51, 1975.
131. Levin VA, Hoffman W, Weinkam J: Pharmacokinetics of BCNU in man: A preliminary study of 20 patients. Cancer Treat Rep 62:1305–1312, 1978.

4. The Role of Intracranial Surgery for the Treatment of Malignant Gliomas

JOSEPH RANSOHOFF

INTRODUCTION

The malignant glioma, grades III and IV, is a cancer of rare biological consistency which until the advent of recent therapeutic advances had a median survival time of 6 months and killed 90% in 18 months [1]. There are 385,000 deaths from cancer each year in the USA of which 50,000 (13%) are associated with central nervous system involvement. It is estimated that of the 50,000 deaths, 8,500 (17%) are caused by primary brain tumors and 5,000 of these can be attributed to the malignant glioma or glioblastoma. The number of deaths from malignant gliomas is about half those reported from leukemia but are, in fact, more common than those from Hodgkin's disease [2]. At the New York University Medical Center in the last 1,000 surgically verified primary supratentorial tumors in adults, 500 represent tumors of glial origin. Eighty percent of glial tumors were in the malignant glioma and glioblastoma class, 10% in the grade I, II and mixed glial groups and 10% represented oligodendrogliomas, ependymomas and other less common forms of primary tumors of neural origin.

The role of intracranial surgery in the management of malignant gliomas has changed considerably in the last decade. Improvements in diagnostic techniques, particularly the advent of CT scanning, the development of programs for improved radiation therapy and the advent of multiple chemotherapies for the treatment of brain tumors have all led to a more positive approach to these neoplasms and pari passu a more aggressive and definitive surgical attitude. There is no longer a real justification for the small craniotomy or "biopsy" followed by no definitive therapy, in a sense, sending the patient home to die. It appears that there is no longer justification for a large external decompression achieved by removing a segment of skull and leaving the dura widely open which did prolong life but resulted in cerebral herniation at the decompression site and often the eventual development of cerebral fungus. Even the more recent practice of the removal of large amounts of normal brain tissue (lobectomy) in order to provide for internal decompression is rarely warranted as preservation of normal brain tissue has been shown to be of increasing importance. As survival times lengthen the long-term effects of radiation therapy and chemotherapy are being found to produce signs and symptoms of progressive cerebral atrophy with the loss of higher integrative functions, hence the need to spare normal tissue.

Surgery in the management of malignant gliomas must be justified on the basis of good surgical principles as well as on sound oncologic concepts. Aggressive tumor resection is unquestionably the most effective method of rapid reduction of tumor burden. In the case of the tumor under consideration, not only is tumor burden reduced but the hypoxic core area is known to contain those cells most resistant to radiation therapy and least available to chemotherapy delivered via the vascular system. This region also contains the necrotic areas which cannot be cleared by the brain, lacking as it does an effective lymphatic system. The surgical aims must include the reduction of the mass effect of the tumor including the shift of brain structures and the relief of increased intracranial pressure. Postoperative neurological deficits should be minimal both in terms of focal and diffuse signs as the patient's functional status post surgery has been shown to have a significant correlation with survival time.

In summary then, intracranial surgery in the management of malignant brain tumors should be based on the neurosurgical principles of elimination of symptoms and the preservation of normal anatomy and oncological principles which include reduction of tumor burden, hypoxic cells and necrotic tissue.

DIAGNOSIS

Malignant astrocytoma has its peak incidence between the ages of 40 and 60 and affects males more commonly than females in a ratio of 3 to 2. While they may occur anywhere in the central nervous system they are most common in the supratentorial white matter. Malignant gliomas arising in the pons, cerebellum and spinal cord constitute less than 5% of these tumors in adults [3]. About 5% are multifocal, showing neither macro or microscopic continuity with other tumor centers. Approximately 2% are truly multicentered with no continuity between tumor centers [4].

Signs and symptoms of intracranial gliomas are about equally divided between phenomenon related to diffuse cerebral dysfunction, focal neurological deficits and convulsive seizures. Complaints related to increased intracranial pressure include headache, obtundation and finally nausea and vomiting. Diffuse signs of cerebral dysfunction which are often subtle at the onset include difficulty with short-term memory, subtle personality changes and such things as lack of attention to personal hygiene. These are often only recognized in retrospect. A frank generalized or focal motor seizure is, of course, a dramatic event and leads to early definitive diagnostic evaluation. The onset of hemiparesis, a visual field defect and even mild speech disturbance are at times missed for several weeks and

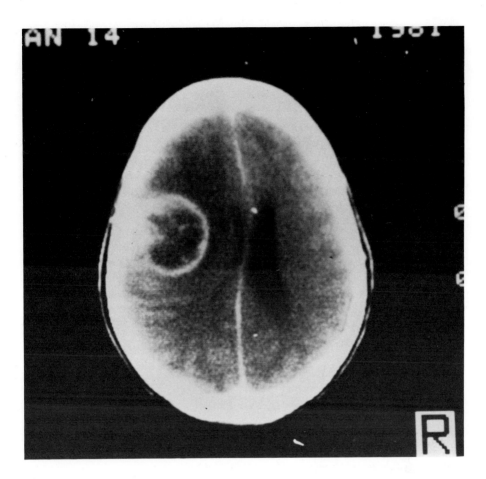

Figure 1. Typical enhanced CT scan of glioblastoma.

in the older age group often attributed to cerebral vascular disease. Depending upon the location of the glioma, which arises in the central white matter, those occurring in the nondominant temporal lobe, frontal lobes, or corpus collosum can often be subtle in terms of the development of symptoms and grow to considerable size before the patient is suspect of harboring a brain tumor.

CT scanning with and without contrast enhancement particularly with the newer high resolution equipment, is relatively definitive in terms of establishing the diagnosis of a single intracranial malignant tumor. A nonenhancing mass lesion surrounded by significant white matter edema with solid or so-called ring (Figure 1) enhancement is seen in malignant gliomas with the differential diagnosis being that of a single intracranial metastatic tumor or on rare occasions, a brain abscess. A single intracranial metastatic tumor cannot be distinguished from a malignant glioma by neuroradiological studies. The radio

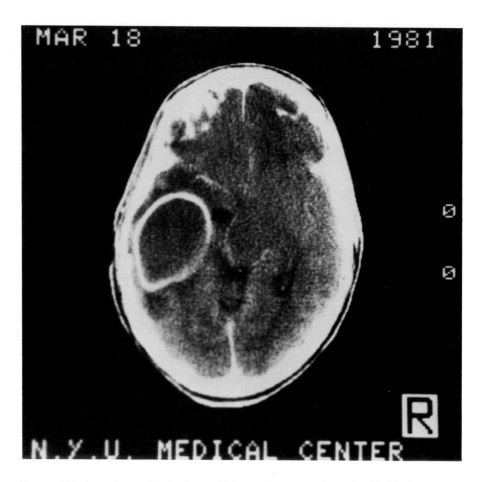

Figure 2. CT enhanced scan of brain abscess. Note smooth contour of capsule with thinning and poor definition medially.

nuclide scan of brain abscess often will demonstrate a ring type pickup in contrast to the solid image seen in the malignant gliomas. The CT ring enhancement in brain abscess often shows a thinning of this ring in the deeper tissues, at times the site of perforation into the ventricular system (Figure 2). Double dose enhancement studies as well as delayed scanning can be useful if multiple lesions are suspected and radio nuclide scanning still remains an important study from two aspects. The Technetium 99 scan can at times demonstrate a multiplicity of lesions not seen on the earlier generation scanners. Secondly, the lower grade gliomas tend to have CT studies showing a nonenhancing mass (Figure 3) and negative brain scans which is important clinical information relative to treatment. Cerebral angiography will also often demonstrate typical neovascularity producing the so-called tumor stain in malignant gliomas and the presence of an

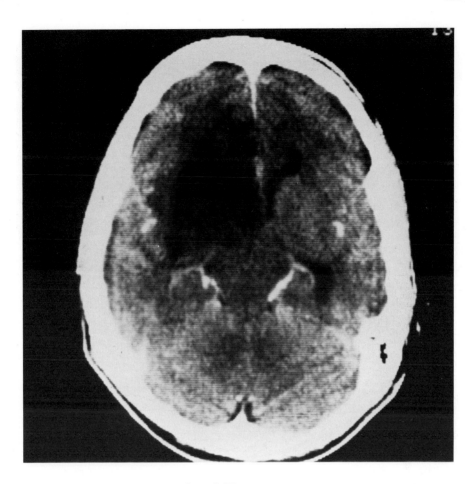

Figure 3. Grade II astrocytoma on enhanced CT scan.

"early draining vein" is seen in malignant tumors but can also be seen with brain abscesses. Cerebral angiography should be selective, i.e., internal carotid injection and external carotid injection via a femoral catheter technique. Extensive supply to an otherwise malignant appearing tumor from the external carotid vasculature is much more likely to represent a meningioma, or other type of tumor (Figures 4 and 5). Lumbar puncture is not indicated and ventriculography and pneumoencephalography have about totally disappeared from the neurodiagnostic armamentarium in the diagnosis of malignant brain tumors. If biopsy is deemed necessary in order to plan therapy in a surgically inaccessible tumor this is best carried out via stereotactic techniques with the aid of CT scanning.

In one study 96% of all clinically symptomatic gliomas were detected by CT scan versus 76% detected by brain scan, 75% by angiogram, and 80% by clinical

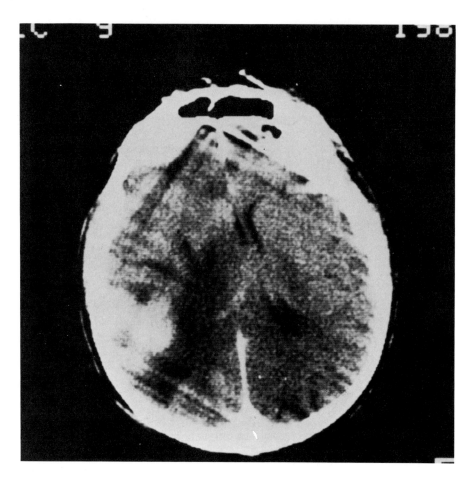

Figure 4. CT of meningioma with edema. Differential diagnosis – glioma.

assessment. Four percent of malignant gliomas were not detected by CT scanning. Rarely false positive scans were noted when artifacts were interpreted as tumors [5]. In another study, the CT scan was able to differentiate among different histological grades of gliomas with reasonable accuracy.

In summary then, a positively enhancing CT scan with neovascularity and early draining veins being demonstrated on selective angiography is almost diagnostic of malignant glioma or a single intracranial metastasis. In the absence of a tumor stain on angiography, a solidly positive radio nuclide scan has almost equal diagnostic significance.

From a systemic point of view, of course, short of subtle and special studies of immunological reactivity the systemic work-up of the patient with malignant glioma should be within normal limits dependent upon the presence of preexisting disease. The sedimentation rate should also be within normal limits.

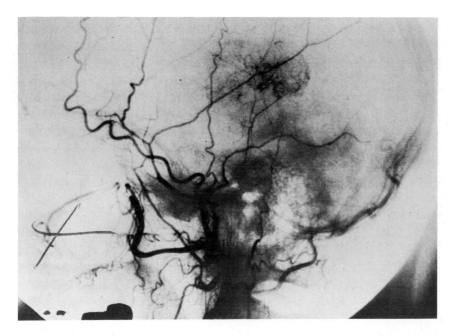

Figure 5. Selective external angiogram demonstrating dural supply to meningioma.

PREOPERATIVE MANAGEMENT AND SURGICAL DECISION MAKING

The preoperative management of the patient with a malignant glioma should include several days of anticonvulsant therapy (Dilantin, 100 mgs., t.i.d. and phenobarbital, 30 mgs., b.i.d., the latter being added in the patient who has preoperative seizures as a part of the presenting symptomatology. Corticosteroids are routinely used preoperatively and can be of great importance in the final decision making process, particularly in patients with significant neurological symptomatology, either obtundation or/and neurological deficit. The patient whose hemiparesis and whose dysphasia clear on large doses of intravenous corticosteroids administered over several days becomes a possible candidate for surgical tumor removal whereas the initial impression would have been contrary to this position. Corticosteroids do not have an oncolitic effect and hence, the response, for example, to one gram of methylprednisolone in divided doses intravenously for several days is based on the reduction of surrounding brain edema rather than a direct effect on the tumor mass itself and indicates that the tumor is not involving necessarily critical areas subserving brain function. The patient who, on the other hand, does not respond with improvement in neurological deficit, all other things being equal, is one in whom surgical intervention is likely not to improve the neurological condition. In conjunction

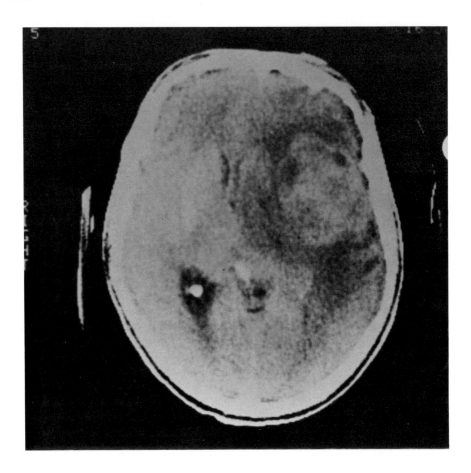

Figure 6. CT of dominant hemisphere glioma.

with the response to corticosteroids, a study of the CT scan and angiography will be important relative to the surgical decision. The CT scan will demonstrate the relationship of the tumor mass to the adjacent ventricular systems and is of greater value than the angiogram which often demonstrates displacement of normal vasculature. In a tumor which on the basis of vascular displacement might be considered to be arising within the thalamus will be seen on CT scan to displace the ventricular system medially and upward particularly the temporal horn, indicating the true and surgically accessible location of the tumor. When, however, a tumor stain on angiography can be seen to surround critical vasculature such as middle cerebral artery and is being fed by thalamo perforating vessels, it can be indicative of the possibility of a poor surgical result (Figure 6 and 7). Extremely hypervascular gliomas, except those located in polar positions, that is, the tip of the frontal lobe or the temporal lobe or occipital lobe, can also represent difficult

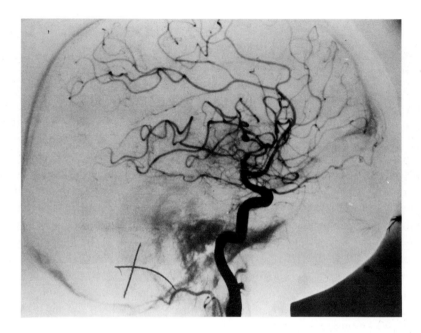

Figure 7. Angiogram demonstrating right carotid and middle cerebral arteries encased in tumor stain.

surgical problems and may sway the surgeon against surgical intervention. Corpus collosum or so-called butterfly gliomas, particularly those arising in the posterior half of the structure are rarely good surgical candidates. Frontal lobe gliomas which appear to be invading the corpus collosum are not necessarily, however, excluded from surgical management. The decision, therefore, for primary surgical intervention is based upon the patient's neurological status with or without the use of high dose corticosteroids, the CT scan and angiographic demonstration of the exact location of the tumor mass, realizing that the goal of primary surgical therapy is that of achieving a radical removal of tumor tissue with preservation of normal brain and restoration or maintenance of good neurological function. Primary surgical intervention generally excludes those patients with tumors located directly within the area subserving speech, those arising clearly within the thalamus on either side, those arising from the corpus collosum, particularly the posterior half of this structure. These patients are generally considered for radiation therapy and chemotherapy as a primary mode of treatment and surgery reserved for the occasional delayed intervention. The criterion for delayed surgery and surgery of recurrent tumors will be discussed in the subsequent section of this chapter.

SURGICAL TECHNIQUE

Once the most important surgical decision has been reached, that is, to carry out primary surgical removal and a careful preoperative planning has been undertaken, all patients are placed on 250 mgs. of Solu-Medrol intravenously the night before surgery and again the following morning. Anticonvulsant dosages are doubled for the 12 hours preoperatively and 500 ccs. of 20% mannitol are given intravenously via a previously placed central line after induction of anesthesia. Whereas the details of the surgical technique are not applicable to this chapter, certain surgical principles will be mentioned.

The scalp incision should, if possible, be placed behind the hair line, however, cosmetic considerations are of secondary importance particularly as high dose radiation therapy produces severe hair loss, usually on a permanent basis. A large free bone flap is developed centered over the area of brain planned for elective excision if the tumor is not seen to be presented on the brain surface. The tumor should be approached from the direction opposite from these areas involving the most critical brain functions except in those situations where the tumor is located mainly at one of the polar areas in the brain where a segment of surrounding edematous brain can be removed as an en bloc resection. Many gliomas have a pseudocapsule of gliogenous tissue and can be removed grossly staying in this plane between tumor and edematous brain. Tumor resection should be carried wherever possible to predetermined limits, whether it be the lateral wall of the ventricle, the medial margin of the tentorium or other clearly identifiable neuroanatomical structures. The operating microscope is used as needed in tumor dissection and is particularly helpful in defining the margin between neoplasm and surrounding edematous or subtle gliomatous capsule. A newly designed instrument, (CUSA) [6], recently commercially available, removes tumor with the use of ultrasound. The tip of this instrument measuring 5 mms. in diameter, vibrates 20,000 per second and emulsifies tumor tissue. It is then removed through the same instrument with the use of irrigation and suction. Tumor removal by this technique has enabled surgeons to dissect neoplastic tissue from adjacent critical brain areas such as speech and motor centers without the production of a significant neurological deficit. This technique produces neither heat nor traction to surrounding tissues and, hence, its particular use for the removal of areas of firm tumor tissue. Once the planned resection has been achieved careful hemostasis is achieved in the tumor bed with the use of the two-point cautery which coagulates small vessels but does require a body ground plate, hence, the current does not pass through adjacent structures and can be used along the brainstem or other critical areas and has proven to be of great benefit in achieving hemostasis and preventing postoperative hematomas. The tumor cavity is then lined with a thin layer of oxidized (absorbable) cotton or gauze to prevent further oozing which might occur in the postoperative period.

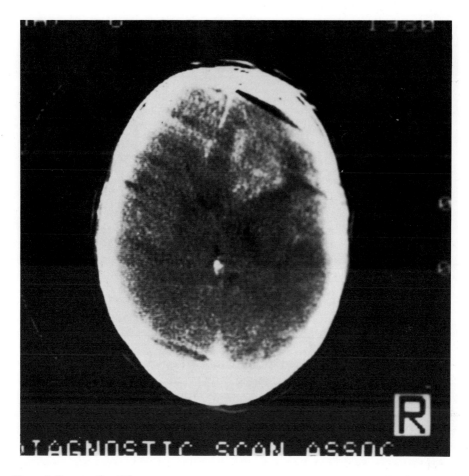

Figure 8. Preoperation CT scan.

The greatest risk of postoperative hemorrhage occurs in those situations where tumor tissue has been incompletely removed as the friable and abnormal capillaries are much more difficult to control than normal brain vasculature. Following hemostasis, the dura is closed as completely as possible and the free bone flap wired in place. There is rarely, if ever, the indication for leaving the dura open and discarding the bone flap. The brain should be well relaxed following the completion of tumor removal and if this is not the case, one must suspect either a deep hemorrhage or inadequate tumor removal and additional tumor must be removed. Only if this is not possible, additional brain is sacrificed in order to achieve adequate relaxation. It has parenthetically, been clearly demonstrated that a so-called external decompression reduces cerebral tissue pressure and invites the occurrence of overwhelming brain edema [7]. In the postoperative period corticosteroids are continued at a high dosage level (1 gram

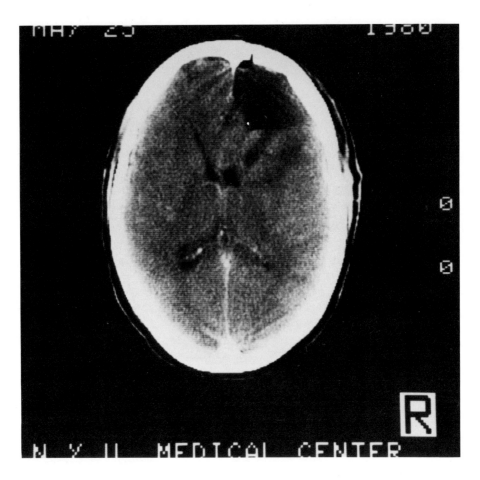

Figure 9. Postoperative CT scan 10 days following 'gross total removal'.

of Methylprednisolone q.d. in divided doses) for three or four days and then gradually tapered over the subsequent week to ten days. A noncontrast CT scan should be obtained early in the postoperative period to document the amount of postoperative brain shift and the presence of any significant hematoma, either in the tumor bed or in the subdural or epidural spaces and the amount of residual tumor (Figures 8 and 9). Anticonvulsants, of course, are continued.

In summary then, the preoperative planning, the operative techniques and the postoperative care must be carried out in a meticulous fashion in order to achieve an accepted morbidity and mortality rate in the conduct of radical glioma surgery. In patients who have been in good condition prior to surgery, the mortality or/and serious morbidity rate should be that of all elective neurosurgery, somewhere in the neighborhood of 7% to 10%. Review of our past 100 patients with glioblastoma whose preoperative Karnofsky Scales were 70 or better revealed a mortality or morbidity rate (K Scale below 60) or 11%. Postoperative radiation therapy and chemotherapy can be initiated within a week to ten days following surgical removal. In patients who have been maintained on corticosteroids for an inordinately long period of time prior to surgery, one can experience the phenomenon of cerebral steroid dependency and a far slower steroid taper may be required, particularly with the use of early postop radiation therapy.

Surgical removal of recurrent gliomas can be considered in patients who have had early satisfactory results and in whom regrowth of tumor can be documented on CT scan. One would hesitate to consider a patient for reoperation short of nine months to a year following the initial surgery and only in those patients in whom meaningful existence has been provided by the initial treatment regime. The tumor, of course, should be strictly localized to its original site and if evidence of extension across the midline to the other hemisphere or into deeper brain structures, including the thalamus or brainstem can be documented on CT scan surgery is not applicable. A far greater number of patients now are being seen with strictly local recurrence of tumor, undoubtedly as a result of the more rigorous use of radiation therapy and chemotherapy than previously had been the case. These tumors at the time of repeat surgery are often very well demarcated from surrounding brain and postoperative CT scan can be seen often to be clean of any signs of gross tumor. In carefully selected patients a significant prolongation of useful life can be achieved, particularly if additional chemotherapy is applicable following the removal of tumor regrowth (Figure 10).

Secondary surgery can at times also be considered, that is, tumor removal in patients who have been deemed to be inoperable in terms of tumor location in the initial evaluation or in whom for other reasons, radiation and chemotherapy has been the initial treatment of choice. If an initial good response has been seen with significant reduction of tumor size and later regrowth is documented, the experience has been similar to that mentioned above in terms of the ease of tumor removal as compared at times to the primary surgical experience. In general, tumor vascularity is significantly reduced particularly in those lesions which were hypervascular on initial angiography and the sharp demarcation between tumor and surrounding tissue makes the consideration of the later secondary surgery in selected cases a reasonable option. Once again, the use of the ultrasonic surgical aspirator has proven to be invaluable in terms of approaching relatively critical brain areas.

The current status of surgical therapy of malignant gliomas can be summarized as follows. With the correlative information available from CT scanning, radio nuclide scanning, cerebral angiography as well as that obtained from the general medical work-up, surgery is rarely indicated to establish the diagnosis of a malignant glioma. When indicated, a carefully controlled biopsy can be obtained through a burr hole needle biopsy, localization being achieved with the use of CT scanning in two planes. Carefully planned and executed surgical approach to

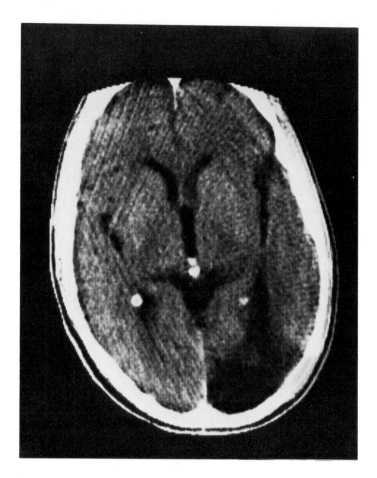

Figure 10. Postoperative scan following removal of recurrent occipital lobe glioma.

these tumors can often achieve a near total removal of grossly recognizable tumor tissue as documented on postoperative CT scans. Mortality and morbidity rates in these highly malignant tumors can be kept at acceptable levels. Clearly, radiation therapy and chemotherapy have made a significant impact on long-term survival in the treatment of this cancer, however, surgery still remains the single most effective method for achieving a rapid reduction of tumor burden and in addition, one achieves a reduction in increased intracranial pressure and relief of neurological signs and symptoms, hence, avoiding the complications of long-term corticosteroid therapy. Finally, following surgery, the other antitumor programs have the best chance for achieving a significant increment of tumor kill [8]. Surgical therapy, therefore, has a distinct role to play in the multidisciplinary approach to the treatment of these difficult and highly aggressive neoplasms.

REFERENCES

1. Walker MD, Alexander E, Hunt W et al: Evaluation of BCNU and/or radiotherapy in the treatment of anaplastic gliomas. J Neurosurg 49:333–343, 1978.
2. Shapiro W: Chemotherapy of nervous system neoplasms in primary intracranial neoplasms. In: Scher J, Ford D (eds). New York, Spectrum Press, 1979.
3. Russell DS, Rubinstein LJ: Pathology of the nervous system, 4th edition. Baltimore, The Williams and Wilkins Company, 1977, pp 147–282.
4. Jellinger K: Glioblastoma multiforme: morphology and biology. Acta Neurochirg 42:5–32, 1978.
5. Ambrose J, Gooding HR, Richardson AE: In: Brain 98:569–582, 1975.
6. Flamm E, Ransohoff J, Wuchinich D et al: Preliminary experience with ultrasonic aspiration in neurosurgery. Neurosurg 2:240–245, 1978.
7. Cooper PR, Rovit RL, Ransohoff J: Hemicraniectomy in the treatment of acute subdural hematoma: a reappraisal. Surg Neorol 5:25–30, 1976.
8. Ransohoff J, Lieberman A: Surgical therapy of primary malignant brain tumors. Clin Neurosurg 25:403–411, 1978.

5. Surgical Management of Endocrine-Active Pituitary Adenomas

CHARLES B. WILSON

In the short span of a single decade, neurosurgeons have assumed an increasingly prominent role in the management of pituitary adenomas. The prevailing practice in this country ten years ago restricted surgical therapy of these adenomas to cases of large suprasellar tumors. Endocrine-active pituitary adenomas were treated either preferentially by some form of irradiation (acromegaly) or indirectly by removal of the end organ, with or without concurrent pituitary irradiation (Cushing's disease); and many adenomas were not treated at all because the nature of the endocrinopathy was not recognized (prolactin-secreting microadenoma). The secretory products of pituitary tumors were of no relevance as an indication for surgery, except in a minority of cases of acromegaly. Neurosurgeons were concerned primarily with accomplishing incomplete removal of large tumors. The objective was decompression of the optic chiasm, usually through a craniotomy, and postoperative radiation therapy was a routine. Surgical intervention was approached with no thought of performing a curative operation.

Over the course of ten years, the treatment of pituitary disorders has changed substantially. Transsphenoidal microsurgery has been applied to the full spectrum of pituitary adenomas, achieving successful therapy, and even cure. Experience attests to the safety and effectiveness of this approach.

The rapid evolution of pituitary microsurgery reflects three principal independent advances. a) Radioimmunoassays for pituitary hormones afforded precise laboratory diagnosis of pituitary hypofunction and hyperfunction, the development of a reliable measurement of prolactin having a particularly dramatic impact. b) Two neuroradiologic procedures, thin-section polytomography and computerized tomography (CT), afforded a degree of diagnostic precision that had been unattainable with pre-existing diagnostic techniques. c) Jules Hardy refined the transsphenoidal approach to the sella turcica by combining the use of a surgical microscope and televised intraoperative fluoroscopy, and in 1969 reported the first successful removal of a pituitary microadenoma. This seminal achievement established the technical feasibility of selectively removing endocrine-active intrasellar adenomas and restoring normal pituitary function.

Between June 1970 and July 1981, 800 pituitary adenomas have been removed

Table 1. Summary of 800 pituitary adenomas (June 1970-July 1981)

Tumor type	Hormone secreted	No. treated	Total (%)	Mortality (%)
Endocrine-active			630 (79)	0 (0)
	Prolactin (PRL)	331		
	Growth hormone (GH)	171		
	Adrenocorticotrophic hormone (ACTH)	126		
	Thyroid-stimulating hormone (TSH)*	2		
Endocrine-inactive			170 (21)	2 (1.2)
Total			800 (100)	2 (0.25)

*One tumor secreting TSH and GH is included under GH.

transsphenoidally at the University of California, San Francisco (UCSF) (Table 1). The 630 endocrine-active adenomas in this series account for 79% of the total. This preponderance of endocrine-active adenomas directly reflects a changing therapeutic preference among endocrinologists, and indirectly reflects the emergence of transsphenoidal microsurgery as a statistically superior treatment for the group of diseases associated with these adenomas.

An important aspect of the increasing acceptance of transsphenoidal microsurgery is the low surgical risk associated with the procedure: in the 630 operations for endocrine-active adenomas, there have been no deaths, and in 170 operations for endocrine-inactive tumors there were two deaths, for an overall operative mortality rate of 0.25% (Table 1).

Post, Jackson, and Reichlin [1] have recently published an excellent monograph that comprises a far more detailed review of the area of pituitary adenomas than I shall attempt to present in this chapter. With no presumption of reviewing all perspectives, and with admitted bias, in the following sections I will consider current views on the behavior, diagnosis, and surgical treatment of pituitary adenomas against the background of my own experience.

1. CLASSIFICATION OF PITUITARY ADENOMAS

The classic histologic categorization of adenomas (on the basis of light microscopy) as chromophobic, eosinophilic, or basophilic has gradually given way to classification according to the hormonal secretion(s) produced by the tumor (Table 2). This scheme, deriving from Landolt [2] and Kovacs [3, 4], correlates with the respective clinical syndromes resulting from proliferation of various stem-cell lines. Conspicuously absent from this scheme is the term 'chromophobic', which was formerly used to designate the largest subdivision of

Table 2. Morphologic classification of pituitary adenomas

Current classification
Growth hormone cell adenoma
Prolactin cell adenoma
Mixed: growth hormone cell – prolactin cell adenoma
Corticotrophic cell adenoma
Thyrotrophic cell adenoma
Gonadotrophic cell adenoma
Undifferentiated cell adenoma
Non-oncocytic (null cell)
Oncocytic (oncocytoma)
Acidophilic stem cell adenoma

predominantly endocrine-inactive adenomas. The term has no meaningful application, as the agranular appearance by light microscopy is misleading: ultrastructural studies have demonstrated secretory granules in 'chromophobic' adenomas, and sensitive radioimmunoassays show that as many as 50% of these produce hormone [5, 6]. The only truly nonsecreting lesions are the oncocytomas, which are thought to be transformed epithelial cells without endocrine potential, and the null cell adenomas recently characterized by Kovacs *et al.* [4].

The seeming paradox, that endocrine-inactive tumors possess secretory granules and cellular machinery for hormone production [2], can be explained by the fact that a tumor is endocrine-active only if it elaborates a biologically active hormone in sufficient quantity to exceed normal levels of the hormone in the blood. Thus, endocrine-inactive tumors can occur: a) when adenoma cells have lost the ability to produce hormone as a result of degeneration or dedifferentiation, b) when a normal hormone is produced in amounts too small to be detected, c) or when an abnormal hormone is produced but not recognized by biologic receptor sites or detected by radioimmunoassay.

It is, however, the local growth characteristics and size of the tumor – irrespective of endocrine activity – that best predict its biologic behavior [5, 7]. Classification on the basis of the degree of sellar destruction (grade) and extrasellar extension (stage) therefore has prognostic value and aids in designing therapy.

An anatomic (radiographic and operative) classification of pituitary adenomas, a modification of Hardy's classification scheme [7, 8], has been described elsewhere (Table 3) [5]. Briefly, grade I tumors are less than 10 mm in diameter and confined entirely within the sella; the sella may be focally expanded, but usually it is intact. Grade II tumors are greater than 10 mm and do not perforate the sellar floor; the sella is almost always enlarged. Grade III tumors focally perforate the dural membrane and cortical bone of the anterior wall and

Table 3. Anatomic (radiographic and operative) classification of pituitary adenomas

Relationship of adenoma to sella and sphenoid sinuses (grade)
  Floor of sella intact
    I : Sella normal or focally expanded; tumor < 10 mm
    II : Sella enlarged; tumor ≥ 10 mm
  Sphenoid
    III : Localized perforation of sellar floor
    IV : Diffuse destruction of sellar floor
  Distant spread
    V Spread via CSF, or blood-borne
Extrasellar extension (stage)
  Suprasellar extension
    0 : None
    A : Occupies cistern
    B : Recesses of 3rd ventricle obliterated
    C : 3rd ventricle grossly displaced
  Parasellar extension
    D*:Intracranial (intradural)
    E : Into or beneath cavernous sinus (extradural)

*Designate anterior (1), middle (2), or posterior (3) fossa.
From Wilson CB: Neurosurgical management of large and invasive pituitary tumors. In: Clinical Management of Pituitary Disorders, Tindall GT, Collins WF (eds), New York, Raven Press, 1979, pp 335 – 342. Adapted with permission.

floor of the sella and extend into the sphenoid sinus or sphenoid bone. Grade IV tumors perforate these structures diffusely. Grade V tumors spread via the cerebrospinal fluid (CSF) or hematogenous routes.

Stage 0 tumors have no suprasellar extension. Stage A tumors occupy the suprasellar cistern but do not deform the third ventricle. Stage B tumors occupy the suprasellar cistern and obliterate only the anterior recess of the third ventricle. Stage C tumors obliterate the anterior recesses of the third ventricle, and deform and elevate its floor. Stage D tumors have intradural, intracranial extension; the subcript 1, 2, or 3 designates extension into the anterior, middle, and posterior fossa, respectively. Stage E tumors invade the cavernous sinus through the lateral dural envelope of the sella.

Gross and microscopic necrosis is characteristic of many pituitary adenomas, irrespective of the endocrine activity, or inactivity, of the neoplastic cells. While necrosis, and occasionally consequent hemorrhage, may have dramatic clinical expression as pituitary apoplexy, much more often regressive changes in an adenoma go unrecognized, and the event, or repeated events, is clinically silent [9]. Spontaneous improvement in the appropriate hypersecretion syndrome or a partially empty sella (an enlarged sella containing an adenoma with intrasellar extension of CSF) [10] are clinical manifestations of tumor necrosis.

2. TREATMENT

Every decision regarding the treatment of pituitary adenomas must have two goals: a) to decompress neural structures (for example, the visual pathways) rapidly and effectively; and b) to normalize the hormonal secretion without producing other endocrine imbalances. For endocrine-active adenomas, these goals can be achieved only if all of the pathological tissue is excised. The patient's condition is not truly ameliorated merely by decreasing the abnormal hormone secretion because the endocrine disease persists, despite a certain reduction in its activity. The treatment modalities available are surgery, radiation therapy, and medical therapy [11].

2.1. Surgery

The operative approaches to pituitary adenomas have been described in detail elsewhere [12]. The selection between the transcranial and the transsphenoidal approach depends on the configuration of the tumor. The transsphenoidal approach is the technique of choice for tumors occupying the sella, whether or not there is sphenoid extension, and also for tumors with vertical suprasellar expansion without lateral extension. The advantages of this approach are that it allows selective excision of tumor so as to preserve normal pituitary function, it entails a low risk of morbidity, and it is well tolerated by debilitated or high-risk patients. It is contraindicated for those tumors with significant lateral parasellar extension or massive suprasellar expansion. A few adenomas are dumbbell-shaped, with a constriction at the diaphragma sella that prevents safe intrasellar delivery of the suprasellar mass; such tumors are better approached transcranially [13]. Only 1% of the pituitary tumors referred to UCSF during the period of this study have required a transcranial approach.

2.1.1. The Transsphenoidal Approach
In our institution, the patient receives nose drops containing Bacitracin for 2 or 3 days before the operation. Systemic antibiotics are no longer considered necessary because almost all cultures obtained from the sinus mucosa removed intraoperatively are either sterile or grow only normal respiratory flora.

The transsphenoidal procedure is performed with the patient in a semi-sitting position. The head is held by pin fixation, and is turned to face the surgeon, who stands at the patient's right. A fluoroscopic image intensifier is positioned at an angle that will provide a collimated lateral view of the sella on a television monitor. If the tumor extends above the sella, air is introduced through a lumbar subarachnoid catheter to fill the third ventricle and the suprasellar cisterns.

An antiseptic scrub solution is used to cleanse the nasal passages and the upper gingival and labial mucosa. If a graft is required, the right lateral thigh is disinfected. The nasal and labial submucosal tissues are infiltrated with epinephrine (1:200,000) in lidocaine (0.5%), after which a horizontal incision is made high on the gingiva between the canine teeth. The soft tissues are lifted upward to expose the lower half of the piriform aperture, which is then enlarged by removing the maxillary rim, both laterally and inferiorly.

A submucosal plane is developed along the nasal floor and septum, and the mucosa is reflected laterally by advancing a self-retaining speculum down the nasal septum to the sphenoidal rostrum. The anterior nasal spine is removed. The cartilaginous part of the nasal septum is either cut at its base and subluxated laterally (so it can be replaced at the end of the operation), or else it is removed. The posterior, osseous part of the septum and the adherent cartilage are removed. The rostrum of the sphenoid is opened with a rongeur or an air drill. The opening into the sinus is enlarged using punches, after which the sphenoid sinus mucosa is reflected, removed, and submitted for culture. Fluoroscopic monitoring and a direct view of the field maintained with the operating microscope (300 mm objective, 12.5 × eyepiece) guide all subsequent maneuvers.

The anterior wall of the sella is perforated with an air drill; then with small punches, bone is removed from the tuberculum sellae superiorly to the floor inferiorly, and laterally to the medial edge of each cavernous sinus. Venous bleeding is controlled with tiny pledgets of oxidized cellulose (Surgicel). Bipolar forceps are used to coagulate the exposed dura around its periphery, and the dura is excised to permit maximal exposure of the sellar contents.

2.1.1.1. Intrasellar Adenomas. The various types of intrasellar adenomas reside in individual, preferential sites throughout the sellar region. These tumors are seldom larger than 2 cm in diameter. Tumors of 5 mm or less in diameter are seldom visible at the exposed surface. If the tumor's location is not evident, either from preoperative CT scans or from inspection of the exposed gland, incisions are extended into the gland. In patients with adenomas secreting growth hormone (GH) and prolactin, vertical incisions are made in the lateral wings. A midline vertical incision may disclose the adenoma in patients who have small, adrenocorticotrophic hormone- (ACTH-) secreting adenomas, but the tumors associated with Cushing's disease may be found occupying any of a number of positions within the sella (described in greater detail in Section 5).

Adenomas are individually distinct, and while their consistency and color may vary, they can be unequivocally distinguished from normal anterior pituitary tissue. Although it may be difficult to differentiate them from the posterior lobe, confirmation by frozen section is decisive. Multiple incisions into the anterior pituitary have not produced detectable impairment of the glandular function.

There is little difficulty involved in exposing larger intrasellar tumors, except for certain prolactin-secreting adenomas that occupy the lateral angle at the base of the dorsum sellae and frequently burrow into the body of the sphenoid bone. Adequate exposure of these tumors requires removal of the ipsilateral sellar floor and subjacent cancellous sphenoid bone. All gross tumor is removed under 16 × 25 × magnification. It is not necessary to obtain frozen sections of grossly normal tissues because there is a distinct boundary between the tumor and the normal gland.

When the tumor is confined to the sella, and if the cavity does not communicate with the subarachnoid space, Gelfoam soaked in absolute alcohol is packed into the cavity repeatedly, for a total exposure to alcohol of 6 to 10 minutes, in order to destroy any remaining microscopic nests of tumor cells. Although the alcohol penetrates the exposed surface of the normal gland, the depth of penetration seems to be negligible, and I have observed no detrimental effect on pituitary function.

The extent to which the remaining cavity is packed varies, depending on the position of the tumor's superior surface in relation to the sellar diaphragm. Fat is used to fill the cavity; the removal of muscle will result in a painful thigh, and the use of fat has proved equally satisfactory. A piece of cartilage carved to cover the sellar opening is slipped inside the dural edges in the case of larger tumors or, in the case of smaller tumors, is placed extradurally beneath the bone edges. After the anesthesiologist has briefly elevated intrathoracic pressure to verify adequate exclusion of CSF, the cartilage graft and surrounding bone are covered with a thin layer of biological adhesive.

2.1.1.2. Adenomas with Suprasellar Extension. The anesthesiologist retains the lumbar spinal subarachnoid catheter in order to control the position of the suprasellar tumor and the diaphragma sellae. The entire surface of the suprasellar capsule must be in the surgeon's field of vision. When the tumor is of modest suprasellar proportions, this presents no problem because the capsule is forced into the sella by the slightly increased intracranial pressure associated with general anesthesia. If the suprasellar capsule does not descend as the intrasellar tumor is removed, however (which is sometimes the case), the anesthesiologist can inject normal saline, in small increments, into the subarachnoid catheter until intracranial pressure forces the suprasellar tumor into the operative field. Occasionally, the capsule falls to the bottom of the excavated sella, obscuring the posterior extension of the tumor above and behind the dorsum sellae. In such a case, withdrawal of CSF will elevate the capsule, affording the surgeon a view of any tumor remaining in a posterior-superior direction. A dental mirror is used to inspect all surfaces of the cavity remaining after the removal of larger tumors and, when used adeptly, can provide an excellent view of otherwise inaccessible areas.

Two or more silver clips are attached to the midportion of the suprasellar capsule; they serve both as a fluoroscopic guide to avoid excessive packing of the sella and as a marker to follow in the postoperative period, to determine the presence of a hematoma at first, and later, tumor recurrence.

Bacitracin solution is used often during the operation to irrigate the operative field. After the final irrigation, the speculum is removed, the sublabial incision is closed with catgut sutures, and both nasal cavities are packed gently with petrolatum gauze coated with Bacitracin ointment. Patients are instructed to use an aqueous nasal ointment for 2 to 3 months.

2.1.1.3. Adenomas with Extension into the Sphenoid Sinus. Surprisingly often, small nodules of tumor perforate the dura and bone and come to lie within the sphenoid sinus (or, rarely, within the body of the sphenoid bone) covered only by mucosa. These small extensions can be removed cleanly and easily. It is difficult, sometimes impossible, however, to remove massive extensions into the sinus completely, and the deficient sellar walls can make the proper orientation problematical.

2.2. Radiation Therapy

Radiation therapy is reserved for lesions that cannot be totally removed by surgery: a) those with extension into or below the cavernous sinus, such as large and invasive adenomas with parasellar extension; and b) those causing Nelson's syndrome. Irradiation is not used as primary therapy for pituitary adenomas at our institution.

3. PROLACTIN-SECRETING PITUITARY ADENOMAS (PROLACTINOMAS)

Prolactin- (PRL-) secreting adenomas, or prolactinomas, constitute the largest group of pituitary adenomas, irrespective of secretory activity, yet their natural history remains uncertain. The behavior and benign clinical manifestations of small prolactinomas have set them apart from the adenomas that produce Cushing's disease and acromegaly, two distinct metabolic entities that have severe and eventually fatal consequences quite independent of the tumor's effects as a mass. Whereas the clinical necessity of treating the patient with either Cushing's disease or acromegaly is unquestioned, the indications for treating the patient harboring a small prolactinoma are controversial; there are those who question the wisdom of treating these patients at all, whether surgically or otherwise.

Unlike adenomas that secrete ACTH or GH, prolactinomas have dissimilar manifestations in men and women; and in contrast to Cushing's disease and acromegaly, in which the presence of a pituitary adenoma can be predicted with virtual certainty on the basis of laboratory results alone, a confident preoperative diagnosis of prolactinoma requires radiographic evidence of an intrasellar tumor unless the patient's PRL values exceed 200 ng/ml. The established effectiveness of bromocriptine has introduced another issue into the controversy regarding these adenomas: whereas long-term medical therapy is relatively ineffective in the treatment of Cushing's disease and acromegaly, therapy with bromocriptine not only corrects the biochemical pathology of prolactinomas, but also has an effect in reducing the tumor's size that is, at times, dramatic.

I cannot resolve these unsettled issues relating to prolactinomas, but observations derived from the histories of patients evaluated and treated at UCSF, supplemented by selected publications in the literature, have led to tentative conclusions that will become evident.

3.1. Clinical Manifestations

Although Forbes *et al.* recognized in 1954 that pituitary adenomas could be the cause of amenorrhea and galactorrhea [14], until recently prolactinomas were not recognized as a frequent cause of secondary amenorrhea and galactorrhea. PRL could not be measured directly until 1963 [15], and polytomography of the sella was not applied to the diagnostic evaluation of amenorrhea and infertility until even later. Clearly evident now is the frequency of hyperprolactinemia in amenorrheic populations; equally evident is the frequency of prolactinomas as the cause of hyperprolactinemia.

Considering first the women in the series of patients treated at UCSF: 80% of women presented with secondary amenorrhea and galactorrhea (either spontaneous or expressible); 10% had primary amenorrhea, and half of these women had galactorrhea; and 10% presented with oligomenorrhea and galactorrhea, with secondary amenorrhea without glactorrhea, or with secondary amenorrhea only.

Among men, a prolactinoma is typically undetected during the initial phase of hyperprolactinemia; they generally receive a correct diagnosis and subsequent treatment only after a large tumor has caused either significant panhypopituitarism or compression and invasion of parasellar structures. In this series, only seven men have been treated because of a symptomatic hyperprolactinemia with otherwise intact pituitary function: two were adolescents with gynecomastia; two were adults with gynecomastia and galactorrhea; two were impotent; and one, a medical student whose hyperprolactinemia was discovered accidentally when he volunteered as a normal blood donor in an endocrinology laboratory, retrospectively admitted to a decrease, but not loss, of libido. Large PRL-secreting adenomas have a slight predilection for women, but the nearly equal frequency of these tumors between sexes contrasts conspicuously with the rarity of recognized microadenomas (in our experience, less than 5%) occurring in men. Presumably, the menstrual cycle and the female breast are quite sensitive to the primary and secondary hormonal effects of a slight excess of PRL, whereas libido and potency in men are less vulnerable targets. Supporting this assumption is the frequency with which symptoms occurred in women who had PRL levels of 50 – 100 ng/ml, as contrasted to that in the men in our series: in six of the seven symptomatic men, the PRL levels were 195 ng/ml or higher. Additional evidence has been obtained by Weiss and Martin (personal communication), who found hyperprolactinemia in only two of 100 impotent men; both men had normal sellas.

Hyperprolactinemia may have extragonadal manifestations, although possibly only in patients normally predisposed to these disorders. A recent, rapid, and often excessive gain of unwanted weight has been a complaint of hyperprolactinemic women with a frequency that suggests a relationship between these two conditions. Correction of hyperprolactinemia by either operation or bromocriptine has been followed in some cases by an impressive weight loss that is not accountable to a change in dietary habits. Equally impressive has been the incidence of emotional lability as an elicited symptom and its reversal following correction of hyperprolactinemia. Whereas obesity and emotional instability may reflect multiple nonendocrine factors, the study recently reported by Klibanski et al. [16], which showed a decreased density of the bone in hyperprolactinemic women, nullifies the assumption that the estrogen deficiency related to hyperprolactinemia is benign. Of the 14 women in their series (mean age, 29 years), 7 had amenorrhea for 2 years or less, and only one had a PRL level higher than 60 ng/ml: hyperprolactinemia that is neither extreme nor long-standing causes significant osteoporosis and its potential morbidity.

3.2. Biology and Pathology

A prolactinoma's capacity to secrete PRL theoretically has no definable relationship to either its rate of growth or its potential for further growth. Despite this theoretical independence of a secretory cell's activity from its capacity to replicate, the level of PRL in blood, taken as an index of secretory activity, has a direct correlation with the size of prolactinomas, exclusive of the bulk contributed by necrosis and cysts.

Necrosis, often with liquification (cyst formation), is a common surgical finding in prolactinomas of all sizes. As a tumor cell population expands (or contracts) at a rate determined by the balance between cell production and cell death, this propensity toward spontaneous necrosis may be related to the prolactinoma's indolent growth. In the series of women with prolactinomas and secondary amenorrhea who have been treated at UCSF, younger patients had a shorter history of menstrual dysfunction, smaller tumors, and lower levels of PRL than their older counterparts. This indisputable correlation of age, duration of amenorrhea, and level of PRL with tumor size, together with the evidence of necrosis in most prolactinomas, supports the conclusions that prolactinomas secrete PRL in proportion to their size, and that their rate of growth is slow. Although we have encountered exceptions to this generalization, the great majority of prolactinomas follow this pattern.

The majority – probably two-thirds – of massive pituitary adenomas secrete PRL; in the past, these tumors were included in the category of chromophobe adenomas. Lundberg *et al.* [17] described 11 patients with extensive and invasive PRL-secreting pituitary adenomas. In most of these patients, the initial symptoms had been either a decrease of libido and potency or amenorrhea and galactorrhea, and there was a lapse of 6 to 36 years between the onset of symptoms and diagnosis. In three cases, growth from an intrasellar, noninvasive adenoma had been documented by x-ray films of the skull several years before the diagnosis was made. Although an unknown proportion of prolactinomas may have a finite capacity for growth and may attain their limit of growth before they exceed the defined size of a microadenoma, there is at present no means for identifying this group of truly benign adenomas.

In the series of patients treated at UCSF, the most commonly encountered mixed adenoma contained PRL- and GH-secreting cell populations. Typically, the women who were acromegalic had amenorrhea and galactorrhea, whereas the acromegalic men had either no clinical expression of hyperprolactinemia or gynecomastia that on rare occasions was accompanied by galactorrhea. In a few women with otherwise typical amenorrhea and galactorrhea and no features of acromegaly, modest elevations of GH were detected unexpectedly. The common association of GH- and PRL-secreting cells in a single adenoma reflects their origin from a single acidophil stem cell, the progenitor of both cell types.

Horvath *et al.* [18] recently reported 15 cases of acidophil stem cell adenomas, most of which were large and locally invasive, that were characterized by a short history of clinical expression, relatively low hormonal activity, and a single cell type with composite morphologic features of GH- and PRL-secreting cells. In this type of adenoma, the acidophil stem cell secretes both hormones, whereas mixed GH-PRL adenomas, which are much more common, contain two distinct cell types.

3.3. Laboratory Evaluation

Reports attempting to distinguish functional hyperprolactinemia from hyperprolactinemia produced by prolactinoma on the basis of laboratory values have contained conflicting conclusions, and a critical analysis of the subject goes beyond the purpose of this review. The test in widest use today determines variations in the PRL level in response to the administration of TRH [19], but in our experience, recently reported by Chang et al. [20], a blunted response to TRH is characteristic of hyperprolactinemic patients with amenorrhea, and does not confirm the presence of an adenoma. A variety of other tests have been used with a similar lack of success. We have had no experience with tests measuring the diurnal variation of PRL secretion, most recently reported by Malarkey et al. [21].

In men whose basal PRL values are above 100 ng/ml, we have so far encountered no difficulty in establishing a prolactinoma as the cause of hyperprolactinemia. In women, hyperprolactinemia with basal values in excess of 200 ng/ml almost invariably indicates a prolactinoma; with some exceptions, basal values between 100 and 200 ng/ml have the same significance. Women with basal levels of 50 – 100 ng/ml, for which the term moderate hyperprolactinemia seems appropriate, present the major problem. Values in this range may have no recognized cause; and in women with moderate hyperprolactinemia that is not a result of medication, pregnancy, or hypothyroidism, a confident prediction of an underlying prolactinoma cannot be based solely on laboratory values. A few patients with moderate hyperprolactinemia have a readily identified intrasellar or superasellar pathology, most commonly an endocrine-inactive pituitary adenoma, infrequently a suprasellar tumor such as a craniopharyngioma [22], that interferes with the pituitary stalk, the 'stalk section effect'.

In our experience, the confident diagnosis of a prolactinoma in a patient with basal PRL levels below 200 ng/ml requires radiographic identification of an intrasellar tumor; and even with unequivocal radiographic demonstration of a pathologic abnormality in the pituitary gland, a few transsphenoidal explorations have revealed either a diffusely enlarged anterior lobe (pituitary 'hyperplasia') or an innocent cyst, usually in the pars intermedia.

3.4. Radiographic Diagnosis

Before CT scanning reached the present level of precision, the diagnosis of a prolactinoma was based on thin-second polytomography of the sella and pneumoencephalography. Even earlier, bilateral carotid angiography was performed routinely, but after a retrospective analysis of 100 intrasellar adenomas indicated that angiograms were of no value, this study was discontinued [23]. Because moderate hyperprolactinemia occasionally may be an unexplained finding in patients with a primary empty sella, and because a partially empty sella is associated with almost one-fourth of intrasellar prolactinomas, pneumoencephalography continued to be performed until the resolution of CT scans permitted identification of empty and partially empty sellas.

In a postmortem study of 100 patients with no history of pituitary disease, polytomograms of excised sphenoid bones were correlated with gross and microscopic pathology of the sella and its contents; the results of this study were reported recently from our Neuroradiology Unit [24]. In nine cases, polytomograms of the sella were considered indicative of a microadenoma, but an adenoma was not found on the examination of serial sections through these sellas. Consequently, a falsely positive diagnosis of a pituitary microadenoma can be expected in 9% of sellar polytomograms from an unselected population, which implies that approximately 10% of abnormal sellas will not contain a clinically significant adenoma. The same investigators [25] determined the accuracy of thin section axial CT scans made with a GE-8800 scanner. A probable diagnosis of adenoma was based on: a) a convex upper border of the pituitary gland (seen in 2% of normal glands); b) height of the gland greater than 7 mm; c) lateral deviation of the pituitary stalk; and d) a focal area of decreased or increased attenuation relative to the normal gland. Using these criteria, the presence of a prolactinoma can be established with a high degree of accuracy in a patient in whom the clinical and laboratory data are consistent with the diagnosis. Adenomas measuring 6 mm or greater can be recognized with confidence, but smaller tumors pose a problem: in approximately 20% of clinically normal patients, high resolution CT scans of the pituitary indicate focal hypodensities measuring 3 mm or greater. Based on pathologic examination of postmortem material, one-half of these focal lesions are cysts of the pars intermedia and the other half are clinically functionless adenomas or areas of focal hyperplasia.

3.5. Prediction of Postoperative Result

The results of transsphenoidal removal of PRL-secreting adenomas in the series of patients treated at UCSF have been reported by Keye *et al.* [26]. Four interrelated factors predicted a successful outcome, defined as the return of menses (Table 4). The patient's age and the duration of amenorrhea were of predictive value, but the more accurate predictive factors were basal PRL and tumor size. The experience of others is in accord with our own [27–29]. Recently, Schlechte *et al.* [28] reported that patients with primary amenorrhea and those with secondary amenorrhea unrelated to the use of oral contraceptives had a poorer outcome than had patients with either post-partum amenorrhea or amenorrhea occurring in close temporal relation with the use of oral contraceptives. In our series, patients with primary amenorrhea have not fared as well as those with secondary amenorrhea, but the patients with primary amenorrhea more frequently had preoperative PRL values greater than 200 ng/ml; we have not observed any differences in outcome related to the use of oral contraceptives, but this factor has not been examined in detail. In Schlechte's series, the mean PRL

Table 4. Return of menses following transsphenoidal removal of a prolactin-secreting adenoma

Age (years)	%
less than 30	92
more than 30	67
Duration of amenorrhea (years)	
less than 2	100
2 – 5	87
5 – 10	78
more than 10	50
Preoperative basal prolactin (ng/ml)	
21 – 50	91
51 – 100	100
101 – 200	100
more than 200	33
Tumor size (cm)	
less than 1	100
1 – 2	94
more than 2	14

in patients with primary amenorrhea was 200 ± 80 ng/ml. In their group of 16 patients with spontaneous amenorrhea, the mean PRL was 180 ± 32 ng/ml, as opposed to mean values of 95 ± 47 ng/ml and 106 ± 15 ng/ml in the groups with post-partum amenorrhea and with amenorrhea in relation to oral contraceptive use, respectively. Adenomas that destroy the sellar floor and those that extend into the sphenoid sinus (grades III and IV) are seldom cured unless the extension is small [29].

3.6. Indications for Operative and Nonoperative Management

The policies currently determining recommended management at UCSF reflect our favorable experience with the microsurgical approach, the assumption that most prolactinomas grow relatively slowly, and the evidence that surgical treatment is less successful in older patients, in those who have had a longer duration of amenorrhea, and in those with higher PRL values and larger tumors. The following recommendations apply for the cases in which a clinical presentation, laboratory data, and a high resolution CT scan, taken together, eliminate any doubt that the patient harbors a PRL-secreting pituitary adenoma.

Operative removal is recommended for all macroadenomas, most of which will be accompanied by PRL values greater than 100 ng/ml. If the tumor is smaller than 2 cm, a curative operation is likely. If the tumor is larger than 2 cm and noninvasive, the outcome may be successful; if it is not, reduction of the tumor's bulk will enhance the effectiveness of subsequent treatment with bromocriptine and irradiation. Sheline's review of irradiation in the treatment of large pituitary adenomas, with and without prior surgery [30], provides the rationale for advising operation for large tumors in which a surgical cure is uncertain or unlikely. Based on evidence from laboratory and clinical trials, irradiation is most effective against the smallest tumors.

More controversial is the management of microadenomas. The indications for operative removal are four:

a) *Desire for pregnancy*. With selective removal of a microadenoma, there is a high probability of restoring normal pituitary and ovarian function and restoring fertility. The small but significant risk of clinically significant growth of the tumor during an induced pregnancy is obviated.

b) *Primary amenorrhea*. Evidence cited earlier suggests that tumors arising at an early age behave more aggressively than those arising later in life; procrastination with medical management seems ill-advised.

c) *Males*. Experience, both personal and published, with the surgical and nonsurgical management of microadenomas in men is limited because a diagnosis is rarely made at this early stage of tumor growth. The probable progression to a macroadenoma and the likelihood of an excellent surgical outcome favor removal.

d) *Personal choice*. Some patients either do not tolerate, or do not wish to take, bromocriptine, and yet are disinclined to do nothing when informed of the tumor's unpredictability and the favorable outcome that can be expected with surgery.

Patients who do not fit into the preceding categories can be observed at regular intervals, the behavior of the tumor being monitored by basal PRL levels and periodic CT scans. I am aware of only one patient whose tumor expanded without a concomitant rise in PRL. Bromocriptine can be prescribed to treat troublesome galactorrhea, and in some patients it can be prescribed in the attempt to counteract excessive weight gain and emotional instability. Unclear at this point is the role of long-term administration of bromocriptine to prevent the now recognized hypoestrogenic osteoporosis that accompanies sustained hyperprolactinemia [16].

In patients for whom nonoperative management is elected at the time of the initial diagnosis, two subsequent developments are considered secondary indications for surgery: a) progressive elevation of PRL levels approaching 200 ng/ml in an untreated patient, or a progressive elevation or PRL values in a patient whose symptoms and hyperprolactinemia were initially controlled by bromocriptine and who is still receiving the drug; and b) enlargement of the tumor, as determined by serial polytomograms or CT scans.

3.7. Pregnancy

Induced pregnancy in a patient known to harbor a prolactinoma carries a small but serious risk [31, 32] of complications related to rapid expansion of the tumor. Prophylactic pituitary irradiation has been used in patients who wish to become pregnant. If a tumor becomes symptomatic during pregnancy, bromocriptine can be administered. We advise against pregnancy for patients known to have a PRL-secreting tumor, and recommend that the administration of bromocriptine for therapeutic purposes should be accompanied by mechanical contraceptive measures. The risk of a serious complication arising during an induced pregnancy seems significantly larger than the risk involved in transsphenoidal surgery. It is on this basis that we advise surgical treatment when pregnancy is desired.

3.8. Results of Transsphenoidal Microsurgery

Hardy [29] has the largest published series of patients treated with transsphenoidal microsurgery, and his excellent results indicate the outcome that can be anticipated in the hands of a skilled and experienced surgeon. In his series of 160 women with PRL-secreting adenomas, cures were obtained in 85 – 90% of all patients with grade I and grade II tumors and preoperative PRL levels up to 200 ng/ml, and in 60% of those with tumors in these grades and PRL levels of 200 – 500 ng/ml. Only one-third of patients with grade III, and none of those with grade IV, tumors were cured by operation. Of 25 patients with PRL levels above 500 ng/ml, only 3 (12%) were cured.

Tucker *et al.* [33] have shown that a normal basal PRL level determined 6 months after operation predicts a cure. Our total series has not been analyzed in detail, but to my knowledge, only one patient who had a normal PRL level 6 weeks postoperatively has had a subsequent recurrence of tumor.

Our early experience has been reported [26, 34]. At present, the data from 100 patients treated subsequently, each of whom has undergone complete endocrine evaluation by our Reproductive Endocrinology Unit, are being analyzed. A preliminary review of this recent series indicates that an excellent result was obtained following removal of grade I and grade II prolactinomas, with cures in 80 – 90% of patients who had favorable predictive factors, the most important being the preoperative basal PRL level.

Our experience with two small prolactinomas occurring in men has been reported [35]. Both had preoperative PRL values below 200 ng/ml and both were cured.

3.9. Radiation Therapy

Sheline [30] described the radiation therapy technique used at this institution (4-ME V linear accelerator, bilateral coronal arc (60°) fields with moving wedge (60°) filters) in reporting the results of radiationtherapy for pituitary tumors. Because the great majority of large chromophobe adenomas that were treated before the PRL assay became available were actually prolactinomas, his experience is relevant to the subject.

Our experience has shown that irradiation involves a very low risk of complications and a very high probability of arresting tumor growth. Nonetheless, PRL levels seldom approach the normal range, even after many years, and present evidence suggests that sterile tumors, viable but incapable of proliferative activity, maintain the capacity to secrete PRL. Delayed, irradiation-induced damage to the adenohypophysis and hypothalamus causes some degree of hypopituitarism in a significant proportion of patients, at least one-third of whom require hormone replacement therapy. We have not used radiation therapy as the primary approach to prolactinomas, but irradiation is indicated for incompletely removed tumors.

3.10. Bromocriptine: Its Role in Diagnosis and Treatment

The dopamine agonist, bromocriptine, may have several sites of action in reducing the normal or abnormal secretion of PRL, but its dominant effect in the patients with a prolactinoma is direct inhibition of the tumor's secretory activity [36]. A second but less predictable effect of bromocriptine is a reduction, often dramatic, in the dimensions of prolactinomas — an effect that we have observed as early as 11 days after treatment was initiated. The subject of prolactinoma regression during treatment with bromocriptine has been reviewed recently by Thorner et al. [37]. The effectiveness of bromocriptine depends on its being administered continuously; with few exceptions the size of the tumor and the PRL level return to, or approach, pretreatment values. We have reported a patient whose prolactinoma metastasized to the cerebellum during continuous, long-term treatment with bromocriptine [38], but with this exception, regrowth of a responsive prolactinoma during bromocriptine therapy has not been reported. The mechanism accounting for the drug's antitumor effect has not been determined. The prompt reduction in tumor volume may reflect a shrinking of individual tumor cells and a decrease in vascularity of the tumor. Although the drug may have a direct oncolytic effect, there have been no reports of long-term administration of bromocriptine resulting in a documented elimination of a tumor. On this basis, it seems unlikely that its oncolytic action is great; the term oncostatic is perhaps more descriptive of its action.

As a diagnostic test to distinguish tumor from other causes of hyperprolactinemia, the administration of bromocriptine has been valueless. As a means of restoring ovulatory menstruation and fertility in women, of reversing impotency in men, and of stopping galactorrhea in both sexes, bromocriptine has been a notable success. As a pharmacologic means of reducing tumor volume, the drug is predictably effective in more than half of large prolactinomas. We have used bromocriptine for this purpose in patients after operation and during and after irradiation. Although we have not used bromocriptine specifically to obtain a preoperative reduction in tumor bulk, this may be a useful measure in certain instances.

4. GROWTH HORMONE-SECRETING PITUITARY ADENOMAS

The signs of a GH-secreting pituitary adenoma develop so gradually that acromegaly may progress unnoticed until the patient's features have become markedly distorted and the tumor has grown beyond the microadenoma stage. Although the disorder varies in the severity of its manifestations, it can produce life-threatening metabolic complications. As many as 89% of untreated acromegalics die by age 60 [39], most frequently from cardiovascular and cerebrovascular disease [40].

There is a continuing controversy regarding which, among the therapeutic modalities available for the treatment of a GH-secreting pituitary tumor, is the most efficacious. Irradiation was first used successfully for the treatment of acromegaly in 1909 [41], and it remains an alternative to surgical intervention. The neurosurgical options include cryohypophysectomy [42–44], the implantation of radioactive sources within the pituitary gland [45, 46], craniotomy [47], and transsphenoidal surgery, either for removal of the entire pituitary gland or for selective removal of adenomatous tissue [5, 7, 13, 44, 48–55].

Despite the significance of clinical improvement obtained by any therapeutic means, the primary factor determining cure of acromegaly is the restoration of normal GH production. The development of a radioimmunoassay for GH has made it possible to quantitate the level of success or failure of a form of therapy, and to compare the relative efficacy of the various modalities that are now used in the treatment of acromegaly.

4.1. Surgical Management

From June 1970 to June 1980, 137 patients underwent transsphenoidal surgery at UCSF for the removal of GH-secreting pituitary adenomas [55]. For one group of patients (Group A), this surgery was the first therapeutic intervention;

the others (Group B) underwent the surgery after having had previous therapeutic intervention. All patients were followed until April 1981.

4.1.1. Preoperative and Postoperative Evaluation

Neuroendocrinologic evaluation of anterior pituitary function included a preoperative and postoperative determination of the GH level in serum, as well as determinations of the levels of serum thyroxine, T_3 uptake, thyroid-stimulating hormone, luteinizing hormone (LH), follicle-stimulating hormone (FSH), serum PRL, and cortisol, and the levels of hydroxycorticosteroids and ketosteroids in urine. Glucose and insulin tolerance tests complete the basic endocrinologic assessment.

All patients were evaluated neuroradiologically by hypocycloidal sellar polytomography in the anterior-posterior and lateral planes. Until recently, pneumoencephalography was performed in all patients, but during the past year this procedure has been discontinued in favor of magnified CT reconstructions of the sella and parasellar regions in the coronal and sagittal planes. Although formerly we performed carotid arteriography routinely, this study is now reserved for patients in whom an aneurysm is suspected from the CT scan.

4.1.2. Operative Procedure

The operative approach is the same as that described earlier in this chapter (cf. Section 2). Because the tumor in acromegalics is characteristically large and readily identified, exploratory incisions through the anterior lobe, which are usually necessary to disclose the often elusive ACTH-secreting tumor, are rarely required. After the intrasellar portion of the tumor is removed, any suprasellar component usually descends into the operative field; if it does not, saline is injected into an indwelling lumbar subarachnoid catheter to encourage downward migration of the tumor. The exact strategy for the surgical resection is determined by the size, consistency, shape, and location of the tumor. The goal is selective, total removal of all adenomatous tissue with preservation of all anterior pituitary gland.

4.1.3. Surgical Results

The surgical results were analyzed considering preoperative and postoperative endocrinologic, neurologic, ophthalmologic, and neuroradiologic data. Cure was defined as clinical regression and restoration of fasting GH levels to normal (less than 10 ng/ml in our laboratory), as determined 6 weeks after surgery. A partial response was defined as clinical improvement and reduction of the GH level to less than 50% of the preoperative value, as determined 6 weeks after

Table 5. Acromegaly: results of transsphenoidal surgery

Result	Group A (%) (102 patients)	Group B (%) (35 patients)
Cure		
surgery alone	80 (78)	26 (74)
surgery + postop irradiation	16 (16)	
total cured	96 (94)	26 (74)
Partial response	6 (6)	2 (6)
Failure	0 (0)	7 (20)

From Baskin DS, Boggan JE, Wilson CB: Transsphenoidal microsurgical removal of growth hormone-secreting pituitary adenomas. A review of 137 cases. J Neurosurg 56:634–641, 1982. Reproduced with permission.

Table 6. Acromegaly: surgical morbidity

Complication	No.	Tumor grade & stage
Group A (8%)		
Meningitis*	2	II-O, II-C
CSF leak*	1	IV-C
Pseudoaneurysm	2	II-E, III-B
Nasal deformity	3	I-A, II-A, IV-B
Group B (14%)		
Meningitis*	2	II-O, III-B
CSF leak*	2	II-C, IV-O
Ophthalmoplegia (temporary)	1	IV-O

*Full recovery with treatment.

From Baskin DS, Boggan JE, Wilson CB: Transphenoidal microsurgical removal of growth hormone-secreting pituitary adenomas. A review of 137 cases. J Neurosurg 56:634–641, 1982. Reproduced with permission.

surgery. Any other result was considered failure. The mean follow-up period was 37.1 months; 100% patient follow-up was achieved.

The results obtained in this series are reported in detail elsewhere [55] (Table 5). Briefly, in Group A, for whom transsphenoidal surgery was the first therapeutic intervention, 78% of patients (80 of 102) were cured by transsphenoidal surgery alone, and an additional 16% (16 of 102) were cured after postoperative irradiation (combined cure rate, 94%). All failures and partial responders had preoperative GH levels greater than 50 ng/ml and suprasellar extension of tumor. In Group B, who underwent the surgery after having had previous therapeutic intervention, 74% of patients (26 of 35) were cured, 6% (2

Table 7. Acromegaly: rate of cure in relation to grade and stage of tumor

Group A				Group B			
Grade	Cure (%)	Stage	Cure (%)	Grade	Cure (%)	Stage	Cure (%)
I	100	0	100	I	no pts	0	81
II	97	A	100	II	82	A	66
III	96	B	82	III	71	B	60
IV	82	C,D,E	80	IV	64	C,D,E	50

From Baskin DS, Boggan JE, Wilson CB: Transsphenoidal microsurgical removal of growth hormone-secreting pituitary adenomas. A review of 137 cases. J Neurosurg 56:634–641, 1982. Reproduced with permission.

of 35) were partial responders and 20% (7 of 35) were failures. Of the patients who had received prior irradiation only, 88% (7 of 8) were cured. All failures and partial responders had preoperative GH levels greater than 40 ng/ml; 56% had suprasellar extension. No patient in this series died of a complication related to surgery; the surgical morbidity rate was 22% (Table 6).

The incidence of new postoperative hypopituitarism in patients treated by transsphenoidal surgery alone (Group A) was 5%; of patients subsequently irradiated, 71% developed hypopituitarism. There was a 66% incidence of hypopituitarism in Group B patients who had received radiation therapy previously. These data indicate that the incidence of postoperative hypopituitarism is considerably greater in patients treated by surgery and irradiation, and support the observation that treatment with radiation therapy often causes hypopituitarism [48, 56, 57].

There was no correlation of outcome with age, sex, preoperative duration of symptoms, or specific symptomatology in either group of patients. The cure rate decreased in relation to the increased grade and stage of the tumor (Table 7). The two factors consistently associated with a partial response to, or failure of, therapy were a high preoperative GH level and the presence of suprasellar extension of tumor. It is reasonable to assume that both of these factors occur in association with aggressive and invasive neoplasms, which are more difficult to remove totally because of their tendency to extend beyond the operative field.

Six patients in this series had come to treatment for gigantism rather than acromegaly. All six had GH levels greater than 40 ng/ml, and five had suprasellar extension; all six of these patients were cured. It is interesting to speculate whether or not patients with gigantism may have a better prognosis than most acromegalics, but this subgroup in our series was too small to warrant statistical evaluation.

The results achieved in this series are consistent with those reported by others, although the cure rate in the group of patients for whom this surgery was the first therapeutic intervention (Group A) is higher than any reported previously. The distinct difference between the results achieved in this group and those in the other patients (Group B) indicates that greater success can be expected in patients who have had no prior therapy. The results suggest also that patients who have had a single prior therapeutic intervention fare better than those who have had either multiple or more diffusely destructive procedures. It appears that prior therapy with only irradiation of the pituitary adenoma does not affect the outcome for patients subsequently treated by transsphenoidal surgery.

Although it has been suggested that a GH level of less than 5 ng/ml is required before the patient can be considered cured, our results indicate that a GH level of less than or equal to 10 ng/ml in the postoperative period correlates well with long-term cure. Of the three patients in our series whose disease recurred after an initial apparent cure, two had postoperative GH levels of less than 5 ng/ml. Among 18 patients with GH levels between 5 and 10 ng/ml, only one has had a recurrence.

4.2. Dynamic Testing of the GH Response

Thirty-nine of the patients in this series were evaluated and followed by the staff of the General Clinical Research Center at UCSF. Using our definition of cure, the cure rate in this special subgroup of patients was 90% (35 of 39). However, only 71% (10 of 14) had a normal GH response to thyrotropin-releasing hormone (TRH), only 64% (25 of 39) had a normal response to oral glucose, and only 57% (8 of 14) had a normal response in both tests postoperatively. The overwhelming majority of patients in this series considered cured by our more simple definition of cure, which is based solely on the return of the GH level to the normal range, have experienced and maintained dramatic resolution in the symptoms and signs of their disease. Based on our evaluation of GH dynamics in the patients who had these tests, a higher failure rate would be expected in our entire series of acromegalic patients than the one observed.

Nine patients had striking acromegalic features but normal preoperative GH levels. In all of these patients, the GH response to dynamic testing suggested the presence of a GH-secreting adenoma; the diagnosis was confirmed at surgery in each case. We conclude that tests of GH dynamics have diagnostic value in the analysis of patients with acromegaly and gigantism, but the prognostic significance of these tests is still unclear.

5. ADRENOCORTICOTROPHIC HORMONE-SECRETING PITUITARY ADENOMAS

In 1932, Cushing, observing basophilic adenomas during autopsy in 6 of 12 patients who had shown clinical manifestations of hypercortisolism [58], implicated a pituitary disorder in the disease that now bears his name. He warned that these small, occult pituitary adenomas might escape detection unless serial sections of the gland were explored. A year later, Naffziger performed a craniotomy on a young woman with Cushing's disease and accomplished incomplete removal of a tumor that produced a dramatic remission of the disease, although it recurred after a year [59]. This was, to my knowledge, the first operation for removal of a corticotrophic adenoma.

Although evidence implicating the anterior pituitary in the pathogenesis of Cushing's disease was undisputed, the intrasellar pathology could not be established by radiographic methods available at the time. This technologic limitation, in combination with the uncertainty regarding the pathogenetic process of the disease, led conservative opinion to argue against treatment by total hypophysectomy. The introduction of cortisone promoted bilateral adrenalectomy as the accepted treatment for Cushing's disease until Hardy's pivotal report in 1969 offered selective removal of pituitary microadenomas via the transsphenoidal aproach as a therapeutic alternative [60].

5.1. Surgical Management

Pituitary adenomas that secrete ACTH have a clinical behavior that is altogether different from that of other hormone-secreting adenomas that are more benign, and they may have a different etiology [61, 62]. Although the adenomas responsible for the disorder have been attributed to a disturbance of the hypothalamic regulation of corticotrophin releasing factors and have been thought to represent hyperplastic or hypertrophic growth rather than autonomous neoplasm [62], results following selective transsphenoidal removal of discrete adenomas argue otherwise.

Between February 1974 and October 1980, 77 patients diagnosed as having Cushing's disease underwent transsphenoidal exploration of the sella turcica at UCSF [63, 64]. Three of the patients had previously undergone unsuccessful pituitary irradiation, and three had recurrent Cushing's disease after an attempted bilateral adrenalectomy.

5.1.1. Preoperative and Postoperative Evaluation
The diagnosis of Cushing's disease can be established if the patient has hypercortisolism, if corticosteroid levels in plasma or urine are not suppressible with

overnight or low-dose dexamethasone administration, and if adrenocortical steroid secretion is suppressed to less than 50% of baseline levels with the administration of high-dose dexamethasone [63, 65, 66]. ACTH levels in plasma are either within the normal range or moderately elevated, measurable levels in the presence of hypercortisolism being consistent with ACTH-induced bilateral adrenocortical hyperplasia [67]. Selective venous sampling for ACTH by catheterization of the inferior petrosal sinus confirms ACTH hypersecretion from a pituitary source; this procedure should be used in any case in which an ectopic source is suspected [68].

During the first years of this series [63], routine anteroposterior and lateral radiographs and hypocyloidal polytomograms of the sella turcica were obtained for all patients [69]; bilateral carotid angiography and pneumoencephalography were performed in the majority, although carotid angiography provided no useful information and was later abandoned, and CT has supplanted pneumoencephalography. Preoperative radiologic evaluation now includes thin-section CT scans of the sella, CT scans of the adrenal glands, and thin-section polytomograms of the sella. Any abnormality detected by polytomography and CT scanning influences the operative plan for sellar exploration, although most patients with Cushing's disease have pituitary adenomas so small that they are not visible on radiologic studies. The decision to explore the sella is based entirely on the endocrinologic diagnosis, and radiographic studies serve only to direct the operative procedure to an observed abnormality of the sella or its contents [64].

For postoperative studies, all steroid replacement is stopped at least 24 hours before testing and baseline levels of plasma ACTH and adrenal corticosteroids are obtained. Anterior pituitary function is assessed by measuring basal levels of serum thyroxine, serum tri-iodothyronine uptake, thyrotropin, LH, FSH, and PRL. Serum GH is measured in response to insulin-induced hypoglycemia [70], and in most patients, stimulation tests with TRH and LH-releasing hormone (250 and 100 μg, respectively, given intravenously) are performed [63].

5.1.2. Operative Procedure

After the diagnosis is established, patients (or in the case of children, their parents) are advised that their disorder is probably caused by a microadenoma, even if none is identified during the course of intrasellar exploration. In the event of a negative exploration, we describe the three therapeutic alternatives – pituitary irradiation, bilateral adrenalectomy, and total hypophysectomy. Adult patients are advised to have a total hypophysectomy if no adenoma is identified; for children and young adults, we recommend radiation therapy rather than hypophysectomy.

If total hypophysectomy is a therapeutic alternative, a lumbar subarachnoid cathether is introduced after the induction of general anesthesia and before the patient is positioned for the transsphenoidal operation. The anesthesiologist is prepared to inject normal saline (without preservative) through the catheter at the surgeon's request, in order to prevent intracranial aspiration of blood and to check the adequacy of sellar closure at the end of the procedure.

The transsphenoidal procedure varies only in minor respects from the technique described by Hardy [7], although because the adenomas causing Cushing's disease are such very small tumors, more minute dissection may be required to expose them than is necessary for the other pituitary adenomas we have described.

If the preoperative tomograms and CT scans indicate a focal abnormality, or if the appearance of the exposed anterior lobe suggests the location of the tumor, that region is explored initially. Direct visualization of adjacent normal structures is required to accomplish complete removal of an identified adenoma, and sometimes this requires excising a wedge of uninvolved, overlying anterior lobe.

If there is no clue to the tumor's location, the initial exploratory incision is made vertically in the midline, and the incision is extended from the upper surface of the anterior lobe to the sellar floor, and to the surface of the posterior lobe. It is important to expose the posterior lobe fully because, as Cushing noted, occasionally an adenoma is found occupying the pars intermedia and invading the posterior lobe [58]. The second and third vertical incisions are made into the lateral wings. Because the adenoma may arise on the surface of the anterior lobe, the next step is to separate the gland from the cavernous sinus and the sellar floor and to inspect the upper surface beneath the arachnoid and sellar diaphragm. The last step is to make a horizontal incision to the plane of the posterior lobe, literally bisecting the anterior lobe, which has now been divided into eight sections. Any disproportionately large section that remains can be divided, but these four incisions (three vertical and one horizontal) approach the limits of technical possibility.

After all abnormal tissue is removed – including any portion of the posterior lobe that is in contact with the adenoma – the cavity is first irrigated with absolute alcohol and then filled with gelatin foam saturated with absolute alcohol. Although there is no evidence that absolute alcohol is effective in destroying residual tumor cells, we have had no reason to abandon this precautionary step. After the cavity is exposed to alcohol for 10 minutes, it is packed either with alcohol-saturated gelatin foam or with the gelatin foam and fat, and the sella is closed with cartilage. If there is a possibility that a CSF leak may develop postoperatively, the sphenoid sinus is filled with fat.

If no tumor is found, and if the patient has consented to hypophysectomy, then total hypophysectomy is performed and all intrasellar tissues are excised. I prefer to use subcutaneous fat for packing the sella, and use a carved piece of nasal septal cartilage to reconstitute the anterior wall of the sella.

On the day of the operation, patients receive exogenous steroid coverage (hydrocortisone, 300 mg in divided doses), which is tapered to maintenance replacement dosages by the 7th to 10th day postoperatively (30 mg of hydrocortisone or 1.0 mg of dexamethasone daily). They are then maintained on replacement hydrocortisone in doses of 20 to 30 mg/day as long as evidence of hypoadrenalism persists. We have not found it necessary to administer mineralocorticoid replacement therapy [63].

Despite the advantages of the transsphenoidal approach in allowing the surgeon to remove a microadenoma selectively, completely, and safely, the operation must be pursued judiciously. Identification and removal of a microadenoma, which we define as tumors that are entirely intrasellar and are less than 10 mm in diameter, require wide exposure of the sella and a bloodless surgical field. As more than one-half of microadenomas are 5 mm or less in diameter, the probability of performing a curative selective adenomectomy is greatly reduced in a case that poses inadequate exposure, anatomical obstacles, or otherwise suboptimal operating conditions. In the face of technical restrictions such as these, abandoning the operation is preferable to exposing the patient to the risks involved in an intrasellar exploration that is compromised.

5.1.3. Surgical Results

Of the 77 patients in this series, 66% were cured by selective adenomectomy, and an additional 10% were cured by total hypophysectomy. In five other cases (7%), the procedure was abandoned without opening the dura because the exposed dura contained formidable venous sinuses. These five cases were technical failures, but with additional experience, and using newly designed forceps with 45° and 90° angled tips, two similar cases were later managed successfully, with curative selective adenomectomy in both instances.

The results in patients with microadenomas were decidedly superior to those in patients with extrasellar adenomas (Table 8). Seven patients who had a total hypophysectomy when an adenoma was not identified during exploration were cured, but at the expense of panhypopituitarism. Selective removal of a microadenoma with preservation of anterior pituitary function was successful in 81% of cases. Age, sex, clinical manifestations of the disease, and laboratory values failed to predict the outcome of operation, and extrasellar extension of the tumor predicted an unfavorable outcome.

Complications of surgery were few. In patients who developed transient diabetes insipidus, the condition resolved within a few weeks. No patient developed CSF rhinorrhea, meningitis, hemorrhage, visual impairment, or oculomotor dysfunction. Of the four patients who were failures in the group with microadenomas, three had persistent hypercortisolism and one had a recurrence after hypercortisolism initially resolved. One of the patients with persisting disease has normal cortisol values and is clinically well 4 years after removal of

Table 8. Cushing's disease: results of pituitary microsurgery*

Result	No.	Percent
Microadenomas		
Corrected with selective removal	48	81
Corrected with total hypophysectomy	7	12
Persistence	3	5
Recurrence	1	2
	59	100
Extrasellar Extension		
Corrected with selective removal	3	23
Corrected with total hypophysectomy	1	8
Persistence	6	46
Recurrence	3	23
	13	100

*Patients with dural venous sinus excluded.
From Wilson CB, Tyrrell JB, Fitzgerald PA, Forsham PH: Surgical management of Cushing's disease. In Hormone-Secreting Pituitary Tumors, Givens HM (ed), Chicago, Year Book Medical Publishers, 1982, pp 199–208. Reproduced with permission.

Table 9. Cushing's disease: postoperative follow-up of tumors with initial correction of hypercortisolism

Follow-up period (years)	No.	Recurrence
Microadenomas		
< 1	15	0
1–2	8	0
2–3	8	0
3–7	18	1
	49	1
Extrasellar tumors		
< 1	1	0
1	2	2
2	2	1
4	1	0
	6	3

From Wilson CB, Tyrrell JB, Fitzgerald PA, Forsham PH; Surgical management of Cushing's disease. In Hormone-Secreting Pituitary Tumors, Givens HM (ed), Chicago, Year Book Medical Publishers, 1982, pp 199–208. Reproduced with permission.

the microadenoma, but dynamic testing has shown that he does not have normal diurnal cycles.

Postoperative follow-up of the patients in this series has identified four patients who developed recurrent Cushing's disease after the initial apparent success of the operation (Table 9). None of the patients treated by total hypophysectomy suffered a recurrence, and 48 of the 49 patients with microadenomas and early postoperative endocrine evaluation indicating cure of the disease have remained well. The excellent prognosis for patients whose disease is initially corrected by selective microadenomectomy is reinforced by our follow-up observations of 26 patients for 2 to 6½ years. After an initially successful removal of extrasellar adenomas, three recurred, all within 2 years after operation. One patient in the series died 6 weeks after operation. Her postoperative course had been uneventful; her death was attributed to myocardial infarction, although postmortem verification was not obtained.

5.2. Pathology

Although insufficient material was obtained for microscopic examination from several patients who were cured by selective adenomectomy, the clinical evidence justifies our assumption of an adenoma [64]. The typical adenoma responsible for Cushing's disease is white or greyish-white and soft, being gelatinous or, in some cases, even liquescent. These adenomas can be minute, the smallest accurately measured tumor having a diameter of 1.5 mm. Failing to find an adenoma after serially sectioning an excised pituitary gland does not exclude the possibility that a microadenoma has simply gone unidentified. For example, as the posterior lobe often excavates the base of the dorsum sellae, small lobules may be either overlooked or aspirated while the surgeon clears the field of blood from the basal dural venous sinuses. The posterior lobe, like the adenoma, is soft and white, adding another possible source of error in obtaining and identifying a specimen. Three tumors identified in our series were confined to the pars intermedia and adjacent posterior lobe. Although none of our adenomas involved the pituitary stalk, theoretically a small adenoma originating in the pars tuberalis could be overlooked in dividing the pituitary stalk.

In a case not included in this series, I identified a 2-mm adenoma on the surface of the anterior lobe lying against the cavernous sinus near the dorsum sellae only as the pituitary gland was being mobilized in preparation for a total hypophysectomy. I suspect that adenomas were similarly overlooked in the two cases in our series in which an adenoma was not proved by pathologic examination after total hypophysectomy. Schnall *et al.* have reported a case of Cushing's disease in which they found diffuse hyperplasia of corticotrophs and no adenoma in the excised pituitary gland [71], but their well documented study is exceptional.

5.3. The Argument for Transsphenoidal Resection as Primary Treatment

Our series of 77 cases [63, 64] confirms that the diagnosis of Cushing's disease accurately predicts the presence of an ACTH-secreting pituitary microadenoma. Our earlier conclusions that these microadenomas frequently present with typical endocrine features in the absence of tomographic abnormalities of the sella turcica, and that hypothalamic abnormalities of ACTH regulation are a consequence of hypercortisolism rather than a manifestation of a primary abnormality of the central nervous system, remain unchanged.

Cushing's disease should be treated initially by transsphenoidal exploration, regardless of the patient's age. In children, the frequently limited pneumatization of the sphenoid sinus presents a technical obstacle to the operation. Nonetheless, our results in children have been excellent; we have successfully corrected the disease without recurrence in each of seven cases. When surgical exploration fails to disclose an ACTH-secreting microadenoma, total hypophysectomy is indicated, except in children and young adults.

The record of success achieved with transsphenoidal adenomectomy is superior to that obtained with either pituitary irradiation or bilateral adrenalectomy. The procedure achieves immediate cessation of hypercortisolism in the great majority of patients; the instances of recurrence are relatively few. The operative risk is low, and there is a decided probability of achieving and maintaining normal pituitary function, although panhypopituitarism may be an unavoidable consequence of correcting hypercortisolism in an adult. Moreover, the potential development of Nelson's syndrome is averted – a factor that cannot be overlooked, considering that the adenomas responsible for this syndrome have a potential for invasiveness and malignant growth that makes them more sinister than other pituitary adenomas. We treat Nelson's syndrome with radical surgery: total hypophysectomy for microadenomas, and radical removal plus irradiation for larger lesions. In our experience, none of 10 patients with macroadenomas causing Nelson's syndrome achieved normal ACTH levels and four died of the effects of tumor, whereas normal ACTH secretion was restored in four of nine patients with microadenomas [72]. The poor results obtained in the treatment of post-adrenalectomy adenomas provides ample argument for performing transsphenoidal surgery when an adenoma is first discovered.

Medical therapy of pituitary adenomas may supercede present forms of surgery and irradiation at some time in the future; but despite intriguing possibilities suggested in recent trials involving the use of bromocriptine [36], the results of medical treatment of Cushing's disease have so far proved disappointing [73]. In short, Cushing's diagnosis was correct, and it will be difficult to improve on the direct form of treatment that he conceived.

6. ENDOCRINE-INACTIVE ADENOMAS

The endocrine-inactive tumors should be mentioned briefly. The patients in our series came to medical attention because of hypopituitarism and involvement of adjacent structures, usually the optic chiasm. The tumors were large, often massive, and operative removal was necessarily incomplete. Recovery of vision was excellent [74]. Both fatalities in this group, the only deaths in the total series, occurred in elderly patients, and no deaths have occurred in the last 500 patients treated.

7. CONCLUSION

Transsphenoidal microsurgery can be advised as a safe and effective primary treatment for pediatric and adult patients with gigantism, acromegaly, and Cushing's disease. Although the behavior of prolactin-secreting microadenomas in patients of both sexes requires further study, it seems clear that the majority of these adenomas grow slowly – but they grow, nonetheless. Transsphenoidal selective removal of the prolactinoma carries a high probability of cure in younger patients who have hyperprolactinemia with PRL levels below 200 ng/ml; and for the patient desiring pregnancy, operative removal of the adenoma is advocated on the basis of a predictably low risk and the likelihood of a successful outcome.

ACKNOWLEDGMENTS

The author is indebted to the physicians who referred the patients in these series for evaluation and treatment, as well as to the physicians, nurses, and staff of the University of California Hospitals who assisted in their therapy. Thanks are also due to Christine Barbera for technical assistance, and to Susan Eastwood for editorial collaboration in the development of this chapter.

REFERENCES

1. Post KD, Jackson IMD, Reichlin S (eds): The Pituitary Adenoma, New York/London, Plenum Medical Book Company, 1980, pp 109–118.
2. Landolt AM: Ultrastructure of human sellar tumors. Correlations of clinical findings and morphology. Acta Neurochir (Suppl 22):1–167, 1975.
3. Kovacs K, Hovrath E, Ezrin E: Pituitary adenomas. Pathol Annu 2:341–382, 1977.
4. Kovacs K, Horvath E, Ryan N, Ezrin C: Null cell adenoma of the human pituitary. Virchows Arch A Path Anat Histol 387:165–174, 1980.

5. Wilson CB: Neurosurgical management of large and invasive pituitary tumors. In: Clinical management of pituitary disorders, Tindall GT, Collins WF (eds), New York, Raven Press, 1979, pp 335 – 342.
6. McCormick WF, Halmi NS: Absence of chromophobe adenomas from a large series of pituitary tumors. Arch Pathol (Chicago), 92:231 – 238, 1971.
7. Hardy J: Transsphenoidal surgery of hypersecreting pituitary tumors. In: Diagnosis and treatment of pituitary tumors, Kohler PO, Ross GT (eds), International Congress Series No 303, New York, American Elsevier, 1973, pp 179 – 194.
8. Hardy J, Vezina JL: Transsphenoidal neurosurgery of intracranial neoplasm. In: Advances in Neurology (vol 15), Thompson RA, Green JR (eds), New York, Raven Press, 1976, pp 261 – 274.
9. Findling JW, Tyrrell JB, Aron DC, Fitzgerald PA, Wilson CB, Forsham PH: Silent pituitary apoplexy: Subclinical infarction of an adrenocorticotropin-producing pituitary adenoma. J Clin Endocrinol Metab 52:95 – 97, 1981.
10. Domingue JN, Wing SD, Wilson CB: Coexisting pituitary adenomas and partially empty sellas. J Neurosurg 48:23 – 28, 1978.
11. Landolt AM: Progress in pituitary adenoma biology – results of research and clinical applications. In: Advances and technical standards in neurosurgery (vol 15), Krayenbuhl H (ed), Wein, Springer-Verlag, 1978, pp 3 – 49.
12. Landolt A, Wilson CB: Tumors of the sella and parasellar area in adults. In Neurological Surgery (edn 2), JR Youmans (ed), WB Saunders, Philadelphia, 1982, pp 3107 – 3162.
13. Wilson CB, Dempsey LC: Transsphenoidal microsurgical removal of 250 pituitary adenomas. J Neurosurg 48:13 – 22, 1978.
14. Forbes AP, Henneman PH, Griswold GC, Albright F: Syndrome characterized by galactorrhea, amenorrhea and low urinary FSH: Comparison with acromegaly and normal lactation. J Clin Endocrinol Metab 14:265 – 271, 1954.
15. Friesen H, Tolis G, Shiu R, Hwang P: Studies on human pituitary prolactin: chemistry, radioceptor assay and clinical significance. In: Human prolactin, Pasteels JL, Robyn C (eds), Amsterdam, Excerpta Medica, 1973, pp 11 – 23.
16. Klibanski A, Neer RM, Beitins IZ, Ridgway EC, Zervas NT, McArthur JW: Decreased bone density in hyperprolactinemic women. N Engl J Med 303:1511 – 1514, 1980.
17. Lundberg PO, Drettner B, Hemmingsson A, Stenkvist B, Wide L: The invasive pituitary adenoma. A prolactin-producing tumor. Arch Neurol 34:742 – 749, 1977.
18. Horvath E, Kovacs K, Singer W, Smyth HS, Killinger DW, Erzin C, Weiss MH: Acidophil stem cell adenoma of the human pituitary: Clinicopathologic analysis of 15 cases. Cancer 47:761 – 771, 1981.
19. Marrs RP, Bertolli SJ, Kletzky OA: The use of thyrotropin-releasing hormone in distinguishing prolactin-secreting pituitary adenoma. Am J Obstet Gynecol 138:620 – 625, 1980.
20. Chang RJ, Keye WR, Monroe SE, Jaffe RB: Prolactin-secreting pituitary adenomas in women. IV. Pituitary function in amenorrhea associated with normal or abnormal serum prolactin and sellar polytomography. J Clin Endocrinol Metab 51:830 – 835, 1980.
21. Malarkey WB, Goodenow TJ, Lanese RR: Diurnal variation of prolactin-secretion differentiates pituitary tumors from the primary empty sella syndrome. Am J Med 69:886 – 890, 1980.
22. Kapcala LP, Molitch ME, Post KD, Biller BJ, Prager RJ, Jackson IMD, Reichlin S: Galactorrhea, oligo/amenorrhea, and hyperprolactinemia in patients with craniopharyngiomas. J Clin Endocrinol Metab 51:789 – 800, 1980.
23. Richmond IL, Newton TH, Wilson CB: Indications for angiography in the preoperative evaluation of patients with prolactin-secreting pituitary adenomas. J Neurosurg 52:378 – 380, 1980.
24. Turski PA, Newton TH, Horton BH; Anatomic-polytomographic correlation of sellar contour. AJNR 1981 (in press).
25. Chambers EF, Turski PA, Newton TH: Regions of low density in the contrast-enhanced pituitary gland: Normal and pathologic processes. Radiology 144:109 – 113, 1982.

26. Keye WR, Chang RJ, Monroe SE, Wilson CB, Jaffe RB: Prolactin-secreting pituitary adenomas in women. II. Menstrual function, pituitary reserves, and prolactin production following microsurgical removal. Am J Obstet Gynecol 134:360–365, 1979.
27. Aubourg PR, Derome PJ, Peillon F, Jedynak CP, Visot A, Le Gentil P, Balagura S, Guiot G: Endocrine outcome after transsphenoidal adenomectomy for prolactinoma: Prolactin levels and tumor size as predicting factors. Surg Neurol 14:141–143, 1980.
28. Schlechte J, Vangilder J, Sherman B: Predictors of the outcome of transsphenoidal surgery for prolactin-secreting pituitary adenomas. J Clin Endocrinol Metab 52:785–789, 1981.
29. Hardy J: Microsurgery of pituitary disorders (Royal College Lecture). Ann R Coll Phys Surg Can 13:294–298, 1980.
30. Sheline GE: Conventional radiation therapy in the treatment of pituitary tumors. In: Clinical management of pituitary disorders, Tindall GT, Collins WF (eds), Raven Press, New York, 1979, pp 287–314.
31. Bergh T, Nillius SJ, Wide L: Clinical course and outcome of pregnancies in amenorrheic women with hyperprolactinaemia and pituitary tumors. Br Med J 1:875, 1978.
32. Shewchuk AB, Adamson GD, Lessard P, Ezrin C: The effect of pregnancy on suspected pituitary adenomas after conservative management of ovulation defects associated with galactorrhea. Am J Obstet Gynecol 136:659, 1980.
33. Tucker H St G, Grubb SR, Wigand JP, Taylon A, Lankford HV, Blackard WG, Becker DP: Galactorrhea-amenorrhea syndrome: Follow-up of forty-five patients after pituitary tumor removal. Ann Intern Med 94:302–307, 1981.
34. Domingue JN, Richmond IL, Wilson CB: Results of surgery in 114 patients with prolactin-secreting pituitary adenomas. Am J Obstet Gynecol 137:102–108, 1980.
35. Pont A, Shelton R, Odell WD, Wilson CB: Prolactin-secreting tumors in men: Surgical cure. Ann Intern Med 91:211–213, 1979.
36. Baskin DS, Wilson CB: Bromocriptine treatment of pituitary adenomas. Neurosurgery 8:741–744, 1981.
37. Thorner MO, Martin WH, Rogol AD, Morris JL, Perryman RL, Conway BP, Howards SS, Wolfman MG, MacLeod RM: Rapid regression of pituitary prolactinomas during bromocriptine treatment. J Clin Endocrinol Metab 51:438–445, 1980.
38. Martin NA, Hales M, Wilson CB: Cerebellar metastasis from a prolactinoma: Occurrence during treatment with bromocriptine. J Neurosurg Case report. 55:615–619, 1981.
39. Evans HM, Briggs JH, Dixon JS: The physiology and chemistry of growth hormone. In: The pituitary gland, vol I: Anterior pituitary, Harris GW, Donovan BT (eds), Berkeley, University of California Press, 1966, pp 439–491.
40. Wright AD, Hill DM, Lowy C, Fraser TR: Mortality in acromegaly. Quart J Med (New Series) 39:1–16, 1970.
41. Gramegna AG: Un cas d'acromégalie traité par la radiothérapie. Rev Neurol 17:15–17, 1909.
42. Adams JE, Seymour RJ, Earll JM, Tuck M, Sparks LL, Forsham PH: Transsphenoidal cryohypophysectomy in acromegaly. Clinical and endocrinological evaluation. J Neurosurg 28:100–104, 1968.
43. DiTullio MV, Rand RW: Efficacy of cryophypophysectomy in the treatment of acromegaly. Evaluation of 54 cases. J Neurosurg 46:1–11, 1977.
44. Kondo A, Handa H, Matsumura H, Makita Y: Operative treatments for acromegaly: Comparison of transsphenoidal cryogenic and microsurgical hypophysectomy. Neurol Med Chir (Tokyo) 19:683–693, 1979.
45. Hibbert J, Shaheen OH: The treatment of acromegaly by yttrium implantation. J Laryngol Otolarynol 91:1–9, 1977.
46. Joplin GF, Cassar J, Doyle FH et al: Treatment of pituitary tumors by interstitial irradiation. In: Pituitary adenomas. Biology, physiopathology and treatment, Derome PJ, Jedynak CP, Peillon

F (eds), Second European Workshop, La Pitié-Sâlpetrière, Paris, Asclepios Publishers, 1980, pp 219 – 232.
47. Cushing H: Intracranial tumors. Notes upon a series of two-thousand verified cases with surgical mortality percentages pertaining thereto. Springfield, III, Charles C Thomas, 1932, 150 pp.
48. Arafah BM, Brodkey JS, Kaufman B, Velasco M, Manni A, Pearson OH: Transsphenoidal microsurgery in the treatment of acromegaly and gigantism. J Clin Endocrinol Metab 50:578 – 585, 1980.
49. Delalande G, Derome PJ, Jedynak CP et al: Transsphenoidal surgery in acromegaly. Results in 102 patients according to the European enquiry. In: Pituitary adenomas. Biology, physiopathology and treatment, Derome PJ, Jedynak CP, Peillon F (eds), Second European Workshop, La Pitié-Sâlpetrière, Paris, Asclepios Publishers, 1980, pp 320 – 321.
50. Garcia-Uria J, del Pozo JM, Bravo G: Functional treatment of acromegaly by transsphenoidal microsurgery. J Neurosurg 49:36 – 40, 1978.
51. Hardy J, Wigser SM: Transsphenoidal surgery of pituitary fossa tumors with televised radiofluoroscopic control. J Neurosurg 23:612 – 619, 1965.
52. Hay ID, Teasdale G, Beastall GH, Ratcliffe JG, McCruden D, Davies DL, Thompson JA: Transsphenoidal surgery for acromegaly: A comparison of cryosurgery and microsurgery. J Endocrinol Suppl 79:44P – 45P, 1978.
53. Laws ER, Piepgras DG, Randall RV, Abboud CF: Neurosurgical management of acromegaly. Results in 82 patients treated between 1972 and 1977. J Neurosurg 50:454 – 461, 1979.
54. U HS, Wilson CB, Tyrrell JB: Transsphenoidal microhypophysectomy in acromegaly. J Neurosurg 47:840 – 852, 1977.
55. Baskin DS, Boggan JE, Wilson CB: Transsphenoidal microsurgical removal of growth hormone-secreting pituitary adenomas. A review of 137 cases. J Neurosurg 56:634 – 641, 1982.
56. Aloia JF, Archambeau JO: Hypopituitarism following pituitary irradiation for acromegaly. Hormone Res 9:201 – 207, 1978.
57. Eastman RC, Gorden P, Roth J: Conventional supervoltage irradiation is an effective treatment for acromegaly. J Clin Endocrinol Metab 48:931 – 940, 1979.
58. Cushing H: The basophil adenomas of the pituitary body and their clinical manifestations (pituitary basophilism). Bull Johns Hopkins Hosp 50:137 – 195, 1932.
59. Lisser H: Hypophysectomy in Cushing's disease. J Nerv Ment Dis 99:727 – 733, 1944.
60. Hardy J: Transsphenoidal microsurgery of the normal and pathological pituitary. Clin Neurosurg 16:185 – 216, 1969.
61. Liddle GW, Shute AM: The evolution of Cushing's syndrome as a clinical entity. Adv Intern Med 15:155 – 175, 1969.
62. Liddle GW: Pathogenesis of glucocorticoid disorders. Am J Med 53:638 – 648, 1972.
63. Tyrrell JB, Brooks RM, Fitzgerald PA, Cofoid PB, Forsham PH, Wilson CB: Cushing's disease: Selective transsphenoidal resection of pituitary microadenomas. N Engl J Med 298:753 – 758, 1978.
64. Wilson CB, Tyrrell JB, Fitzgerald PA, Forsham PH: Cushing's disease: Surgical management. In: Hormone-Secreting Pituitary Tumors, Givens JR (ed), Chicago, Year Book Medical Publishers, 1982, pp 199 – 208.
65. Pavlatos FCh, Smilo RP, Forsham PH: A rapid screening test for Cushing's syndrome. JAMA 193:720 – 723, 1965.
66. Liddle GW: Tests of pituitary-adrenal suppressibility in the diagnosis of Cushing's syndrome. J Clin Endocrinol Metab 20:1539 – 1560, 1960.
67. Besser GM, Edwards CRW: Cushing's syndrome. Clin Endocrinol Metab 1:451 – 490, 1972.
68. Findling JW, Aron DC, Tyrrell JB, Shinsako JH, Fitzgerald PA, Norman D, Wilson CB, Forsham PH: Selective venous sampling for ACTH in Cushing's syndrome. Ann Int Med 94:647 – 652, 1981.

69. Weinstein M, Tyrrell B, Newton TH: The sella turcica in Nelson's syndrome. Radiology 118:363–365, 1976.
70. Tyrrell JB, Wiener-Kronish J, Lorenzi M, Brook RM, Forsham PH: Cushing's disease: growth hormone response to hypoglycemia after correction of hypercortisolism. J Clin Endocrinol Metab 44:218–221, 1977.
71. Schnall AM, Kovacs K, Brodkey JS, Pearson OH: Pituitary Cushing's disease without adenoma. Acta Endocrinol (Copenh) 94:297–303, 1980.
72. Wilson CB, Wright DC, Gutin PH: Endocrine-active pituitary adenomas and their treatment by transsphenoidal microsurgery. Presented at the meeting of the Texas Association of Neurological Surgeons, San Antonio, Texas, May 3–6, 1979.
73. Lamberts SWJ, Klijn JGM, de Quijada M, Timmermans HAT, Uitterlinden P, de Jong FH, Birkenhäger JC: The mechanism of the suppressive action of bromocriptine on adrenocorticotropin secretion in patients with Cushing's disease and Nelson's syndrome. J Clin Endocrinol Metab 51:307–311, 1980.
74. Wilson CB, Dempsey LC: Neuro-ophthalmology and transsphenoidal removal of pituitary adenomas. In: Neuro-Ophthalmology update, Smith JL (ed), Masson, New York, 1977, pp 221–226.

6. Immunologic Considerations of Patients with Brain Tumors

M.S. MAHALEY Jr. and G. YANCEY GILLESPIE

INTRODUCTION

Continued interest in the immunologic considerations of patients with malignant gliomas stems from the frustration that clinicians face in current concepts of management coupled with the attractiveness of the hypothetical, selective destruction of malignant cells by elements of the immune system. Our frustration with current therapies is exemplified by the fact that the most effective combinations of surgery, radiotherapy, and chemotherapy for malignant gliomas only increases the median survival time from 17 weeks for patients untreated postoperatively to a little over a year for optimally treated patients [1]. The attractiveness of mobilizing the immune system to control neoplasia has undoubtedly been promoted by the recognized cellular immune responses that all too successfully result in transplantation rejections together with the humoral (antibody) immune mechanisms responsible for specific and timely neutralization and destruction of microbial invaders.

However, the immediate application of immunologic manipulations to the control of intracranial neoplasms has also been an exercise in frustration related to several practical considerations. To date, no glioma-specific antigens have been identified conclusively; therefore, the development of vaccines specific for this disease or even the prospects of an immune reaction directed specifically or nonspecifically against this neoplastic process are still speculative.

Equally confounding is the remarkable degree of heterogeneity of malignant gliomas that has been recognized more recently [2, 3]. Not only has conventional light microscopic histological description of these tissues varied considerably within the neuro-ectodermal origins of this disease but more specific differences of morphological and biochemical nature have now been described, not only from patient to patient within a given tumor type but also from one population of cells to another within a given individual's tumor.

If malignant gliomas behave in the face of therapy in some way similar to that of other human cancers, then combination approaches to therapy will be needed. Surgery will undoubtedly remain an important ingredient for purposes of precise histological identification, of immediate debulking of a significant portion of the mass, and of making tissue available for investigative analysis. Radiotherapy has certainly proven itself as an important adjunctive form of therapy, with damage to normal brain remaining the single most important limiting dose factor to be considered [4]. Chemotherapy has emerged as a useful further adjunct to postoperative management, although one is limited at the present time by the number of effective drugs available and the uncertainty as to the best route and timing of administration. Toxicity to the lungs, liver, kidneys and bone marrow in addition to that to the central nervous system has emerged as a significant dose limiting factor with chemicals [5].

The role that immunotherapy may play within this scenario is, as yet, undefined. Since considerable investigation has now been made into the nature and functional capacity of the immune system in patients with malignant gliomas, it would seem appropriate to review these findings as well as the current status of pilot efforts in the evaluation of immunotherapeutic approaches to this disease.

IMMUNOCOMPETENCE OF PATIENTS WITH BRAIN TUMORS

In the last few years, considerable information has been forthcoming regarding the immune status of patients with primary intracranial tumors. *Cellular immune mechanisms*, which may be responsible for delayed hypersensitivity-type reactions, graft versus host reactions, allograft reactions, lysis of virus infected target cells, and probably cytotoxic effects upon tumor cells, have been most exhaustively studied. At the time of surgical diagnosis, it has been shown that patients with malignant gliomas generally suffer impaired delayed hypersensitivity responses to such common skin tests as tuberculin, candida, trichophyton, mumps and streptokinase [6, 7, 8]. This anergy also manifests itself as a diminished blastogenic response by peripheral blood leukocytes that have been stimulated *in vitro* with mitogens or in mixed lymphocyte cultures [7, 9, 10]. Furthermore, patients with malignant brain tumors have a depression of their circulating lymphocytes, particularly T cells, that persists throughout the course of their disease [8, 11].

It is uncertain when, during the course of tumor progression, these cellular immune capabilities become impaired, but it is apparent that these defects are significantly worse in those cases with the most anaplastic morphology. Although the degree of cellular immune depression soon after surgery does not predict anticipated survival time, a *further* reduction in skin test responses serially in a given patient later in time does correlate with regrowth of the intracranial neoplasm [12]. The role suppressor cells may play in the overall immunological integrity of the brain tumor patient is currently debated [13, 14]. The practicality of stimulating an effective cellular immune response in brain tumor patients must be tempered by information provided by the studies just reviewed.

General and specific *humoral immune responsiveness* by patients with malignant glioma has been less extensively studied. Mahaley *et al.* [8] followed serum antibody titers to tetanus and influenza in glioblastoma patients who were challenged serially after surgery. Although their antibody levels rose initially, subsequent serial challenge injections failed to increase their antibody titers. Despite bimonthly boosting, most patients suffered a progressive deterioration in responsiveness that closely paralleled their worsening clinical status. The nature of this developing serologic anergy has not been elucidated.

It has been noted that serum immunoglobulin levels of brain tumor patients are essentially normal [15] with one exception; namely, there is somewhat elevated serum IgM level observed most often in those patients with gliomas of the most malignant histologic grade [8]. However, elevation in IgM levels has also been reported in 15 patients with meningiomas [16]. While the actual number of circulating B-cells does not appear to be diminished [17], the ability to mobilize an effective antibody immune response to a new foreign antigen (keyhole limpet hemocyanin) appears to be significantly impaired in patients with malignant gliomas (Mahaley, unpublished). This would suggest that there may be a basic impairment in the glioma patient's ability to recognize any new antigens, including tumor antigens, in a way that would effectively mobilize an antibody response. Finally, from the serum of patients with malignant gliomas, a soluble blocking factor has been described that (a) prevents *in vitro* blastogenesis of T-cells from normal individuals [7] as well as those from tumor-bearing patients, and (b) blocks the cytotoxicity of peripheral blood leukocytes from glioma patients towards glioma cells in culture [18].

Unfortunately the degree of normality of these various immune mechanisms as determined peri-operatively cannot serve to predict the anticipated life expectancy in the individual patient. However, subsequent serial decline in immune competence, as measured by these parameters, has adumbrated the onset of further clinical worsening on the part of the patient. Brooks *et al.* [12] formulated a predictive equation based on a combination of non-tumor specific lymphocyte assays that, when monitored serially, can signal tumor recurrence 4 to 9 weeks *prior* to the patient's clinical deterioration. This study re-emphasized the utility of scrially monitoring the immune status of cancer patients.

Another possible correlation of immune status and survival time has been suggested by deTribolet and coworkers [19, 20] who quantified circulating immune complexes pre- and postoperatively in patients with intracranial tumors. Elevated levels of circulating immune complexes, measured by the radiolabeled Clq-binding assay [21], were detected significantly more often preoperatively in patients with grade III/IV gliomas than in patients with low grade [I/II] gliomas. Moreover, malignant glioma patients with elevated immune complex levels had significantly shorter survival times than did those patients with nonelevated immune complex levels.

Thus, it would seem important to continue to monitor general immune competence parameters at regular intervals. In particular, one should attempt to document any changes that might occur with exploratory efforts at immunotherapy.

EVIDENCE OF AN ANTI-TUMOR RESPONSE BY GLIOMA PATIENTS

Despite what appears to be a progressive, systemic anergy in patients with malignant gliomas, a number of lines of direct and circumstantial evidence exist for an immune response to these neoplasms. Perivascular mononuclear cell infiltration, a hallmark of cell-mediated immune responses to foreign tissues, has been observed in histologic sections of malignant gliomas sampled at autopsy [22] or biopsied at surgery [23]. Its incidence and extent has been correlated positively with length of patient survival [24, 25, 26]. T lymphocytes have been shown to compose up to 50% of these inflammatory mononuclear cells [27] and a considerable number of macrophages have been identified in malignant glioma tissues [28, 29]. The inability of most glioma patients to mount or maintain a substantial mobilization of inflammatory mononuclear cells at the tumor site may be a manifestation of the macrophage migratory dysfunction previously described for cancer patients [30]. This migratory dysfunction does not appear to extend to neutrophils, however. Kay et al. [31] reported that neutrophils from 5 patients with varying grades of malignant glioma exhibited chemotaxis toward chemotactic stimuli that was similar in magnitude to that of 30 healthy individuals.

In addition to intratumoral immune-related cells, immunoglobulin has been detected *in situ* on both glioblastoma cells [32] and bound to Fc-receptor bearing cells (presumably macrophages) present within CNS tumors [33]. In light of the observations Aarli *et al.* [34] that normal human IgC binds to myelin sheaths, glia and neurons, however, it may be presumptuous to ascribe this to a specific immune response. Despite this *caveat*, the carefully controlled studies of Wood *et al.* [33] would suggest that many, but not all, CNS malignancies contain immunoglobulin associated predominantly with infiltrating Fc-receptor positive cells and, less commonly, with Fc-receptor negative tumor cells. In no instance was immunoglobulin associated with the membranes of both types of cells. Thus, it would appear that there is morphologic evidence, in some patients at least, of an attempt to respond immunologically to their tumors.

Further evidence of immune reactivity has been provided through the application of *in vitro* techniques to quantify immune system function(s) directed towards malignant glioma. *In vitro* blast transformation of patients' peripheral blood mononuclear cells has been demonstrated following co-cultivation with autologous glioma cells [35, 36]. This has been taken to indicate an antigen-recognition event by immune memory cells. In a somewhat analogous fashion, Sheikh *et al.* [37] demonstrated that adherence of leukocytes from 80% of glioma patients could be inhibited by exposure to extracts of glioma tissue. Further analysis of the specificity of this phenomenon [38] has suggested that it was a glial tumor-associated response. However, blastogenic transformation or leukocyte adherence inhibition in response to antigen could possibly be observed for lymphoid cells that have either effector (i.e., tumoricidal) or regulatory (i.e., suppression) function. The stimulation of a cell-mediated immune response disadvantageous to the tumor-bearing host may explain why this mechanism is ineffective. Indeed, where *in vitro* cytotoxic activity by patient lymphocytes has been sought against either autochthonous or allogeneic cultured tumor cells, in well-controlled studies, there has been only occasional evidence of a tumor-specific or tumor-associated effect. Several studies have shown that some glioma patients possess peripheral blood lymphocytes or serum antibodies capable of specific cytotoxicity directed against cultured, autologous glioma cells [18, 39]. However, patients' sera were also found to contain 'blocking factors' that could prevent the cytotoxic effect of patients' lymphocytes on glioma target cells [6, 18, 39]. Whether or not this 'blocking effect' is another manifestation of a soluble substance in patients' sera that can inhibit lymphocyte transformation *in vitro* [6,10] remains uncertain. Moreover, *in vitro* cytotoxicity studies have suggested that the natural occurrence of cell-mediated cytotoxicity by brain tumor patients' leukocytes is infrequent and can be demonstrated *in vitro* only when a relatively large number (10^5) of effector cells are exposed to a much smaller number (10^2) of potential tumor target cells [18].

The search for anti-glioma antibodies in sera from patients with primary intracranial neoplasms has been conducted by a number of investigators using a variety of assays that depend upon either primary or secondary antibody interactions. Assays that rely solely on the primary union of antibody with its specific antigen as detected by immunofluorescence, radioimmunoassay or mixed cell hemadsorption have been utilized to suggest the presence of tumor-specific or tumor-associated cell surface antigens on human glioma cells. Likewise, assays that require a secondary interaction of the antibody to produce precipatation of the antigen or tumor cytotoxicity through the action of complement or lymphoreticular cell elements have purported to demonstrate an antitumor effect of glioma patients' sera.

However, the results obtained thus far, by all of these techniques, appear conflicting and almost impossible to interpret due to the great variability in the percentage of positive reactions that have been reported. A good example of this can be found in efforts reported from three different laboratories using essentially similar techniques. Kornblith *et al.* [40, 41] employed both complement-dependent (CDC) and antibody-dependent cell mediated cytotoxicity (ADCC) assays in screening serum samples from a large number of patients. Using a single glioblastoma cell line, LM, they reported that greater than 70% of the sera reacted positively by the CDC assay. In a second series, none of the glioma patients' sera tested had anti-glioma activity by the ADCC assay. In contrast, Woosley et al. [18] found that a small, but statistically significant, percentage (20%) of patients reacted to autologous glioma cells, in preference to autologous normal cells, by either the CDC assay, or the ADCC assay. In an attempt to resolve these differences, Martin-Achard et al. [42] tested sera from a large number of patients with primary brain tumors against 8 glioblastoma cell lines using both techniques. Their findings were not consonant with the previous report of tumor-specific antibody responses by the majority of patients with malignant glioma. Only a small percentage of significant antibody-mediated tumor cytotoxicity tests were obtained, and the majority of those patients tests could be prevented by absorption of the sera with platelets and histologically unrelated tumor cells. Further studies by Kornblith et al. [43, 44] were performed on 25 glioma patients' sera and autologous cells. They reported complement-dependent cytotoxic antibodies in 44% of these cases. Moreover, absorption experiments with cultured fibroblasts, white blood cells, erythrocytes, platelets and normal brain homogenates were reported to indicate that the principal antibody activities were directed against glioma-specific cell surface antigens. Thus, the occurrence of CDC or ADCC reactions, when similar reactions to autologous normal cells are excluded, would appear to be an infrequent although definite *in vitro* phenomenon.

ATTEMPTS TO BOOST IMMUNOLOGIC COMPETENCE

Two preliminary studies have been carried out attempting to boost general immune competence in patients with malignant gliomas. Since neither of these studies included a control population of patients, both must be considered to be pilot efforts. DeCarvalho [45] utilized BCG inoculation, with no results suggestive of either an alteration in immune reactivity or clinical response. Selker et al. [46] evaluated the administration of C. parvum and, likewise, did not demonstrate any change in immune competence or any definitive clinical response to treatment. Mahaley et al., [47] utilized levamisole in a controlled, prospective, randomized study of patients who all had surgical resection of a significant portion of their tumor followed by a course of radiotherapy and BCNU chemotherapy. Levamisole was administered at the beginning of adjunctive, postoperative therapy of those patients randomized to receive it. In these clinical studies, the addition of levamisole immunostimulation therapy to the most effective conventional treatment for patients with anaplastic gliomas failed to improve subsequent serial immune profiles and did not alter the survival times of those patients compared to those randomized not to receive levamisole. It would appear, therefore, that this addition of levamisole immunostimulation to conventional therapy for patients with malignant gliomas not only fails to improve overall survival time but also does not significantly boost general immune competence. A similar protocol of therapy evaluated concurrently in the avian sarcoma virus (ASV) induced glioma model in F344 rats also did not show any evidence of further prolongation of anticipated survival [47]. Therefore, up to the present time, these limited trials have failed to identify an immune stimulant or immune adjuvant that by itself is effective clinically in patients with malignant gliomas. However, it is quite possible that adjuvants of this type may be useful in combination with other forms of immunologic intervention. Obviously, other immunostimulants such as OK-432 and thymosin [48] remain to be more thoroughly evaluated as single agents.

ADOPTIVE IMMUNOTHERAPY

Attempts to transfer immune competence to patients with cancer include adoptive transfer of cellular elements of the immune system. Autologous leukocytes were first infused directly into the intracranial cavity in 3 patients with malignant gliomas by Trouillas and Lapras in 1970 [49]. Takakura et al. [50] infused peripheral white blood cells into the intracranial space of malignant brain tumors of 10 patients with recurrences. Although increased survival of patients receiving these intratumoral infusions was reported, these non-randomized studies were not confined to one tumor type and were not otherwise cell-controlled. Thus, interpretation of these results is difficult. Young et al. [51] injected autologous leukocytes directly into the tumor bed of 18 patients with recurrent *glioblastoma multiforme*. This non-random, uncontrolled study resulted in a clinical improvement in at least 9 patients with relatively long term benefits implied in at least 4 of these. Interest in this approach continues [52].

There has been a paucity of immune studies carried out on these patients, and it is reasonable to conclude that there has been no evidence of a significant immune reaction on the part of patients so treated nor of any proven therapeutic effectiveness of these treatments.

PASSIVE IMMUNOTHERAPY

In vitro studies have demonstrated the presence of antibodies in the sera of patients with brain tumors capable of adherence to the surfaces of autologous and homologous glioma cells from tissue culture [36, 53]. It has been suggested that IgM may be the most effective in complement fixation on glioma surface membranes and that IgG may inhibit this reaction [54]. Complement mediated antibody fixation onto glioma cells *in vivo* might certainly be a basis for serotherapy.

The administration of a xenogeneic (rabbit) antiglioma antibody preparation to patients with malignant gliomas was first reported by Mahaley et al. in 1965 [55]. This pilot study was carried out primarily to evaluate the expression of brain tumor antigens *in vivo* and the localization potential for these putative antiglioma antibodies. Although these investigations demonstrated that there were likely distinctive glioma antigens and that xenogeneic antibody raised against these antigens could be demonstrated to localize within glioma tissue, clinical application of serotherapy has not been pursued vigorously. Of course, at the time this approach had its own limitations with reference to the requirement for extensive absorptions to ensure specificity as well as the heterologous nature of the antiglioma antibody preparation. With the emerging technology for production of monoclonal specific antibodies in quantities sufficient for exhaustive study, the role that serotherapy may play in the management of malignant gliomas undoubtedly will be re-evaluated and more completely investigated. Thus far, several categories of antigens have been identified on human glioma cells by monoclonal antibodies raised by different investigators. A neuroectodermal (possibly oncofetal) antigen has been demonstrated to be shared by melanoma, glioma and neuroblastoma cells [56–61] but not by any other nongliogenous cell types. The second class of monoclonal antibody defined antigen is a 'glioma-mesenchymal extracellular matrix' antigen recently described by Bourdon et al. [62]. Finally, monoclonal antibodies have been raised against what appear to be 'common glioma-associated' antigens expressed by a majority of malignant gliomas [60]. As more of these monoclonal antibody reagents become available, they will be invaluable tools in probing the extraordinary heterogeneity of malignant gliomas.

ACTIVE IMMUNOTHERAPY

The very first efforts at immunotherapy in patients with malignant gliomas involved active immunization, utilizing the patient's own tumor injected subcutaneously in a living state [63]. Although the tumor cells grew locally, there was neither enhancement nor shortening of the patient's survival time over what would have been predicted. Morever, these investigators did not detect any complement-fixing serum antibodies directed against autologous glioma cells. In studies carried out a year later, Grace [64] reported using the same approach in six patients, with similar results. Although patient survival times were not significantly extended, there was a suggestion that an immunologic reaction had occurred in some of the patients. For example, regression of local implants was observed in four patients, two of whom subsequently developed positive delayed hypersensitivity skin reactions to extracts of their tumors. Trouillas and Lapras [49] implanted glioma cells with or without Freund's complete adjuvant, in a series of twenty patients. They also evaluated immunization with glioma cell extracts in Freund's complete adjuvant at sites distant to the implant sites. Tumor growth appeared to be inhibited at the implant sites, and patients developed positive intradermal delayed hypersensitivity reactions to glioma extracts. Unfortunately, this study included a small number of patients with insufficient immunological assays, and it was uncontrolled insofar as survival times were concerned. Trouillas [65] later reported immunodiffusion precipitation reactions between patients' sera and autologous glioma extracts after intradermal immunization of 14 individuals with autologous glioma extracts incorporated in Freund's adjuvant. The post-immunization sera did not cross react with extracts of meningioma or normal adult brain but did have positive reactions with fetal brain extracts. Febvre *et al.* [66] also reported delayed hypersensitivity skin reactions to lyophilyzed glioma cell lines in patients following immunization to their own gliomas. No reactions occurred to brain tissue itself, and pre-immunization skin tests were all negative. In a prospective, randomized study, Bloom *et al.* [67] injected irradiated, autologous glioma cells subcutaneously in twenty-seven patients, 10 of whom received multiple injections. However, these investigators were unable to demonstrate any improvement in survival or development of positive delayed hypersensitivity reactions to autologous glioma extracts. In this study, immune stimulants or adjuvants were not used, and no attempt was made to select candidates for immunotherapy who still had demonstrable general immunologic reactivity.

Based upon well-founded immunological considerations, as well as review of some of the cases mentioned above, there is a legitimate concern about the development of allergic encephalomyelitis in individuals immunized with any material from the brain (including a neoplasm), particularly when associated with concomitant (such as Freund's or BCG) inoculation [68]. This has led to some caution in pursuing active immunization vigorously in patients with brain tumors.

A pilot study has just been completed at the University of North Carolina, where patients with histologic proven anaplastic gliomas underwent immune screening after surgery. Twenty patients found to have a normal or slightly less than normal degree of general immune competence were offered adjunctive immunotherapy with either of two tissue cultured glioma cell lines (lethally irradiated just prior to inoculation) mixed with a BCG cell wall preparation, with subsequent monthly inoculations of cultured glioma cells only. Each patient also received a full course of cobalt therapy to the head (6000 rads over 6 weeks) and bimonthly chemotherapy with BCNU (80 mg/M^2/day i.v. $\times$ 3). Serial immune studies including skin testing with the glioma cell preparation were carried out bimonthly. The twenty patients were accrued from October 1979 to November

1980. Most of these patients are still living and continue to be followed with the results of their serial immune studies yet to be compiled. From our preliminary analyses, it has become apparent that this form of active immunization can result in the production of a significant quantity of serum antibody directed against the immunizing glioma cells. However, approximately 90% of the antibody titer [69] reflects an immune reaction to fetal bovine serum proteins incorporated into the immunizing tumor cells during their tissue culture growth prior to inoculation [70]. A further significant amount of antibody produced is directed against HLA histocompatibility antigens. However, once serum antibodies against fetal bovine serum and HLA antigens have been absorbed, there remains in some patients a low, but significant, titer of serum antibody reactive against the original glioma cell preparation. This reactivity may represent antibodies directed against distinctive antigenic features of the glioma cell itself.

FUTURE AREAS OF CLINICAL INVESTIGATION

When one considers the variety of cellular composition of the usual anaplastic glioma – morphologic, biochemical, and metabolic heterogeneity – it is easier to understand why radiotherapy and chemotherapy kill only a portion of the neoplastic population. Radiotherapy delivered to the whole brain will never yield an irrepairable lethal effect upon *every* tumor cell, and certain normal brain cells will always suffer damage (witnessed by mental changes, evidence of atrophy, and delayed vascular ischemic changes). Chemotherapy also is dependent upon the same target cell sensitivity in one form or another, and the ultimate sensitivity of the brain itself has been illustrated by the retinopathy and encephalopathy occurring after high dose regional vascular infusion of BCNU [71]. If there are indeed distinctive or specific antigens associated with human glioma cells, immunological therapeutic approaches would be a rational basis for control of this neoplasm, provided all growing or potentially growing cells contain this antigen (unshared with normal tissues) and provided the antigen could serve as a focus for attack by the immune system upon the cell. If the antigenicity of glioma cells is weak, then it may be possible to artificially amplify their antigenic properties by chemical modification. Stavrou *et al.* [72] have demonstrated that dimethylated and triphenylated rat glioma cells elicited a stronger antibody response when tested against native glioma cells than did unconjugated rat glioma cells.

In the last few years, investigators have begun to recognize in most patients with malignant gliomas at the time of surgical diagnosis an immune profile that includes: (a) impaired delayed hypersensitivity reactions to recall skin test antigens; (b) low numbers of circulating lymphocytes, particularly T-cells; (c) impaired ability of lymphocytes to respond to mitogens; (d) elevated serum IgM levels; and (e) poor antibody production in response to repeated boosting with tetanus toxoid or influenza vaccine or to a new antigen, keyhole limpet hemocyanin. An unanswered question remains as to the relevance of these findings to the potential of eliciting an effective immune reaction by the patient to the glioma.

One of the few encouraging reports of immunotherapy in humans with malignant gliomas has been that of results obtained by Trouillas and Lapras [49], who actively immunized patients with *glioblastoma multiforme* using subcutaneous live autologous tumor cells combined with Freund's complete adjuvant. A potential hazard of such immunization is a growth of the subcutaneous glioma implants with subsequent metastasis. If non-viable cells could be substituted for living cells in active immunotherapy, this serious complication would be avoided. An important question is whether non-viable tumor cells would be as effective as living cells, since the antigenicity of the killed cells may be altered and their ability to elicit an anti-tumor response thereby decreased. On the other hand, cryptic antigenic sites on the tumor cells may be exposed during the process of rendering the cells non-viable; indeed, radiation has been shown to increase immunogenicity of mouse lymphoma cells [73] and to cause the release from sarcoma cells of an antigenic component that is capable of activating a cellular immune response against the tumor [74]. In experimental specific and non-specific immunotherapy studies, the efficacy of viable versus radiation-killed (10,000 rads) allogeneic tumor cells was compared in the immunotherapy of ASV-induced brain tumors in rats [47]. No difference in outcome was found. In fact, for the first time, combined forms of immunotherapy *alone* were found to effectively prolong the survival time of ASV glioma bearing rats. When BCNU chemotherapy was added to this combination of immunotherapeutic agents, there was further significant prolongation of survival.

In the future, multi-modality approaches, such as the pilot study recently undertaken at the University of North Carolina, would seem to be the most ethically enlightened approach to these patients. It continues to be important to monitor immune reactions of any patients undergoing immunotherapy trials. Indeed, immunotherapeutic intervention without such monitoring can be poorly justified at this time. Very careful appraisals of possible adverse reactions to such treatments are, likewise, strongly indicated. Such neurological changes may be rather subtle, reflected only in alterations in the patient's mental status and cognitive functions, which may be recognized only by appropriately timed CT scanning, consistent neurological examinations and interview by personnel continuously involved in each patient's care. As an example, Trouillas [64] noted one case of a confusion syndrome and inflammatory response in a patient receiving immunotherapy. For the most part, the only side effect of immunotherapy has been induration at the injection site(s) as a result of granuloma-like tissue reaction. There is little substitute for adequately studied nervous system tissues at the time of death, whether one is considering possible toxic effects of the nervous system from radiotherapy, chemotherapy or immunotherapy. The cause of death needs to be ascertained in as many of these patients as possible and the relation of persistent intracranial tumor to death of the patient must be reconciled. A significant impact into this disease will certainly require the enlightened understanding of referring physicians and a concerted attack mounted jointly by neurosurgeons, neuro-oncologists, and basic scientists working together and learning from each other.

REFERENCES

1. Walker MD, Green SB, Byar DP, Alexander E Jr, Batzdorf U, Brooks WH, Hunt WE, MacCarty CS, Mahaley MS Jr, Mealey J Jr, Owens G, Ransohoff J II, Robertson JT, Shapiro WR, Smith KR Jr, Wilson CB, Strike TA: Randomized comparisons of radiotherapy and nitrosoureas for the treatment of malignant glioma after surgery. N Eng J Med 303:1323–1329, 1980.
2. Bigner DD, Bigner SH, Ponten J, Westermark B, Mahaley MS Jr, Ruoslahti E, Herschman H, Eng LF, Wikstrand CJ: Heterogeneity of genotypic and phenotypic characteristics of fifteen permanent cell lines derived from human gliomas. J Neuropath Exp Neurol 40(3):201–229, 1981.
3. Bullard DE, Bigner SH, Bigner DD: The morphologic response of cell lines derived from human gliomas to dibutyryl adenosine 3':5'-cyclic monophosphate. J Neuropath Exp Neurol 49(2):230–246, 1981.
4. Burger PC, Mahaley MS Jr, Dudka L, Vogel FS: The morphologic effects of radiation administered therapeutically for intracranial gliomas. Cancer 44:1256–1272, 1979.
5. Aronin PA, Mahaley MS Jr, Rudnick SA, Dudka L, Donohue J, Selker R, Moore P: Predicting BCNU pulmonary toxicity in patients with malignant gliomas. An assessment of risk factors. N Eng J Med 303:183–188, 1980.
6. Brooks WH, Netsky MG, Normansell DE, Horowitz DA: Depressed cell-mediated immunity in patients with primary intracranial tumors. J Exp Med 136:1631–1647, 1972.
7. Brooks WH, Caldwell HD, Mortara RH: Immune responses in patients with gliomas. Surg Neurol 2:419–423, 1974.
8. Mahaley MS Jr, Brooks WH, Roszman TL, Bigner DD, Dudka L, Richardson S: Immunobiology of primary intracranial tumors. I. Studies of the cellular and humoral general immune competence of brain tumor patients. J Neurosurg 46:467–476, 1977.
9. Thomas DGT, Lannigan CB, Behan PO: Impaired cell-mediated immunity in human brain tumours. Lancet 1:1389, 1975.
10. Young HF, Sakalas R, Kaplan AM: Inhibition of cell-mediated immunity in patients with brain tumor. Surg Neurol 5:19–23, 1976.
11. Brooks WH, Roszman TL, Rogers AS: Impairment of rosette-forming T lymphocytes in patients with primary intracranial neoplasms. Cancer 37:1869–1873, 1976.
12. Brooks WH, Latta RB, Mahaley MS, Roszman TL, Dudka L, Skaggs C: Immunobiology of primary intracranial tumors. Part 5: Correlation of a lymphocyte index and clinical status. J Neurosurg 54:331–337, 1981.
13. Roszman TL, Brooks WH, Elliott LH: Immunobiology of primary intracranial tumors. VI. Suppressor cell function and lectin-binding lymphocyte subpopulations in patients with cerebral tumors. Cancer: In press, 1981.
14. Braun DP, Penn RD, Flannery AM, Harris JE: Immunoregulatory cell function in peripheral blood leukocytes of brain cancer patients. Personal communication of a preprint.
15. Weiss JF, Morantz RA, Bradley WP, Chretien PB: Serum acute-phase proteins and immunoglobulins in patients with gliomas. Cancer Res 39:542–544, 1979.

16. Tokumaru T, Catalano LW: Elevation of serum immunoglobulin M(IgM) level in patients with brain tumors. Surg Neurol 4:17–21, 1975.
17. Brooks WH, Roszman TL, Mahaley MS, Woosley RE: Immunobiology of primary intracranial tumours. II. Analysis of lymphocyte subpopulations in patients with primary brain tumours. Clin Exp Immunol 29:61–66, 1977.
18. Woosley RE, Mahaley MS Jr, Mahaley JL, Miller GM, Brooks WH: Immunobiology of primary intracranial tumors. 3. Microcytotoxicity assays of specific immune responses of brain tumors patients. J Neurosurg 47:871–885, 1977.
19. deTribolet N, Martin-Achard A, Louis JA: Circulating immune complexes in patients with gliomas. Acta Neurochir (Suppl) 28:437–474, 1979.
20. Martin-Achard A, de Tribolet N, Louis JA, Zander E: Immune complexes associated with brain tumors: Correlation with prognosis. Surg Neurol 13:161–163, 1980.
21. Heier HE, Carpentier N, Lange G, Lambert PH, Godel T: Circulating immune complexes in patients with malignant lymphomas and solid tumors. Int J Cancer 20:887–894, 1977.
22. Ridley A, Cavanagh JB: Lymphocytic infiltration in gliomas: Evidence of possible host resistance. Brain 94:117–124, 1971.
23. Takeuchi J, Barnard RO: Perivascular lymphocytic cuffing in astrocytomas. Acta Neuropath (Berl) 35:265–271, 1976.
24. DiLorenzo N, Palma L, Nicole S: Lymphocytic infiltration in long-survival glioblastoma: Possible host's resistance. Acta Neurochir 39:27–33, 1977.
25. Palma L, DiLorenzo N, Guidetti B: Lymphocytic infiltrates in primary glioblastomas and recidivous gliomas. J Neurosurg 49:854–861, 1978.
26. Brooks WH, Markesbery WR, Gupta GD, Roszman TL: Relationship of lymphocyte invasion and survival of brain tumor patients. Ann Neurol 4:219–224, 1978.
27. Stavrou D, Anzil AP, Weidenbach W, Rodt H: Immunofluorescence study of lymphocytic infiltration in gliomas. Identification of T lymphocytes. J Neurol Sci 33:275–282, 1977.
28. Wood GW, Morantz RA: Immunohistologic evaluation of the lymphoreticular infiltrate of human central nervous system tumors. J Natl Cancer Inst 62:485–491, 1979.
29. Morantz RA, Wood GW, Foster M, Clark M, Gollahon K: Macrophages in experimental and human brain tumors. Part 2: Studies of the macrophage content of human brain tumors. J Neurosurg 50:305–311, 1979.
30. Synderman R, Pike MC: Pathophysiological aspects of leukocyte chemotaxis: Identification of a specific chemotactic factor binding site on human granulocytes and defects of macrophage function associated with neoplasia. In: Leukocyte chemotaxis: Methods physiology and clinical implications, Gallin JI, Quie PG (eds), New York, Raven Press, 1978, p 357.
31. Kay NE, Murray KJ, Douglas SD: Neutrophil chemotaxis in cerebral astrocytoma. Surg Neurol 8:255–257, 1977.
32. Tabuchi K, Kirsch WM: Detection of IgG on glioblastoma cell surface in vivo. Acta Neurochir 43:93–100, 1978.
33. Wood GW, Morantz RA, Tilzer SA, Gollahon KA: Immunoglobulin bound in vivo to Fc receptor-positive cells in human central nervous system tumors. J Natl Cancer Inst 64:411–418, 1980.
34. Aarli JA, Aparicio SR, Lumsden CE, Tonder O: Binding of normal human IgG to myelin sheaths, glia and neurons. Immunology 28:171–185, 1975.
35. Ciembroniewicz J, Kolar O: Tissue culture studies of glioblastoma multiforme. Acta Cytol 13:42–49, 1969.
36. Egger AE: Autoradiographic and fluorescence antibody studies of the human host immune response to gliomas. Neurology (Minneap) 22:246–250, 1972.
37. Sheikh KMA, Apuzzo MLJ, Weiss MH: Specific cellular immune responses in patients with malignant gliomas. Cancer Res 39:1733–1738, 1979.
38. Apuzzo MLJ, Sheikh KMA, Weiss MH, Heiden JS, Kurze T: The utilization of native glioma an-

tigens in the assessment of specific cellular and humoral immune responses in malignant glioma patients. Acta Neurochir 55:180–200, 1981.
39. Kumar S, Taylor G, Steward JK, Waghe MA, Morris-Jones P: Cell-mediated immunity and blocking factors in patients with tumours of the central nervous system. Int J Cancer 12:194–205, 1973.
40. Kornblith PL, Dohan FC, Wood W, Whitman BO: Human astrocytomas: Serum-mediated immunologic response. Cancer 33:1512–1519, 1974.
41. Quindlen EA, Dohan FC, Kornblith PL: Improved assay for cytotoxic antiglioma antibody. Surg Forum 25:464–466, 1974.
42. Martin-Achard A, Diserens A-C, de Tribolet N, Carrel S: Evaluation of the humoral response of glioma patients to a possible common tumor-associated antigen(s). Int J Cancer 25:219–224, 1980.
43. Coakham HB, Kornblith PL: The humoral immune response of patients to their gliomas. Acta Neurochir (Suppl) 28:475–479, 1979.
44. Coakham HB, Kornblith PL, Quindlen EA, Pollock LA, Wood WC, Hartnett LC: Autologous humoral response to human gliomas and analysis of certain cell surface antigens: In vitro study with the use of microcytotoxicity and immune adherence assyas. J Natl Cancer Inst 64(2):223–233, 1980.
45. DeCarvalho S, Kaufman A, Pineda A: Adjuvant chemo-immunotherapy in central nervous system tumors. In: Adjuvant therapy of cancer, Salmon SE, Josen SE (eds), Amsterdam: Elsevier/North Holland Biomedical Press, 1977, pp 495–502.
46. Selker RG, Wolmark N, Fisher B, Moore P: Preliminary observations on the use of Corynebacterium parvum in patients with primary intracranial tumors: Effect on intracranial pressure. J Surg Oncol 10:299–303, 1978.
47. Mahaley MS Jr, Steinbok P, Aronin P, Dudka L, Zinn D: Immunobiology of primary intracranial tumors. V. Levamisole as an immune stimulant in patients and in the ASV glioma model. J Neurosurg 54:220–227, 1981.
48. Ommaya AK, Reed J, Walters CL, Meeker WR, Weiss JF: Thymosin for brain tumor therapy. A phase I trial in patients with malignant gliomas. In: Abstracts of the AANS Meeting. (Poster Section), 1981, p 90.
49. Trouillas P, Lapras CL: Immunotherapie active des tumeurs cerebrales. Neuro-Chirurgie 16:143–170, 1970.
50. Takakura K, Miki Y, Kubo O, Ogawa N, Matsutani M, Sano K: Adjuvant immunotherapy for malignant brain tumors. Jap J Clin Oncol 12:109–120, 1972.
51. Young HF, Kaplan AM, Regelson W: Immunotherapy with autologous white cell infusions ('lymphocytes') in the treatment of recurrent glioblastoma multiforme. Cancer 40:1037–1044, 1977.
52. Neuwelt EA, Clark K, Kirkpatrick JB, Toben H: Clinical studies of intrathecal autologous lymphocyte infusions in patients with malignant gliomas: A toxicity study. Ann Neurol 4:307–312, 1978.
53. Roda JE, Heredero JJ, Villarejo FJ, Roda JM: Tumoural antigens on experimental and human glioblastoma. Acta Neurochir 53:187–204, 1980.
54. Garson JA, Quindlen EA, Kornblith PL: Complement fixation by IgM and IgG autoantibodies on cultured human glial cells. J Neurosurg 55:19–26, 1981.
55. Mahaley MS Jr, Mahaley JL, Day ED: The localization of radio-antibodies in human brain tumors. II. Radioautography. Cancer Res 25:779–793, 1965.
56. Kennett RH, Gilbert FM: Hybrid myelomas producing antibodies against a human neuroblastoma antigen present on fetal brain. Science 203:1120–1121, 1978.
57. Carrel S, Accolla RS, Carmagnola AL, Mach J-P: Demonstration of human melanoma associated antigen(s) by monoclonal antibodies. In: Protides of biological fluids, Peeters H (ed), vol 27, Oxford, Pergamon Press, 1980, pp 505–509.

58. Herlyn M, Clark WH Jr, Mastrangelo MJ, Guerry D IV, Elder D, La Rossa O, Hamilton R, Bondi E, Tuthill R, Steplewski Z, Koprowski H: Specific reactivity of monoclonal anti-melanoma antibodies. Cancer Res 40:3602–3609, 1980.
59. Imai K, Ng A-K, Ferrone S: Characterization of monoclonal antibodies to human melanoma-associated antigens. J Nat Cancer Inst 66(3):489–496, 1981.
60. Schnegg JF, Diserens AC, Carrel S, Accolla RS, de Tribolet N: Human glioma-associated antigens detected by monoclonal antibodies. Cancer Res 41:1209–1213, 1981.
61. Seeger RC, Rosenblatt HM, Imai K, Ferrone S: Common antigenic determinants on human melanoma, glioma, neuroblastoma and sarcoma cells defined with monoclonal antibodies. Cancer Res 41:2714–2717, 1981.
62. Bourdon MA, Wikstrand CJ, Pegram CN, Bigner DD: A glioma-mesenchymal extracellular matrix (GMEM) antigen defined by monoclonal antibody. Fed Proc 40:821 (Abstract 3358), 1981.
63. Bloom WH, Carstaris KC, Crompton MR, McKissock W: Autologous glioma transplantation. Lancet 2:77–78, 1960.
64. Grace JT Jr, Perese DM, Metzgar RS, Sasabe T, Holdridge B: Tumor autograft responses in patients with glioblastoma multiforme. J Neurosurg 18:159–167, 1961.
65. Trouillas P: Immunologie et immunotherapie des tumeurs cerebrales. Neurologique 128:23–34, 1973.
66. Febvre H, Maunoury R, Constans JP, Trouillas P: Reactions d'hypersensibilite retardee avec des lignees de cellules tumorales cultivees in vitro chez des malades porteurs de tumeurs cerebrales malignes. Int J Cancer 10:221–232, 1972.
67. Bloom HJ, Peckham MJ, Richardson AE, Alexander PA, Payne PM: Glioblastoma multiforme: A controlled trial to assess the value of specific active immunotherapy in patients treated by radical surgery and radiotherapy. Brit J Cancer 27:253–267, 1973.
68. Bigner DD, Pitts OM, Wikstrand CJ: Induction of lethal experimental allergic encephalomyelitis in nonhuman primates and guinea pigs with human glioblastoma multiforme tissue. J Neurosurg 55:32–42, 1981.
69. Gillespie RP, Mahaley MS Jr, Gillespie GY: Unpublished observations.
70. Kerbel RS, Blakeslee D: Rapid adsorption of a foetal calf serum component by mammalian cells in culture. A potential source of artefacts in studies of antisera to cell-specific antigens. Immunology 31:881–891, 1976.
71. Grimson BS, Mahaley MS Jr, Dubey HD, Dudka L: Ophthalmic and central nervous system complications following intra-carotid BCNU (Carmustine). Accepted for publication. J Clin Neuroophthal, 1981.
71. Stavrou D, Hulten M, Anzil AP, Bilzer T: The humoral antibody response of rats immunized with chemically modified syngeneic brain cells and glioma cells. Int J Cancer 26:629–637, 1980.
73. Maruyama Y: Dose-dependent recognition of irradiated isogeneic mouse lymphoma cells: Study by terminal dilution assay. Int J Cancer 3:593–602, 1968.
74. Sato I, Nio Y, Abe M: In vitro production of an anti-tumor agent by reticuloendothelial cells. Gann 59:273–280, 1968.

7. Corticosteroids: Their Effect on Primary and Metastatic Brain Tumors

ROBERT G. SELKER

1. INTRODUCTION

During the past fifty years, there have been a number of technical and scientific advances in the treatment of primary malignant and metastatic lesions of the brain. From the surgeon's point of view, modern anesthesiology, magnification, intracranial dehydrating agents, lasers, and ultrasonic tissue resectors have been in the forefront of progress. From the radiologist's vantage point, computer scan localization (C.T.), linear accelerators, implantable sources of irradiation, and the tolerance of normal brain tissue to 6000 or more RADS, has led the way toward greater survival periods. For the oncologist, the availability of new drugs, bone marrow transplants and their success with nonsolid tumor malignancies have provided an almost unlimited horizon of potential cure. From the brain tumor patient's perspective, however, no one single technological advance has contributed as much to his or her overall well being and continued function as has the introduction of steroids. From the despair created by persistent headache and vomiting to the incapacitation of progressive deficit, steroids have seemingly reversed the inevitable and created a sense of well being. Even though long-term steroid administration can be a 'double-edged sword', few patients would exchange the reduced or symptom free interlude.

Historically, adrenal extracts were introduced in the 1930's primarily for the treatment of Addison's Disease. Soon after, cortisone was isolated from the extract initiating an unprecedented era of pharmacologic and physiologic investigation hastened by the demands of war. During the 1940's Prados [1] and Grenell [2] demonstrated rather conclusively the effectiveness of ACTH, adrenal-cortical extract and cortisol in reversing or inhibiting the formation of cerebral edema in the experimental animal. In 1952, Ingraham and Matson [3] reported their experiences with the use of intraoperative cortisone in children with lesions of the pituitary. They and others [4, 5], alluded to the possible use of cortisone in the prevention of cerebral edema in the human. However, it was not until the late 1950's and early 60's that Galicich and French [6], and others [7 – 13], employed steroids specifically for that purpose. Their findings in the glioma and metastatic tumor patient were dramatic.

Thus the stage was set for the introduction of a modality whose use today is considered to be standard therapy [14 – 23]. A modality whose long-term effects are both a blessing and a curse; and whose potential relative to other developments in the treatment of intracranial tumor patients is yet to be fully appreciated. One can only speculate as to the role of steroids in conjunction with implanted sources of irradiation, 'tagged' (cell specific) antimetabolites, or as a provider of a symptom free interval during hyperthermia, essential amino acid deprivation, tumor-specific vaccines, and future developments in genetic engineering.

2. PHYSIOLOGY AND PHARMACOLOGY

The adrenal cortex produces aldosterone (mineral corticoid) and cortisol (glucocorticoid); the latter being anti-inflammatory with a prominent effect on the metabolism of carbohydrate and protein. Cortisol, with its major portion bound to serum proteins, is secreted in a cyclic fashion during the 24 hours of a day. Morning values are the highest, whereas late evening secretion falls dramatically [24, 25]. Although unmetabolized cortisol is excreted in small amounts in the urine, most is processed in the liver where it is converted into conjugated metabolites and excreted in the urine as ketogenic steroids. Except in cases of adrenal neoplasm and the exogenous intake of steroid products, cortisol production is regulated by anterior pituitary secretion of adrenocorticotrophic hormone (ACTH), which in turn is controlled by a 'releasing factor' secreted by the hypothalamus. Dysfunction of the anterior lobe of the pituitary or hypothalamus from whatever cause effects ACTH secretion, cortisol production, and urinary 17-hydroxycorticoid levels. This chain of events can be monitored and is the basis for clinical endocrine evaluations.

As noted with plasma cortisol levels, pituitary ACTH secretion is governed by a circadian schedule and is interrelated with normal sleep patterns. This diurinal variation is an important feature in later discussions of dose schedules and administration. It is apparent that exogenous supplies of steroid will reduce adrenal glucocorticoid production and in turn decrease the secretion of 'releasing factor' from the hypothalamus and ultimately ACTH production by the pituitary. With diminished supplies of glucocorticoid the converse would be true, and the need for increased production of ACTH would be signaled to the pituitary (Figure 1).

Long-term administration of exogenous glucocorticoids will result in depression of the hypothalamic-pituitary-adrenal axis. The degree and rapidity of occurrence is in part dependent upon the product selected, the dose levels employed, the time of day of administration and total duration of therapy. If steroid dependent patients are subjected to stressful situations, (i.e., surgery, intercurrent infection, etc.), steroid supplementation will be required. Target organ

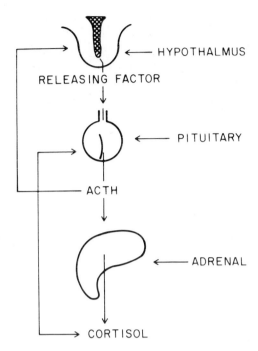

Figure 1. Schematic diagram of hypothalamic-pituitary-adrenal axis 'feedback loop'.

response following acute withdrawl of long-term exogenously supplied steroid is poor, and may be responsible for clinical deterioration independent of the degree of intracranial palliation.

Chemically, adrenal steroids are composed of three 6-carbon hexane rings and one-5 carbon pentane ring. Hydrocortisone and other 'anti-inflammatory' steroids are referred to as 'C_{21} steroids' when an attached 2-carbon chain is present at position 17 (Figure 2). The 'hydroxy' group at position 17 has led to the commonly used terminology, '17-hydroxycorticoids'. C_{21} steroids whose actions effect gluconeogenesis are referred to as 'glucocorticoids'. Synthetic analogs of the hormone cortisol have substitutions in the steroid nucleus which enhance their anti-inflammatory activity, while at the same time decreasing their mineral corticoid effect. These chemical realignments, however, do not effect their normal cortisol-like catabolic actions.

The plasma half-life of cortisol, defined as the time required for a 50% degradation of the initial concentration, should not be confused with its tissue or biologic half-life activity. The biologic half-life constitutes a similar percent degradation, but is measured as *tissue* anti-inflammatory effect. Synthetic analogs may not be detectable in the plasma 100 – 200 minutes after intravenous administration. Nevertheless, their tissue anti-inflammatory action may persist for hours, their metabolic action apparent for days. Therefore, steroid prepara-

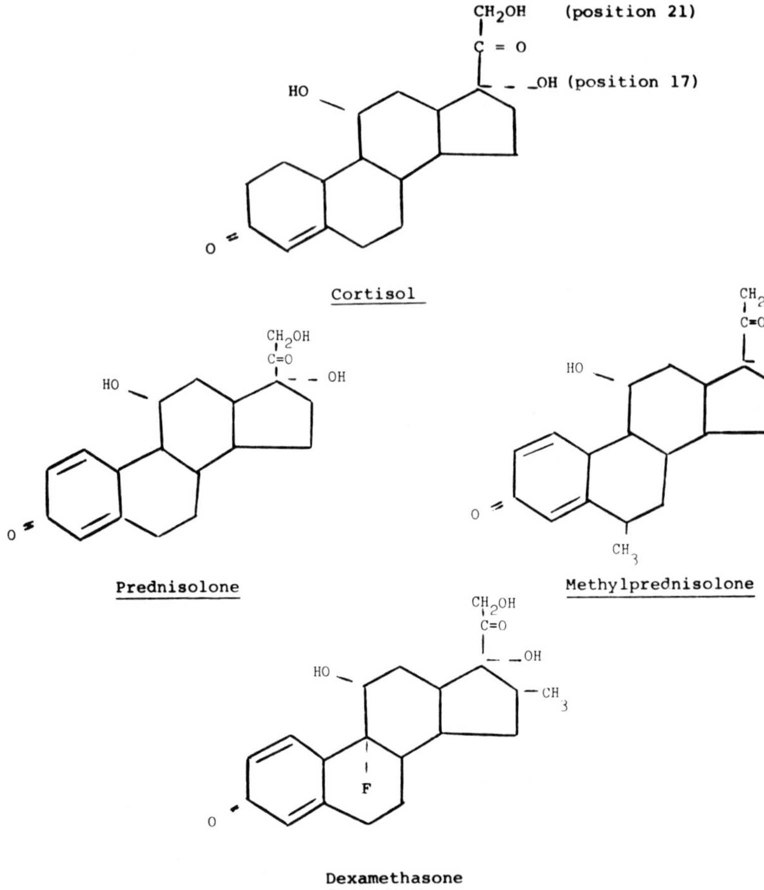

Figure 2. Chemical configuration of cortisol and its synthetic analogs.

tions with prolonged tissue or biologic half-lives are prone to the more rapid production of secondary clinical changes and target organ suppression. Knowledge of these differences are important when considering dose-time relationships in tumor patients expected to require long-term steroid administration. Biologic half-lives for hydrocortisone and cortisol are about 12 hours (short-acting group), prednisone and methylprednisolone 12 – 36 hours (intermediate acting), and dexamethasone 36 – 54 hours (long-acting) [24] (Table 1). All gucocorticoids are 90% bound to serum proteins. The synthetic analogs, however, are loosely bound and diffuse freely into body tissues, perhaps explaining the side effects seen in patients on seemingly low doses [26]. Metabolically, glucocorticoids result in increased liver gluconeogenesis, affect carbohydrate and lipid metabolism as well as protein synthesis. This altered metabolism is clinically apparent in peripheral muscle, on patterns of peripheral glucose utilization and in

Table 1. Comparison of plasma and biological half-lives of natural and synthetic steroid preparations

Classification	Plasma half life (min)	Biological half life (hours)
Short acting		
Cortisol	90	8 – 12
Intermediate		
Prednisone		
Methylprednisolone	200	12 – 32
Long acting		
Dexamethasone	300	36 – 54

various osseous portions of the human skeleton. The commonly seen truncal redistrubution of fatty tissue and the 'moon face' are common manifestations of this altered state.

Steroids possess their anti-inflammatory properties as a result of actions on capillary permeability and vasoconstrictor response to norepinephrine [27]. A subsequent decrease of protein 'leakage' into the inflamed area lessens the normally apparent swelling, while the passage of leukocytes and macrophages into the damaged area is said to be impeded by a steroid induced decrease in local endothelial cell 'sticking' [28, 29]. Even if 'trapped' at the site, macrophages are inhibited in their ability to engulf and process antigens [24]. In short-term clinical use, these effects appear to be minimal. In long-term administration, however, the effects can be quite profound. Concurrent with acute steroid use is a rapid but transient rise in neutrophils, a decrease in lymphocytes and changes in other blood components [30]. The rise in neutrophils is presumed to result from an increased discharge of cells from bone marrow stores and a decrease in removal of cells from the circulation. A similar response, however, has been seen with reduction of steroid dose administration.

Two major groups of lymphocytes. (B & T cells) are responsible for humoral and cell mediated immunity. The latter group (T cell) is derived from precursor cells in bone marrow and processed in the thymus. B cells, on the other hand, arise directly from marrow stem cells and harbor immunoglobulins which react with foreign antigens. Whereas B lymphocytes may be affected by very high dose steroid administration, T lymphocytes are thought to be more readily challenged and impeded.

3. MECHANISMS OF ACTION AND EFFECTS OF STEROID ADMINISTRATION

Therapy with systemic glucocorticoids in the brain tumor patient has become an accepted and widely practiced method of symptom palliation. In the minds of most, steroids 'treat the edema' commonly encountered in this patient population, taking advantage of their low mineral corticoid, high anti-inflammatory properties. Many physicians, however, are unaware of the other attributes ascribed (correctly or incorrectly) to steroid administration; attributes thought to be responsible for the positive clinical patient response. In all probability, the benefits of steroids are not restricted to one mechanism alone, but related to an overall series of 'effects' contributing to the well known objective clinical response.

3.1. Antiedema Effect

At the outset it is important to recognize the probable existence of two forms of cerebral edema as described by Klatso [31]; cytotoxic and vasogenic. Cytotoxic edema is formed primarily by direct cellular (astrocytic) injury with little or no real change in vascular permeability. It is associated with disturbances in cellular metabolism and ion transport, occurs primarily but not exclusively in white matter, and is attended by minimal extravasation of fluid to the extracellular space [32–34]. Vasogenic edema on the other hand, is a reflection of vascular injury and increased permeability with subsequent white matter swelling. It is thought to be the predominant (but not exclusive) type of edema associated with intracranial tumors [31, 35]. This is especially true in the postoperative period where some component of cytogenic edema is thought to be present.

Characterized as 'hydrostatic extravasation' of protein containing fluid into an enlarged extracellular space, vasogenic edema is accompanied by distinct changes in sodium and potassium ion concentrations and transport, as well as overall water content [1, 35–37]. The volume of this hydrostatic 'leakage' from damaged vessels can at times be quantitated with changes in the mean systemic arterial pressure (MAP) (increased MAP pressure being conducive to increased edema) [35]. In all probability it is the vasogenic edema, even though not pure in its existence in the brain tumor patient, that responds most beneficially to steroid administration. In a study of gray and white matter surgical specimens from operated brain tumor patients, Long et al., described changes compatible with both vasogenic and cytotoxic edema [38]. The study substantiated the favorable clinical response to steroid therapy. Others have demonstrated both the protective and therapeutic value of steroids in the animal tumor model, relating it again to a correction in the sodium, potassium and water content of the edematous tissue, a decrease in vascular permeability and subsequent extravasation of high protein containing fluid [1, 2, 39–44].

3.2. Effects on Cerebrospinal Fluid and Cerebral Blood Flow

While it is generally believed that major fluid changes occur in brain tissue water, a number of investigators have reported a significant reduction in CSF production (as high as 50%) when glucocorticoids were administered to the experimental animal [45 – 47]. To my knowledge, however, measurement of CSF flow in man following glucocorticoid administration has not been reported. Weiss and Nulsen [45] have considered this change to occur as a result of secretion rather than absorption of CSF, based on their experimental work in which they created a minus 300 mm pressure hydrostatic differential between venous and CSF spaces. They allude to a further consideration of the effect of glucocorticoids on brain volume which may contribute to a decrease in the flow of CSF. This concept remains interesting, but requires further clinical delineation.

Clinically, reduction of intracranial pressure can be objectively recorded following the administration of steroids [6, 19, 23, 48]; a finding generally attributed to the resolution of accumulated extracellular water. As a result, previously reduced regional cerebral blood flow and impaired autoregulation in edematous areas surrounding an intracranial tumor are also improved [49 – 51].

Although a decrease in CSF secretion can be experimentally documented, in all probability resolution of peritumoral edema and subsequent normalization of autoregulation and blood flow creates the major impact on the recorded reduction of intracranial pressure.

3.3. Steroids as a Diuretic Agent

Almost every pharmacologic agent used clinically to reduce intracranial pressure is accompanied by a diuresis (urea, mannitol, glycerol, hypertonic glucose, furosemide, etc.) [35, 52 – 54]. Glucocorticoids, although not recognized as a diuretic, do induce a diuretic response when administered to brain tumor patients.

The significance of the renal loss of body fluid upon the water content of edematous brain remains unclear [53, 55, 56]. Mannitol, urea, and other so called 'hyperosmolar' agents exert their effect by osmotic force, water depleting the extracellular fluid spaces and subsequent clearance through an obligatory renal loss. Furosemide produces body water loss by inhibiting renal tubular resorption and hence, a secondary reduction in brain water content; a concept questioned by a number of investigators who invoke the possibility of a direct action on brain water [57, 58]. Glucocorticoids on the other hand, may exert their diuretic effect by reducing the secretion of aldosterone or its (aldosterone) ability to promote renal tubular cation exchange causing a secondary increase in the glomerular filtration rate [55]. Whereas the diuretic effect of steroids seems to be variable and independent of the clinical response seen in the brain tumor patient, the combination of a diuretic such as furosemide and a steroid, has an effect greater than the effect of either preparation when used individually [35]. The exact relationship of the combination to the mobilization of extracellular brain water remains obscure, but seems to relate to a steroid resolution of deranged water transport, vascular permeability, and electrolyte transfer with the obligatory loss of body fluid related to the diuretic.

3.4. Direct Effect of Steroids on Tumor Tissue

The rather prompt remission of symptoms seen with steroid administration in a patient with an intracranial mass lesion is best related to resolution of peritumoral edema. Nevertheless, a number of reports have appeared regarding a long-term steroid-induced direct antineoplastic effect [42, 44, 59–72]. Braunschweiger *et al.* [73] employing steroids in an experimental setting, described temporary tumor stasis in C_3H mammary carcinoma. Their findings suggested a reversible block in G_1 without a change in the growth fraction. Clinically, however, observations of patient improvement on long-term steroid therapy have not resulted in statistically significant increases in survival time (BTSG personal communication).

Other suggested mechanisms of steroid influence on tumor growth involve alterations in glucose metabolism, decreased rates of protein and/or DNA synthesis, and a steroid-induced reduction in tumor oxygen consumption [74]. Sherbert *et al.* [63] reported tissue culture evidence of tumor growth inhibition when steroids (in high concentrations) were added to the culture media. In an interesting followup these investigators described *accentuated* growth of similarly treated cells following removal from steroid treated patients; a finding not observed in cells from non steroid treated patients. Further, there appears to be some form of tissue selectivity in the response to steroid with perhaps some predilection for mesenchymal tissue [75].

Not all experimental tumor lines demonstrate a uniform steroid response [76–78]. In the cultured glioma cell for instance, higher concentrations of steroid produce inhibition of growth. Where resistant cells exist, normal growth patterns are resumed after a brief pause [62]. In a transplantable mouse ependymoma and mixed rat glioma Wright *et al* [65] demonstrated steroids to not only completely retard the growth, but transform cellular histology to a less malignant form.

The overall direct neoplastic effect of steroid administration in tumor patients remains largely undetermined. In clinical trials, statistically valid evidence of prolonged survival times has not been forthcoming and tissue culture evidence of inhibition may be misleading. Claims of rapid deterioration and death following withdrawl of steroid should not be mistaken for evidence of reduced cytotoxic effect and resumed tumor growth. Rather, this phenomenon should be more properly viewed as a physiologic dependence upon an exogenously delivered pharmacologic compound.

3.5. Effects of Diagnostic Studies

Surgical localization of a tumor mass is dependent upon a variety of diagnostic imaging. Whereas air contrast displays mass effect alone, radioisotope and contrast enhanced computer scanning is, to a degree, dependent upon changes in the blood-brain-barrier permeability. Hence, the many reports of a change in the image or 'pickup' of intravascularly injected materials before and after steroid administration [19, 56, 79 – 83]. Consistent with its effect on vascular permeability, steroids will alter the uptake of imaging compounds and may incorrectly then be used as evidence for, or against a tumor 'response' to a particular drug or treatment program. Not only may confusion exist with the administration of steroids, but changes in dose level may alter as well the objective measurements of a lesion as seen with isotope scans. Certainly, density readings in contrast enhanced computerized scanning fall victim to the same misinterpretation, as do changes in tumor 'blush' and mass effect as demonstrated by angiography.

Presumably this alteration in imaging is related to the resolution of peritumoral edema and decreased vascular permeability. Brain tumor capillaries lack firm endothelial junctions and are said to contain 'pores or fenestrations' allowing variable passage of materials to the extracellular space [84]. Changes in the passage of contrast materials via this mechanism may be responsible for the variation seen in scans performed immediately after contrast injection, and those performed 30, 60, or 90 minutes later.

4. CURRENT CLINICAL USES OF STEROID IN THE MANAGEMENT OF INTRACRANIAL TUMORS: ALTERNATIVE SCHEDULES OF ADMINISTRATION

From a surgical point of view, the use of steroids in the perioperative period is well established. The oncologist and radiotherapist employ the benefit of steroid therapy during various phases of the patient's treatment and many reports have appeared relative to the beneficial effects [6, 10, 17, 19 – 21, 23, 38, 53, 66, 85, 86]. The current trend is for acute perioperative use, a reasonably rapid tapering, and then complete withdrawl until evidence of recurrence or interim symptomatology requires its reinstitution.

Drug dose levels, originally low by today's standards, seem to have reached the point that if '. . .a little is good, a lot is better'. To a degree that concept may be true. In our own experience, prompt resolution of neurologic deficit can readily be appreciated at 'normal' dose levels. However, recurrence of symptoms may soon be forthcoming and found again to be easily reversible by a modest increase in the dose level. Following a stepwise methodical increase in dosage over a reasonably protracted period of time, 'Maxi' dose levels are achieved. Each increase is accompanied by clinical improvement, but each time quantitatively less improvement than the previous interval and over a shorter period of time. At some point in the patient's course, steroid increase does not equate with clinical improvement. In other words, no additional clinical benefit can be seen to accrue with a relative increase in dose level. Presumably conventional edema is not the cause and tumor mass, hydrocephalus or some other factor has intervened. The difference between dose levels of steroid at initiation of therapy and the dose level beyond which clinical improvement does not occur has, in our institution, been termed the 'therapeutic gap'; a concept in agreement with other published data [12]. In general the more rapidly (in time parameter) the 'gap' is narrowed, the poorer the overall prognosis. Conversely, the slower the narrowing of the 'gap' the less likely is rapid tumor repopulation. As a clinical test, the greater the requirement for higher dose levels of steroid without resolution of the neurologic deficit, the greater the probability of parenchymal involvement and the less likely is peritumoral edema to be the causative factor. It is also then less likely for cytoreductive surgery to reverse the deficit. This is especially true in our experience with the recurrent glioblastoma patient in whom a 'second look' effort is being entertained. Resolution of neurologic impairment with steroid administration or increased dose levels is a positive factor in the selection of candidates for second look surgical consideration. Narrowing of the 'therapeutic gap' may be commensurate with an estimated decrease in brain compliance.

The use of long-term steroid administration after needle biopsy and radiation of a primary glioma, attests to the effect of steroid 'chemical decompression' [21], and to the inevitable peak of the 'therapeutic gap'. Nevertheless, in those patients harboring lesions not amenable to cytoreductive surgical efforts (primary or metastatic), steroids provide significant short term palliation of symptoms [81].

4.1. Variations in Time-dose Administration

4.1.2. Smaller-more frequent Doses. Steroids are most commonly administered in divided doses, (usually every 6 hours) commensurate with their bioavailability and half-life (Figure 3). As noted a peak of plasma availability occurs at approximately 2 hours after oral administration and immediately after a short intravenous infusion or bolus. Thereafter, serum levels drop precipitously over the next 4–6 hours. Clinical improvement is maintained with repeated administration (i.e., every 6 hours), regardless of the peak and valley configuration of plasma availability. The obvious question relates to the drug level required for a clinical response; the peak, or a lesser level. To test this response, a group of patients judged to eventually require long-term steroid administration were subjected to a change in time and dose schedules. The change resulted in an overall decrease in 24-hour cumulative dose while still maintaining a stable neurologic status (i.e., 16 mg of methylprednisolone q 6 hours, to 8 mg q 4 hours, or 64 mg/24 hours vs 48 mg/24 hours). Aside from the night time inconvenience, each

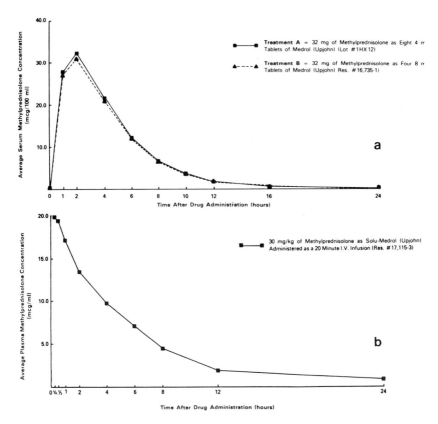

Figure 3. Bioavailability assay of single oral and intravenous administration of Methylprednisolone. (Reproduced with permission of Upjohn Pharmaceuticals).

maintained their improvement, indicating at that point at least a beneficial effect from a stable blood level at a substantially lower overall 24 hour total dose. The response to a similar dose decrease on a 6 hour schedule was not as beneficial. In time, as with the 6 hour schedule, each required increases commensurate with increasing edema and presumably increasing intracranial pressure. Nevertheless, a more stable but lower level of drug availability may be of value (Figure 4).

4.1.3. Alternate-Day Therapy (ADT). The use of long-term steroids on a daily basis produces a multitude of physiologic, endocrinologic and skeletal problems. One finds it difficult, however, to withhold a medication (steroid) with the known potential for clinical improvement. To accomplish each of these objectives, (best neurologic status with least secondary effects) a perusal of the literature led to the concept of using 'alternate-day steroid therapy' [88–96]. The rationale for ADT relates to the extended tissue anti-inflammatory half-life of the synthetic analogs and the possibilities of avoiding pituitary-adrenal sup-

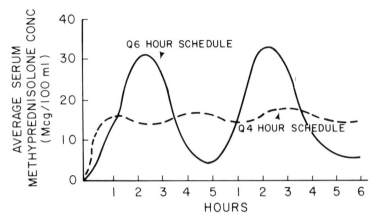

Figure 4. Simulated bioavailability curves for change of time dose relationship from Q 6 hour to Q 4 hour administration (64 mg/24 hours vs 48 mg/24 hours).

pression. ADT permits reasonably normal pituitary-adrenal secretory activity on the 'off' day. The program is intended for use in the patient *after* the perioperative period, or in a patient whose symptoms reappear during radiation therapy and in the patient demonstrating recurrent signs weeks or months after their primary therapy. It is *not* feasible for use in our experience, in patients who are already on long-term chronic administration. Further, it must be cautioned that choice of product is important. Preparations with long biologic and tissue half-lives will produce pituitary-adrenal suppression regardless of the every-other-day administration. Consequently, for this group of patients we utilize methylprednisolone exclusively (Figure 5).

Drug dose levels in the ADT program vary. In general, one calculates the total dose level as if the patient were to be placed on an every 8 hour daily schedule; for instance, 16 mg of methylprednisolone every 8 hours. The total daily dose would be 48 mg (16 mg × 3 = 48). The total then is doubled or tripled (48 × 2 = 96; or 48 × 3 = 144) and is given as a single administration every other morning at approximately 8 am; the time of the highest level of normal endogenous cortisol secretion. If symptoms persist, the every-other-day dosage is increased by increments of 16 mg until the desired effect is achieved or until the trial is declared a failure. When the desired effect has been achieved, the dose is similarly reduced by 16 mg until such time as the minimal effective dose has been determined. One hundred to 300 or more mg/every other day is not unusual and should not deter one from a trial of ADT.

Our experience with ADT has been positive to the extent that *no* patient's initial post surgical steroid schedule is on an every-day basis. It does not preclude the eventual need for every day divided dose administration, but it has postponed the inevitable complications which occur with long-term steroid need. It should

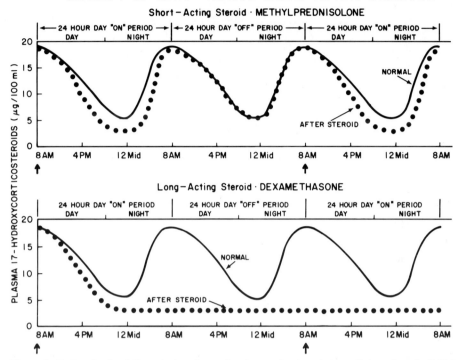

Figure 5. Rationale for 'alternate-day therapy' using a short or intermediate acting (half-life) preparation. (Reproduced with permission of Upjohn Pharmaceutical).

not, in our opinion, be used in the immediate operative period, nor in patients who have been on daily steroids for an extended period of time.

Further efforts to forestall the use of daily steroids include the 'off' day use of diuretics such as furosemide or diamox; again a concept previously noted in the literature [35]. Although perhaps synergistic when used simultaneously, their use in this setting (i.e., on alternate days), seems to be complementary.

4.1.4. Sustained Release Injection. Patient compliance relative to the need for multiple pill ingestion and gastrointestinal complaints, the need for short-term palliation during the period of radiation therapy or to satisfy the need for outpatient dose tapering can be accomplished with the use of IM methylprednisolone (Depo-Medrol[R])*. Utilizing doses of 80–160 mg/injection (80 mg/single site injection) bioavailability studies (Figure 6) indicate useful levels for approximately a week. This form of administration is recommended for short-term or interim needs. We do not recommend this format for long-term application.

*Upjohn Pharmaceuticals.

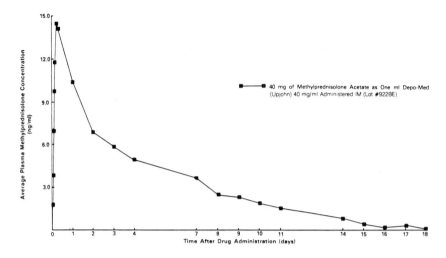

Figure 6. Bioavailability curve for single injection of IM Methylprednisolone (Depo-Medrol^R). (Reproduced with permission of Upjohn Pharmaceuticals).

5. COMPLICATIONS OF STEROID USE

Complications related to long-term steroid use are many and have been described extensively in the literature [12, 24, 30, 56, 97 – 110]. Some are commonplace, others somewhat more esoteric.

5.1. Abnormal Glucose Metabolism

Controlled diabetic patients placed on steroids demonstrate an expected glycosuria and increase in insulin requirements. Similarly, previously unrecognized 'subclinical' diabetics suddenly manifest polyuria, glycosuria, and increased blood sugar levels requiring insulin coverage during the period of steroid administration. Enhanced gluconeogenesis, glycogen deposition in the liver and increased synthesis of hepatic enzymes combine to produce an increased hepatic glucose production. There is some information to suggest a decrease in the activity of insulin receptors and a subsequent decrease in peripheral glucose utilization [111]. For the most part, the augmented abnormality of glucose metabolism is steroid dependent and resolves spontaneously following cessation of drug administration. In the most extreme and unusual circumstances hyperosmolar nonketotic coma has been reported [109].

5.2. Gastrointestinal Abnormalities

There has long been a controversy regarding the ulcerogenic potential of long-term steroid administration [112]. Much of the literature concerning the use of steroids in benign diseases attests to a 'steroid ulcer potential', but the incidence of major bleeding secondary to such ulceration is varied and primarily associated with the use of non steroid antiarthritic drugs (salicylates). In lesions of the central nervous system, the known potential for gastric ulceration irrespective of steroid use, further confuses the issue. Several features help distinguish newly developed steroid ulceration from preexisting gastrointestinal pathology. First, steroid induced lesions are more apt to be gastric than duodenal and lack the usual male preponderence. Secondly, steroid ulceration demonstrates little surrounding inflammatory tissue and is reasonably asymptomatic until a problem occurs (usually hemorrhage). Ulceration of this variety does not produce the usual form of obstruction or perforation.

The mechanism for steroid ulceration is uncertain. There is conflicting evidence to impugn an induced increase in gastric acid secretion. There is, however, considerable experimental evidence to suspect a defect in both the qualitative and quantitative aspects of mucous production by the stomach [99, 113, 114, 155], as well as a modification of mucosal cell renewal leading to an 'altered mucosal defense mechanism' [99]. Although steroids may not in and of themselves produce ulceration, they appear to enhance the potential created by other factors. Therefore, we routinely use anticholinergics and antacid preparations during the short-term, high-dose perioperative period. In the long-term application of steroids, antacids alone seem to suffice. We have abandoned the use of Cimetidine in our brain tumor population because of the detrimental interaction between it and nitrosourea therapy [116].

Although not clearly apparent in our patient population, the occurrence of steroid-induced acute pancreatitis has been reported [117, 118]. Thought to be due to an alteration of pancreatic secretions, it is mentioned here because of its inherent confusion (in the acute state) with peptic ulcer disease.

5.3. Involvement of Muscle and Skin

Observations of patients with Cushing's disease leaves little doubt of the occurrence of myopathy with increased endogenous steroid secretion. Similar observations in patients on long-term administration of exogenous steroid confirm the now familiar finding of proximal muscle weakness (primarily lower limb), thinning of the overlying skin and muscle atrophy. Clinically, this is manifested by weakness associated with fatigue, inability to climb stairs or rise from a chair, and difficulty with such simple movements as getting in out of bed [24, 98, 104, 108, 119–123]. Unrelated to age, sex, or dosage, myopathy in our experience is related to the overall length of therapy. Although reported to promptly improve after cessation of administration, our observations of myopathic patients indicate resolution to be a very slow process (if ever); especially in the patient whose normal course is one of continued debilitation.

Experimentally, the process may be related to decreased substrate oxidation, proportional to the population of white glycolytic fibers in the muscle mass and alteration in the functional activity of muscle mitochondria with inhibition of muscle phosphorylase activation [107]. The degree of interference with oxidative metabolism is most severe with longer acting steroid preparations or when 9-alfa fluorinated compounds are employed [121, 124]. Electromyographic studies reveal decreased motor unit potentials, a lower than normal amplitude on volition and as a rule, no evidence of fibrillation. Nerve conduction studies remain normal presumably excluding direct peripheral nerve involvement.

Microscopically, muscle biopsies from experimental animals reveal changes in muscle fiber size most pronounced in animals receiving long-acting preparations or compounds with a 9-alfa fluorine configuration [108]. Aside from differences in muscle fiber thickness, no evidence of degenerative changes are noted within the muscle.

Changes in the skin and its support structures are varied and common. Stria formation with thinning of the epidermis, alterations in wound healing, acne, easy bruisability, loss of collagen and elastic fibers, and prominence of subcutaneous vascular channels are all manifestations of this change. Vitamin A has been reported to have a beneficial effect [125].

5.4. Skeletal Abnormalities

Most adult patients requiring long-term steroid therapy develop osteoporosis. In the child, long-term administration is associated with retarded skeletal growth occurring primarily because of inhibition of linear bone growth and changes in the epiphyseal closure time [24]. Abnormalities in calcium and nitrogen balance secondary to reduced absorption and associated secondary hyperparathyroidism, altered tissue catabolism and matrix formation, contribute to a decrease in new bone formation and replacement. There may be subtle impairment of growth hormone [24, 97, 98]. Clinical evidence of compression fractures are not uncommon in the brain tumor patient on long-term steroid therapy. In patients with metastatic disease, the possibilities of steroid induced compression fracture must be carefully considered in the differential diagnosis.

5.5. Drug Interactions

Drug interactions involving steroids are well documented in the literature [22, 24, 56, 79, 102, 105, 106, 127 – 128]. Of major importance is the interaction with anticonvulsant medications. Diphenylhydantoin, phenobarbital and diazepam, all have been shown to decrease the half-life of glucocorticoids, and increase the rate of excretion of their metabolites. Most of the currently used anticonvulsants are potent stimulators of hepatic enzyme systems causing an increase in hepatic steroid metabolism. Resulting lower blood levels of the active form of the steroid may cause neurologic deterioration related to a relatively decreased effective dose level.

Steroids are known to increase the rate of salicylate metabolism. Commonly used in the brain tumor patient for control of febrile episodes, increased dose levels of salicylate may be required when used concurrently with exogenous steroid.

5.6. Effects on Blood Elements in the Immune System

The exogenous administration of steroids creates changes in peripheral white cell counts and their immune components. Characteristically, one finds neutrophilia, lymphocytopenia, monocytopenia and a decrease in circulating eosinophils [30, 129 – 133]. The selective cellular decrease rather than being a cell loss, probably occurs as a result of inhibited bone marrow release and/or a sequestering of cells in other than vascular compartments. Steroid-induced inhibition of glucose uptake by lymphocytes may result in a moderate cell lysis [24]. It is presumed that short-term steroid administration causes little or no effect on phagocytosis, however, there may be some inhibition in accumulation of cells at an inflammatory site [30]. This accumulation phenomenon is less apparent with the utilization of alternate-day therapy.

Glucocorticoids may interfere with other phases of the normal inflammatory response; antigen processing, macrophage activation and migration [24]. As described before, changed capillary permeability, less edema formation in the area of inflammation, and less endothelial 'sticking' of white cells all contribute to the decreased host response.

Of the two types of lymphocytes B & T (the former from bone marrow stem cells containing specific immunoglobulins, the latter from bone marrow precursor cells processed in the thymus and responsible for cell mediated immunity), the T lymphocytes are more vulnerable to the adverse effect of steroid administration. Hence, less depression of humoral antibody production and the obvious alteration in cell mediated immunity.

From a clinical point of view, most of the literature relative to steroid administration and infection deals with the nonsurgical patient and disease states such as rheumatoid arthritis, collagen diseases and asthmatics [30, 101, 125, 134, 136 – 139]. Whereas preexisting infection such as tuberculosis and the enhanced possibility of an opportunistic infection have been elaborated upon by others [24, 110], little information is available concerning the surgical risks in a long-term steroid patient harboring a malignancy in whom some future surgical effort may be needed [140]. Our own experience would tend to indicate no major inflammatory problems to occur. Delayed wound healing and superficial 'stitch' like problems have been encountered in a 'second look' surgical group, but major infections involving the central nervous system have not been noted. This has been the experience of the Brain Tumor Study Group as well when high-dose, long-term steroid was employed in a randomized trial (BTSG personal communication). We do have major concern for long-term steroid patients with the potential for infection; i.e., those with implanted foreign bodies (shunts) and patients with the need for acute high-dose intermediate term steroid administration where a nidus of infection preexists. In the latter example, it seems mandatory to remove the origin if possible.

Although difficult to establish statistically, there does not appear to be a major increase in infection in the surgical patient on long-term steroid. However, chronic hypoproteinemia, bone marrow depression from chemotherapy, and chronic cachexia all predispose to the risk and would, in that group of patients, approach the statistically significant increased incidence reported in the renal transplant literature [139].

5.7. Drug 'Dependency'

It is uncertain that 'dependency', which by definition and usage connotes addiction, is the proper term. Perhaps terms such as 'replacement' or 'suppresion' would be more descriptive of the drug need in patients on long-term steroid therapy encompassing the perioperative period and extending beyond the period of radiation. Even though the demonstrated need for such therapy may have passed, (i.e., C.T. evidence of edema, mass effect, midline shift, etc.), attempts at withdrawl result in fatigue, depression, nausea, headache, and a diminished conscious level. This occurs primarily in those patients placed initially on the long-acting preparations. Regardless, acute 'withdrawl' of steroid creates untoward side effects. Our experience indicates that slowly lowered doses over many weeks may ultimately permit reduction of dose. The best therapy is to anticipate the 'dependence' and initially utilize a short-acting preparation. As soon as feasible institute the alternate-day dose schedule of administration. Most likely, the 'dependence' is a reflection of chronic pituitary adrenal suppression. The use of ACTH may worsen the course because of the 'feedback loop', leading to pituitary depression.

5.8. Miscellaneous Complications

A number of other less recognized problems are known to occur with steroid administration; glaucoma with increases in intraocular pressure in as high as 40% of the patients [141 – 143] psychiatric complications (as high as 30% of the patients) [144], cataract formation [97], increased urinary uric acid and creatine excretion, exaggeration of abnormal menstrual cycles, fetal and placental malformations [145], development of secondary neoplasia (primarily lymphoma) [100, 146], thromboembolism [98], and pseudotumor cerebri [97].

6. SUMMARY

The use of glucocorticoids in the management of patients with intracranial space occupying tumors has become the standard of practice in most every neurosurgical center. Beneficial in its effect on peritumoral edema, membrane permeability and realignment of normal intra and extracellular cation and anion relationships, steroids nevertheless retain the potential for secondary complications.

It is important to recognize the need to restrict dose levels to the lowest point commensurate with functional needs. To this end, all patients requiring chronic steroid administration beyond the operative period should be given a trial of alternate-day therapy. Further, there is significant rationale to suggest the initial administration of steroid be with one of the short-acting preparations. The use of such a preparation is mandatory if the alternate-day program is to be successful.

REFERENCES

1. Prados M, Strawger RL, Feindel W: Studies on cerebral edema. II Reaction of the brain to exposure to air; physiologic changes. Arch Neurol Psychiat 54:290, 1945.
2. Grenell RG, McCawley EL: Central nervous system resistance. III The effects of adrenal cortical substance on the central nervous system. J Neurosurg 4:508–518, 1947.
3. Ingraham FD, Matson DD, McLaurin RL: Cortisone and ACTH as an adjunct to the surgery of craniopharyngiomas. New Eng J Med 246:1952.
4. Russek HI, Johnson BL, Russek AS: Cortisone in the immediate therapy of apoplectic stroke. J Am Geriat Soc 2:216, 1954.
5. Roberts HT: Supportive adrenocortical therapy in acute and subacute cerebrovascular accidents with particular reference to brain stem involvement. J Am Geriat Soc 6:686, 1968.
6. Galicich JH, French LA: Use of dexamethasone in the treatment of cerebral edema resulting from brain tumors and brain surgery. Am Pract Diag Treat 12:169–174, 1961.
7. Galicich JH, French LA, Melby JC: Use of dexamethasone in the treatment of cerebral edema associated with brain tumors. Lancet 81:46–53, 1961.
8. French LA, Galicich JH: The use of steroids for control of cerebral edema. Clin Neurosurg 10:212–223, 1964.
9. French LA: The use of steroids in the treatment of cerebral edema. Bull N.Y. Acad Med 42:301–311, 1966.
10. Kofman S, Garvin JS, Nagamani D: Treatment of cerebral metastases from breast carcinoma with presnisolone. JAMA 163:1473–1476, 1957.
11. King DF, Moon WJ, Brown N: Corticosteroid drugs in the management of primary and secondary malignant cerebral tumors. Med J Australia 2:787–881, 1965.
12. Renaudin J, Fewer D, Wilson CB, Boldrey E, Calogero J, Enot J: Dose dependency of decadron in patients with partially excised brain tumors. J Neurosurg 39:302–305, 1973.
13. Beks JWF, Doorenbos H, Walstra GJM: Clinical experiences with steroids in neurosurgical patients. In: Steroids and Brain Edema. Reulen HJ, Schurmann K (ed) Springer-Verlag N.Y. Heidelberg, Berlin, 1972, pp 233–238.
14. Ruderman NB, Hall TC: Use of glucocorticoids in the palliative treatment of metastatic brain tumors. Cancer 18:298–306, 1965.

15. Smith RA, Smith WA: Steroid therapy of cerebral edema. J Med Assoc GA 56:324–328, 1967.
16. Maxwell RE, Long DM, French LA: The clinical effects of a synthetic glucocorticoid used for brain edema in the practice of neurosurgery. In: Steroids in Brain Edema. Reulen HJ, Schurmann K. (ed) Springer-Verlag N.Y., Heidelberg, Berlin, 1972, pp 219–232.
17. Bernard-Weil E, David M: Preoperative hormonal hormonal treatment in cases of cerebral tumor. J Neurosurg 20:841–848, 1963.
18. Einhorn LH, Burgess MA, Vallejos C, Bodey GP, Gutterman J, Mavligit G, Hersh E, Luce J, Frei E: Prognostic correlations and response to treatment in advanced metastatic malignant melanoma. Ca Res 34:1995–2004, 1974.
19. Weinstein JD, Toy FJ, Jaffe ME, Goldberg HI: The effect of dexamethasone on brain edema in patients with metastatic brain tumors. Neurol 23:121–129, 1972.
20. Kramer S, Henderickson F, Zelen M, Schotz W: Therapeutic trials in the management of metastatic brain tumors by different time/dose fraction schemes of radiation therapy. National Cancer Institute Monograph 46:213–220, 1976.
21. Marshall LF, Langfitt TW: Needle biopsy high-dose corticosteroids and radiotherapy in the treatment of malignant glial tumors. National Cancer Institute Monograph 46:157–160, 1976.
22. Gutin PH: Corticosteroid therapy in patients with brain tumors. National Cancer Institute Monograph 46:151–156, 1976.
23. Miller JD, Sakalas R, Ward JD, Young HF, Adams WE, Vries JK, Becker DP: Methylprednisolone treatment in patients with brain tumors. Neurosurgery 1(2):114–117, 1977.
24. Swartz SL, Dluhy RG: Corticosteroids: Clinical pharmacology and therapeutic use. Drugs 16:238–255, 1978.
25. Dluhy RG, Lauler DP, Thorn GW: Pharmacology and chemistry of adrenal glucocorticoids. Med Clin North Amer 57:1155–1165, 1973.
26. Dluhy RG, Newmark RS, Lauter DP, Thorn GW: Pharmacology and chemistry of adrenal glucocorticoids. In: Steroid Therapy. Azarnoff (ed), Saunders, Philadelphia 1975, pp 1–14.
27. Zweifach BW, Schorr E, Black MM: The influence of adrenal cortex on behavior of terminal vascular bed. Ann N.Y. Scad Sci 56:626–633, 1953.
28. Allison F, Adcock MH: Failure of pretreatment with glucocorticoids to midify the phagocytic and bacterial capacity of human leukocytes for encapsulated type I pneumococcus. J Bacteriol 89:1256–1261, 1965.
29. Glaser RJ, Berry JW, Loeb LH, et al: The effect of cortisone in streptococcal lymphadenitis and pneumonia. J Lab Clin Med 28:363–373, 1951.
30. Fauci AS, Dale DC, Balow J: Glucocorticosteroid therapy: Mechanisms of action and clinical consideration. Ann Int Med 84:304–315, 1976.
31. Klatzo I, Neuropathological aspects of brain edema. J Neuropath Exp Neurol 26:1–14, 1967.
32. Luse SA, Harris B: Electron microscopy of the brain in experimental edema. J Neurosurg 17:439–446, 1960.
33. Pappius HM, Gulati DR: Water and electrolyte content of cerebral tissues in experimentally induced edema. Acta Neuropathologica 2:451–460, 1963.
34. Syvertsen A, Haughton VM, Williams AL: The computed tomographic appearance of the normal pituitary gland and pituitary microadenomas. Radiology 133:385–391, 1979.
35. Reulen HJ: Vasogenic brain edema: New aspects in its formation, resolution and therapy. Br J Anesth 48:741–752, 1976.
36. Reulen HJ, Medzihradsky F, Enzenbach R, Marguth F, Brendel W: Electrolytes, fluids and energy metabolism in human cerebral edema. Arch Neurol 21:517–525, 1969.
37. Bennett HS: The concepts of membrane flow and membrane vesiculation as mechanisms for active transport and ion pumping. J Biophys and Biochem Cytol 2(4):99–103, 1956.
38. Long DM, Hartmann JF, French LA: The response of human cerebral edema to glucosteroid administration: An electron microscopic study. Neurology 16:521–528, 1966.

39. Grenell RG, Mendelson J: Adrenal steroids and cerebrovascular permeability – preliminary note. Bull Sch Med Univ Md 39:56–58, 1954.
40. Rovit RL, Hagan R: Steroids and cerebral edema: The effects of glucocorticoids on abnormal capillary permeability following cerebral injury in cats. J Neuropath Exp Neurol 27:277–299, 1968.
41. Scheinberg LC, Herzog I, Taylor JM, et al: Cerebral edema in brain tumors: Ultrastructural and biochemical studies. Ann NY Acad Sci 159:509–532, 1969.
42. Gurcay O, Wilson C, Barker M et al: Corticosteroid effect on transplantable rat glioma. Arch Neurol 24:266–269, 1971.
43. Gurcay O, Wilson CB, Eliason J: The effect of methylprednisolone acetate on transplantable rat glioma. Surg Forum 21:437–439, 1970.
44. Shapiro WR, Posner JB, Corticosteroid hormones: Effects in an experimental brain tumor. Arch Neurol 30:217–221, 1974.
45. Weiss MH, Nulsen FE: The effect of glucocorticoids on CSF flow in dogs. J Neurosurg 32:452–458, 1970.
46. Sato O: The effect of dexamethasone on cerebral spinal fluid production rate in the dog. Brain Nerve 19:49–56, 1967.
47. Garcia-Bengochea F: Cortisone and the cerebrospinal fluid of noncastrated cats. The Amer Surgeon 31(2):123–125, 1965.
48. Kullberg G, West KA: Influence of corticosteroids on the ventricular fluid pressure. Acta Neurol Scand (Suppl 13)41:445–452, 1965.
49. Hadjidimos A, Steingass U, Fischer F, Reulen HJ, Schurmann K: The effects of dexamethasone on rCBF and cerebral vasomotor response in brain tumors. Europ Neurol 10:25–30, 1973.
50. Brock M, Hadjidimos A, Deruaz J, Schurmann K: Regional cerebral blood flow and vascular reactivity in cases of brain tumor. In: Brain and Blood Flow, Pitman, London, 1971, pp 281–284.
51. Reulen HJ, Hadjidimos A, Schurmann K: The effect of dexamethasone on water and electrolyte content and on rCBF in perifocal brain edema in man. In: Steroids and Brain Edema, Reulen JH, Schurmann K (ed) Springer-Verlag, NY, Heidelberg, Berlin, 1972.
52. Cottrell JE, Robustelli S, Post K, Turndorf H: Furosemide and mannitol induced changes in intracranial pressure and serum osmolality and electrolytes. Anesthesiology 47:28–30, 1977.
53. Shenkin HA, Gutterman P: The analysis of body water compartments in postoperative craniotomy patients. Part 3: The effects of dexamethasone. J Neurosurg 31:400–407, 1969.
54. Langfitt T: Possible mechanisms of action of hypertonic urea in reducing intracranial pressure. Neurology 11:196–209, 1961.
55. Liddle GW: Effects of antiinflammatory steroids on electrolyte metabolism. Ann NY Acad Sci 82:854–867, 1959.
56. Gutin PH: Corticosteroid therapy in patients with cerebral tumors: Benefits, mechanisms, problems, practicalities. Seminars in Oncol 2(1):49–56, 1975.
57. Clasen R, Cooke P, Pandolfi S, Carnecki G, Bryar G: Hypertonic urea in experimental cerebral edema. Arch Neurol 12:424, 1965.
58. Schurmann K, Reulen HJ, Hadjidimos A: The influence of dexamethasone and diuretics on perifocal cerebral edema in brain tumors. J Neurosurg Sci 17:68, 1973.
59. Bernard-Weil E, Landau-Ferey J, Ancri D, et al: Clinical effects of combined vasopressin-corticosteroid therapy in patients with recurrent grade III Astrocytomas. Neurochirurgia 15:127–134, 1972.
60. Arpels C, Southam B, Southam C: Effects of steroid on human cell cultures: Sustaining effect of hydrocortisone. Proc Soc Exp Biol and Med 115:102–106, 1964.
61. Kline I, Leighton J, Belkin M, Orr H: Some observations on the response of four established human cell strains to hydrocortisone in tissue culture. Cancer Research 17:780–784, 1957.

62. Mealey J, Chen T, Schanz G: Effects of dexamethasone and methylprednisolone on cell cultures of human glioblastomas. J Neurosurg 34:324–334, 1971.
63. Sherbet GV, Lakshmi MS, Hadded SK: Does dexamethasone inhibit the growth of human gliomas. J Neurosurg 47:864–870, 1977.
64. Chen TT, Mealey J: Effect of corticosteroid on protein and nucleic acid synthesis in human glial tumor cells. Cancer Research 33:1721–1723, 1973.
65. Wright LR, Shaumba B, Keller J: The effect of glucocorticosteroids on growth and metabolism of experimental glial tumors. J Neurosurg 30:140–145, 1969.
66. Kotsilimbas DG, Meyer L, Berson M, Taylor JM, Scheinberg LC: Corticosteroid effect on intracranial melanoma and associated cerebral edema. Neurol 17(3):223–226, 1967.
67. Pomeroy TC: Studies on the mechanisms of cortisone induced metastases of transplantable mouse tumors. Cancer Research 14:201–204, 1954.
68. Shklar G: Cortisone and hamster buccal pouch carcinogenesis. Cancer Research 26:2461–2463, 1966.
69. Bruztowicz RJ: The effect of cortisone on the growth of transplanted ependymomas in mice. Mayo Clin 26:121–128, 1951.
70. Toolan HW: Growth of human tumors in cortisone treated laboratory animals: The possibility of obtaining permanently transplantable human tumors. Cancer Research 13:389–394, 1953.
71. Guner M, Freshney RI, Morgan D, Freshney G, Thomas DCT, Graham D: Effects of dexamethasone and betamethasone in vitro cultures from human astrocytomas. Br J Ca 35:439–447, 1977.
72. Grossfield H, Ragan C: Action of hydrocortisone on cells in tissue culture. Proc Soc Exp Biol Med 86:63–68, 1954.
73. Braunschweiger PG, Stragand JJ, Schiffer LM: The effect of methylprednisolone on cell proliferation in C3H/HeJ spontaneous mammary tumors. (Submitted for publication Cancer Research).
74. Sherbet GV, Lakshmi MS: The surface properties of some human intracranial tumor cell lines in relation to their malignancy. Oncology 29:335–347, 1974.
75. Cornman I: Selective damage to fibroblasts by desoxycorticosterone in cultures of mixed tissues. Science 13:37–39, 1951.
76. Geiger RS, Dingwall JA, Andrus W: The effect of cortisone on the growth of adult and embryonic tissues in vitro. Am J M Sci 231:427–436, 1956.
77. Leighton J, Kline I, Orr HC: Transformation of normal human fibroblasts into histologically malignant tissue in vitro. Science 123:502–503, 1956.
78. Moore AE, Southam C, Sternberg S: Neoplastic changes developing in epithelial cell lines derived from normal persons. Science 124:127–129, 1956.
79. Crocker EF, Zimmerman RA, Phelps ME, Kuhl DE: The effect of steroids on the extravascular distribution of radiographic contrast material and technetium pertechnetate in brain tumors as determined by computed tomography. Radiology 119:471–474, 1976.
80. Fletcher JW, George EA, Henry RE, Donati RM: Brain scans, dexamethasone therapy, and brain tumors. JAMA 232(12):1261–1263, 1975.
81. Marty R, Cain ML: Effects of corticosteroid (dexamethasone) administration on the brain scan. Radiology 107:117–121, 1973.
82. Harris AB: Steroids and blood-brain-barrier alterations in sodium acetrizoate injury. Arch Neurol 17:282–297, 1967.
83. Verdura J, Brown H, White RJ: Use of adrenal steroids in cerebral metastasis: Case report of improvement documented by angiography. Ohio State Med J 50:693–694, 1963.
84. Hirano A, Matusi T: Vascular structures in brain tumors. Human Pathol. 6:611–621, 1975.
85. Matsuoka S, Arakaki Y, Numaguchi K, Uneo S: The effect of dexamethasone on electroencephalograms in patients with brain tumors. With specific reference to topographic computer display of delta activity. J Neurosurg 48:601–607, 1978.

86. Galicich JH, French LA, Melby JC: Use of dexamethasone in treatment of cerebral edema associated with brain tumors. Lancet 81:46–53, 1961.
87. Coakham H: Surface antigen(s) common to human astrocytoma cells. Nature 250:328–330, 1974.
88. Miller J, Stubbs SS: Alternate day methylprednisolone therapy in asthma. Annls of Allergy 28:482–485, 1970.
89. Harter JG, Reddy WJ, Thorn GW: Studies on an intermittent corticosteroid dosage regimen. New Eng J Med 269:591–596, 1963.
90. Harter JG: Corticosteroids: Their physiologic use in allergic disease. N.Y. State J Med 827–840, 1966.
91. MacGregor R, Sheagren J, Lipsett M, Wolff S: Alternate-day prednisone therapy, evaluation of delayed hypersensitivity responses, control of disease and steroid side effects. New Eng J Med 260:1431–1527, 1969.
92. Easton J, Busser R, Heimlich E: Adrenal responsiveness of asthmatic children after long-term alternate day prednisone. J Allerg 43:171, 1969.
93. Schultz S, Newhause R, Russo J: Alternate-day steroid regimen in the treatment of ocular disease. Br J Ophthal 52:461–463, 1968.
94. Ackerman G: Alternate-day corticosteroid therapy in lupus nephritis. Clin Res 16:377, 1968.
95. Falliers CJ, Chai H, Molk L, Bane H, de A. Cardoso R: Pulmonary and adrenal effects of alternate day corticosteroid therapy. J of Allergy and Clin Immunol 49:156–166, 1972.
96. Martin M, Gaboardi F, Podolsky S, Raiti S, Calcgno P: Intermittent steroid therapy: Its effect on hypothalamic-pituitary-adrenal function and the response of plasma growth hormone and insulin to stimulation. New Eng J Med 279; 273, 1968.
97. Dujovne CA, Azarnoff DL: Clinical complications of corticosteroid therapy. Symp Steroid Therapy. Med Clin North Amer 57(4):1331–1342, 1973.
98. Bond WS: Toxic reactions and side effects of glucocorticoids in man. Am J Hosp Pharm 34:479–485, 1977.
99. Fenster LF: The ulcerogenic potential of glucocorticoids and possible prophylactic measure. Med Clin North Amer 57(5): 1289–1294, 1973.
100. Lipsmeyer EA: Development of malignant cerebral lymphoma in a patient with systemic lupus erythematosus treated with immunosuppression. Arth and Rheum 15(2):183–186, 1972.
101. Brothers JR, Olson G; Golk HC: Enhancement of infection by corticosteroids experimental clarification. Surg Forum 24:30–32, 1973.
102. Vincent FM: Phenytoin/desamethasone interaction. Lancet, June 24, p 1360, 1978.
103. Rish BL, Caveness WF: Relation of prophylactic medication to the occurrence of early seizures following craniocerebral trauma. J Neurosurg 38:155–158, 1973.
104. Golding DN, Begg TB: Dexamethasone myopathy. Br Med J 2:1129–1130, 1960.
105. Haque N, Thrasher K, Werk E, Knowles H, Sholiton L: Studies on dexamethasone metabolism in man: Effect of dipheylhydantoin. J Clin Endo & Metab 34:44–50, 1972.
106. Stjernholm M, Katz F: Effects of diphenylhydantoin phenobarbital and diazepam on the metabolism of methylprednisolone and its sodium succinate. JCE & M 41(5):887, 1975.
107. Koski CL, Fifenberick DH, Max SR: Oxidative metabolism of skeletal muscle in steroid atrophy. Arch Neurol 31:407–410, 1974.
108. Faludi G, Mills LC, Chayes ZW: Effect of steroids on muscle. Acta Endocrinologica 45:68–79, 1964.
109. Corcoran FH, Granatir RF, Schlang HA: Hyperglycemic hyperosmolar nonketotic coma associated with corticosteroid therapy. J Fla Med Assn 58:38–39, 1971.
110. Brennan MJ; Corticosteroids in the treatment of solid tumors. Symposium Steroid Therapy. Med Clin of North Amer 57(4):1225–1239, 1973.
111. Kahn CR, Goldfine ID, Neville DM, Roth J, Garrison M, Bates R: Insulin receptor defect: A major factor in the insulin resistance of glucocorticoid excess. Endocrinology 93:168, 1973.

112. Conn HO, Blitzer BL: Nonassociation of adrenocorticosteroid therapy and peptic ulcer. New Eng J Med 294(9):473–479, 1976.
113. Desbaillets L, Menguy R: Inhibition of gastric mucous secretion by ACTH. An experimental study. Amer J Diag Dis 12:582, 1967.
114. Menguy R, Masters YF: Effect of cortisone on mucoprotein secretion by gastric antrum of dogs: phathogenesis of steroid ulcer. Surgery 54:19, 1963.
115. Sun DCH: Effect of corticotropin on gastric acid, pepsin, and mucus secretion in dogs with fistulas. Amer J Dig Dis 14:107, 1969.
116. Selker RG, Moore PM, LoDolce D: Bone marrow depression with Cimetidine plus Carmustine. New Eng J Med 299(15):834, 1978.
117. Carone FA, Liebon A: Acute pancreatic lesions in patients treated with ACTH and adrenal corticoids. New Eng J Med 257:690–697, 1957.
118. Riemenschneider TA, Wilson JF, Vernier RL: Glucocorticoid induced pancreatitis in children. Pediatrics 41:428–436, 1968.
119. Kligman AM, Leyden JJ: Adverse effects of fluorinated steroids applied to the face. J Am Med Assoc 229:60–63, 1974.
120. Harman JB: Muscular wasting in corticosteroid therapy. Lancet 1:887, 1959.
121. MacLean K, Schurr PH: Reversible amyotrophy complicating treatment with fluorocorticosone. Lancet 1:701–703, 1959.
122. Perkoff GT, Silber T, Tyler FH et al: Studies in disorders of muscle: XII. Myopathy due to the administration of therapeutic amounts of 17-hydroxycorticosteroid. Am J Med 26:891–898, 1959.
123. Syme JR: Muscle wasting complicating treatment with dexamethasone. Med J Aust 2:420–421, 1960.
124. Walsh G, DeVivo D, Olson W: Histochemical and ultrastructural changes in rat muscle: Occurrence following adrenal corticotrophic hormone, glucocorticoids, and starvation. Arch Neurol 24:83–93, 1971.
125. Hunt TK, Ehrlich MA, Garcia JA, et al: Effects of vitamin A on reversing the inhibitory effect of cortisone on healing of open wounds in animals and man. Ann Surg 170:633–640, 1969.
126. Werk EE, Choi Y, Sholiton L et al: Interference in the effect of dexamethasone by diphenylhydantoin New Eng J Med 281:32–34, 1969.
127. Burstein S, Klaiber E: Phenobarbital-induced increased 6-B hydroxycortisol excretion: Clue to its significance in human urine. J Clin Endocrin Metab 25:293–296, 1965.
128. Graham G, Champion G, Day P, Paull P: Patterns of plasma concentrations and urinary excretion of salicylate in rheumatoid arthritis. Clin Pharmacol & Therapeut 22:410–420, 1977.
129. Sneiderman CA, Wilson JM: Effects of corticosteroids on complement and the neutrophilic polymorphonuclear leukocyte. Transplantation Proc 7:41–47, 1975.
130. Fauci AS: Corticosteroids and circulating lymphocytes. Transplantation Proc 7:37–40, 1975.
131. Rinehart JJ, Sagone AL, Balcerzak SP et al: Effects of corticosteroid therapy on human monocyte function. New Eng J Med 292:236–241, 1975.
132. Hudson G: The marrow reserve of eosinophils, effect of corticoid hormones on the foreign protein response. Br J Hematol 10:122–130, 1964.
133. Claman HN: Corticosteroids and lymphoid cells. New Eng J Med 287:388–395, 1972.
134. Kirkpatrick CH: Steroid therapy of allergic diseases. Med Clin North Am 57:1309–1320, 1973.
135. Germuth FG: Role of adrenocortical steroids in infection, immunity and hypersensitivity. Pharmacol Rev 8:1–24, 1956.
136. Staples PJ, Gerding DN, Decker JL et al: Incidence of infection in systemic lupus erythematosus. Arthritis Rheum 17:1–10, 1974.
137. McAllen MK, Kochanowski SJ, Shaw KM: Steroid aerosols in asthma: as assessment of

betamethasone valerate and a 12-month study of patients on maintenance treatment. Br Med J 1:171 – 175, 1974.
138. Perper RJ, Sanda M, Chinea G et al: Leukocyte chemotaxis in vivo. II. Analysis of the selective inhibition of neutrophil or mononuclear cell accumulation. J Lab Clin Med 84:394 – 406, 1974.
139. Myerowitz RL, Medevios AA, O'Brien TF: Bacterial infection in renal homotransplant recipients: A study of fifty-three bacteremic episodes. Am J Med 53:308 – 314, 1972.
140. Lieberman A, LeBrun Y, Glass P, Goodgold A, Lux W, Wise A, Ransohoff J: Use of high-dose corticosteroids in patients with inoperable brain tumors. J Neurol Neurosurg Psychiat 40:678 – 682, 1977.
141. Burde R, Becker B: Corticosteroid-induced glaucoma and cataracts in contact lens wearers. JAMA 213:2075 – 2077, 1970.
142. Frenkel M: Blindness due to steroid-induced glaucoma. Illinois Med J 135:160 – 163, 1969.
143. Giles CL: The ocular complications of steroid therapy. Mich Med 66:298 – 301, 1967.
144. Ritchie E: Toxic psychosis under cortisone and corticotripin. J Ment Sci 102:830 – 837, 1956.
145. Warrell DW, Taylor R: Outcome for fetus of mothers receiving prednisolone during pregnancy. Lancet 1:117 – 118, 1968.
146. Schneck DS, Penn I: De novo brain tumours in renal-transplant recipients. Lancet 1:983 – 986, 1971.

8. Applied Radiophysics for Neuro-Oncology

HARRY R. KATZ and ROBERT L. GOODMAN

1. THE ELECTROMAGNETIC SPECTRUM

X-rays and gamma rays are part of a spectrum of electromagnetic radiation that includes radio waves, visible light, and the ionizing radiation used in clinical radiotherapy. All types of radiation in the electromagnetic spectrum are characterized by a frequency (f) and wavelength (λ). The wavelength determines the properties or characteristics of the wave. All electromagnetic waves travel at the speed of light in a vacuum, and the f and λ for any type of electromagnetic radiation are interrelated by the equation:

$$\nu\lambda = C = 3 \times 10^8 \text{ M/sec.}$$

For most purposes, electromagnetic radiation can be considered as a continuous wave. For the purposes of radiation therapy physics, however, it is necessary that electromagnetic radiation be considered as discrete 'packages' or quanta of energy, also called photons. The amount of energy a photon of electromagnetic radiation can carry is dependent upon the wavelength of the radiation. As the wavelength becomes shorter, the energy carried by each photon becomes larger. It is the high energy carried by each photon of electromagnetic radiation in the x-ray range of the electromagnetic spectrum that is responsible for the biological effects of the irradiation. Electromagnetic radiation is considered to be ionizing if the energy of the photon is more than 124 eV [1]. In addition to the uniform velocity and quantum nature of electromagnetic radiation, other qualities that are characteristic of electromagnetic radiation include:
a. Travel in a straight line in free space.
b. Loss of intensity when passing through matter by attenuation (absorption and scatter).
c. Obeying the inverse square law in free space. The intensity of any radiation emanating from a source varies inversely as the square of the distance from the source [2].

The designation 'x-rays' and 'gamma rays' does not imply any quantitative or qualitative differences between these types of radiation. Rather, it describes the different origins of the radiation. X-rays are electromagnetic radiation produced outside the nucleus of the atom, in a device such as an x-ray tube, which accelerates electrons in an electrical current and then stops the electrons in a target of high density material, such as tungsten. Part of the energy of the electrons is, thus, converted into photons of electromagnetic radiation called x-rays. Gamma rays are electromagnetic radiation produced when an unstable nucleus of a radioactive isotope, i.e. radium, spontaneously decays, giving off energy in the form of electromagnetic radiation [3].

2. THE INTERACTION OF RADIATION WITH MATTER: INDIRECTLY IONIZING RADIATION

The ionizing x-rays and gamma rays used in radiation therapy do not produce direct biological damage themselves, but produce it indirectly through the production of charged particles, such as electrons. Hence, they are referred to as indirectly ionizing radiation. The mechanisms whereby X and gamma radiation produce charged particles (the directly ionizing radiation) depend upon the energy of the photons of X and gamma radiation.

Photons may interact with matter by one or more of several methods. These methods involve the interactions of photons with the orbital electrons of the atoms of matter being irradiated. For the purposes of radiation interactions, electrons are considered either 'free' or 'bound'. An electron is considered 'free' if its binding energy is small compared to the energy of the incoming photon with which it interacts. Since organic matter is composed of substances such as hydrogen, oxygen, and carbon, which are of low atomic number, the binding energies of these atoms' electrons are small, in the range of 500 electron volts (eV). As the energies of photons of X and gamma irradiation used in clinical radiotherapy are in the range of 100 KeV to many MeV, the energies of the electrons are, therefore, small compared to these photon energies, and all electrons are considered to be 'free'.

Methods by which photons interact with matter include:

a. The Photo-Electric Effect

In this type of interaction, an incident photon strikes an atom and ejects one of the 'bound' electrons in the inner shells (the K-L-M-N electrons) from the atom. The electron, ejected from the atom, is called a 'photo-electron'. It has an energy equal to the energy of the incident photon minus the binding energy holding the electron in place. The ejection of the photo-electron results in the atom becoming an unstable ion. An outer shell electron moves to the vacant space in the inner shell to establish equilibrium within the atom. As this occurs, a characteristic radiation is emitted from the atom. The probability of the photo-electric effect taking place is highly dependent upon the atomic number of the material being irradiated. It increases with higher atomic number and with increasing energy of the photons. Interaction by the photo-electric effect in tissue composed primarily of water is only noticeable in the very low energy range of 10 KeV to 100 KeV. It is responsible for the detail in diagnostic x-rays but plays practically no role at the megavoltage energies used in clinical radiotherapy [1].

b. Compton Scattering

Compton scattering is the predominant mode of interaction between photons and atoms in clinical radiotherapy (100 KeV − 10 MeV). In this interaction, an incident photon collides with a 'free' orbital electron, transferring some of its energy to the electron, which is then ejected from the atom. The incident photon scatters with a reduced energy and an increased wavelength. With increasing energy of the incident photon, proportionately more of its energy is transferred to the recoil electron and less is retained by the scattered photon, resulting in more energy being possessed by the electron to do biological damage. Absorption of energy by the Compton process is independent of the atomic number of the absorbing material. Therefore, bone and soft tissues absorb radiation equally gram for gram by this process in the megavoltage range of energies in clinical radiotherapy. This accounts for the lack of detail in radiographs taken with megavoltage radiotherapy equipment.

c. Pair Production

This type of interaction between photons and matter begins when the energy of the incident photon exceeds 1.02 MeV (as with Cobalt-60). When an incident photon of 1.02 MeV or greater approaches the nucleus of an atom, the photon may disappear. In its place an electron and a positron (a positive electron) appear. In this case mass has been produced from energy without a change in the net electric charge, as the positron and the electron are of equal and opposite charge. The positron, like the electron, can travel through the irradiated matter, producing ions. It can also collide with a 'free' electron to recombine into 2 photons each of 0.511 MeV, equivalent to the 1.02 MeV of the initial photon energy. This process, called 'annihilation', is the conversion of mass into energy. Pair production as a means of interaction between photons and matter increases in frequency as the photon energy increases beyond 1.02 MeV, but its overall importance in body tissues is low compared with the Compton effect below 10 MeV. Unlike Compton scatter, where bone and soft tissue absorb energy equally, in pair production bone aborbs twice as much energy gram for gram as soft tissue [1−3].

3. INTERACTION OF RADIATION WITH MATTER: DIRECTLY IONIZING RADIATION, CHARGED PARTICLES, AND LET

In addition to the indirectly ionizing X and gamma radiation which produce directly ionizing electrons in matter, there are a variety of charged particles used in radiotherapy which are themselves directly ionizing. When these charged particles pass through matter, they collide with the atoms of matter along their path, leaving behind a 'track' of ionized atoms and molecules. The number of ionized atoms and molecules that a charged particle produces as it traverses any material is a function of the size and the charge of the particle and its energy at any given point along its track. As the energy of the charged particle is decreasing as it travels along it track, the particle slows down. The probability of its interacting with adjacent atoms of matter increases. The unit which describes the transfer of energy from a charged particle to the atoms of matter with which it collides along its track is the linear energy transfer (LET). This is expressed in units of energy per length of track in KeV/u. The greater the charge of the particle and the smaller its velocity, the greater is its LET [1]. LET values for X and gamma radiation are in the range of 0.3 to 2 KeV/u for ^{60}Co and 250 KV x-rays respectively. Neutrons of 14 MeV energy have LET values of 3–30 KeV/u, and neutrons of 2.5 MeV have LET values of 15–80 KeV/u.

LET has important implications in clinical radiation therapy that will be discussed in a later section. The charged particles used in clinical radiotherapy include:

a. Electrons

Electrons are negatively charged particles of small mass. Their range of penetration in tissue is limited, providing a useful method of treating tumors to a limited depth in tissue without irradiating the normal tissue that lies deep to the tumor.

b. Neutrons

Neutrons are not charged (they are electrically neutral), and do not produce direct ionization. They do not interact with electrons and, therefore, penetrate to a greater depth in tissue than do charged particles. The penetration of neutrons is similar to that of ^{60}Co irradiation. Neutrons interact with the nuclei of atoms, predominately hydrogen, in the absorbing material and produce recoil protons, alpha particles, and other heavy nuclear fragments [3].

c. Protons

Protons are positively charged particles of the same mass as the neutron (1800 times that of an electron) and are useful in radiotherapy because of their characteristic deposition of their peak energy in a narrow band of tissue at a controllable depth. This characteristic will be discussed in a later section.

d. Negative Pi-Mesons (Pions)

Pi-mesons are particles intermediate in size between an electron and proton. Their mass is 273 times that of an electron. They are felt to represent the force holding the protons and neutrons together in the nucleus of an atom. They are produced by bombarding a target such as beryllium with very high energy (400 – 800 MeV) protons, producing positive, negative and neutral pi-mesons. Only the negative pi-mesons are used in clinical radiation therapy. They, like protons, deposit their peak energy in a characteristic band of tissue at the end of their path.

4. INTERACTIONS OF IONIZING RADIATION WITH LIVING MATTER

The cells of animal tissue consist of approximately 70 – 80% water. Therefore, the interaction of ionizing radiation with molecules of water is the most well known action studied in living tissue [1, 3]. When ionizing radiation interacts with a water molecule, the molecule may become ionized as in the equation:

$$H_2O \xrightarrow{h\nu} H_2O^+ + e^- \rightarrow H_2O^+ + e^-_{aq}$$

An ion is an atom which has become electrically charged after losing an electron. The H_2O^+ is called an 'ion radical'. A 'free radical' is a very reactive species which contains an unpaired electron in the outer shell. The e^-_{aq} are electrons trapped among water molecules in the cell and called 'solvated electrons'. As H_2O^+ is both charged and has an unpaired electron, it is an 'ion radical'. It decays to form free radicals which are no longer charged but still have an unpaired electron:

$$H_2O^+ \rightarrow H^+ + \cdot OH$$

The $\cdot$ OH is highly reactive free radical which can diffuse through the cell up to a distance of 20 Å and strike a DNA molecule, causing the breaks in DNA that are the ultimate effect of ionizing radiation on the living cell. Approximately 75% of the radiation damage to DNA in mammalian cells is due to the $\cdot$ OH radical [3]. Water molecules may also be excited directly by ionizing radiation to produce an 'excited' water molecule (H_2O^*) which dissociates into the hydroxyl radical ($\cdot$ OH) and hydrogen radical (H $\cdot$):

$$H_2O \xrightarrow{h\nu} H_2O^* \rightarrow H \cdot + \cdot OH$$

In summary, the processes of interaction of ionizing radiation with water molecules in the cell produces 3 reactive species; the e^-_{aq}, $\cdot$ OH, and H $\cdot$, in the proportions of 45%, 45%, and 10% respectively [1]. Present evidence indicates that the · OH radical is responsible for the major damage to the cell [4].

5. MODE OF CELL KILLING BY IONIZING RADIATION

There is a growing body of evidence that the DNA in the nucleus of the cell is the most important 'target' of radiation damage involved in cell killing [4]. Ionizing radiation acts on DNA to produce breaks in one or both of the DNA strands, or alterations in the base sequences. There is also support for the theory that radiation effects on the cell membrane contribute significantly to cell damage [3].

When cells receive very large doses of radiation (tens of thousands of rads) or if they are very sensitive to moderate doses of radiation (small lymphocytes, oocytes), they will begin to degenerate during the resting phase of their cell cycle even before undergoing the next mitosis. This is call 'interphase death'. A cell receiving a lethal but not excessively large dose of radiation (about 200 rads) will continue to function and may subsequently divide one or more times, but will eventually lose the capacity to divide. This is called 'mitotic death' and represents the predominant mode of mammalian cell death after irradiation [4].

For proliferating cells, such as bone marrow stem cells, this loss of capacity for continued proliferation through cell division is called 'reproductive death'. For fully differentiated cells that do not divide, such as nerve, muscle, or secretory cells, this end point of reproductive death is not appropriate. Cell death would, in these cases, be defined as the loss of the cells' specific function, such as impulse conduction in the neuron or contractility in the muscle cell [3]. A surviving cell which has retained its capacity to divide indefinitely and produce a colony (clone) is a 'clonogenic cell'. The ability of mammalian cells to survive various doses of ionizing radiation is represented graphically as a curve (Figure 1). The dose is plotted on a linear scale and the surviving fraction of cells on a logarithmic scale. The curve for mammalian cells irradiated with X or gamma radiation consists of an initial curving portion, the 'shoulder', followed by a straight line portion, the 'slope' (Figure 1, curve b). The slope represents the dose required to reduce the number of clonogenic cells to 37% (1/e) of their initial number. This dose is called the D_o or 'mean lethal dose'. The extrapolation number 'n' is a measure of the width of the shoulder and indicates the cells' ability to repair sublethal damage (damage which can be repaired and will not cause the reproductive death of the cell). The extrapolation number varies widely among different mammalian cell types. An alternative way of expressing the size of the initial shoulder portion of the curve is the quantity D_q or the 'quasithreshold dose' (Figure 1b).

Mitotically active cells, whether tumor or normal tissue cells, have similar sensitivities to radiation, with a D_o of about 140 rads [4, 5]. Some highly active

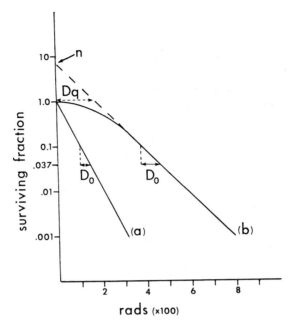

Figure 1. Mammalian cell survival curves. a) Survival curve for cells irradiated with densely ionizing radiation; b) Survival curve for cells irradiated with sparsely ionizing radiation. D_o = mean lethal dose; D_q = quasithreshold dose; n = extrapolation number.

dividing cells such as bone marrow stem cells and small lymphocytes are more sensitive, having a D_o of about 95 rads [4]. The 'shoulder' on the curve implies that radiation damage may be 'accumulated' before a cell undergoes mitotic death [4]. If a dose of radiation is split into two smaller doses and given to a cell population, separated by an interval of time, the number of cells which survive after the two 'split doses' are delivered will be greater than the number of cells which would survive the dose delivered in one fraction. This is due to a reappearance of the shoulder on the curve for the second dose of radiation which 'wastes' some of the dose. This wastage of dose is a necessary consequence of fractionation in clinical practice, which is done in order to minimize the late effects on normal tissue.

The effects of fractionated irradiation on tumors and on normal tissue are summarized in the four "R's" of radiobiology:

a. Repair of Sublethal Damage

As indicated by the shoulder on the survival curve, repair does not vary significantly between tumors and normal tissue. There is some evidence, however, that hypoxia, which, if extreme, can interfere with repair of sublethal damage, and may make tumor cells slightly less able to repair sublethal damage than normal tissue [5]. The time required to repair sublethal damage is generally less than one hour [3]. Cells which have repaired sublethal damage behave in the same way after subsequent doses of radiation as cells which had not been previously irradiated.

b. Repopulation between Fractions

If the cell cycle time of normal tissues were short compared to that of tumors, normal tissue would be restored by cell repopulation between radiation fractions while tumor cells would not be able to repopulate. However, there does not appear to be a significant difference between tumor and normal tissue cell cycles. Some normal tissues do show a markedly increased proliferation rate after radiation depletion (i.e. bone marrow stem cells). This phenomenon permits recovery of normal tissue following a course of fractionated irradiation [5].

c. Reoxygenation

Tumors have been shown to contain approximately 15% hypoxic cells. These cells are more resistant to radiation damage than oxygenated cells. After a dose of radiation kills the oxygenated cells in a tumor, the remaining hypoxic cells will form a higher percentage of the total number of surviving cells in the tumor — almost 100% [5]. As the killed oxygenated cells are removed, more oxygen is available to perfuse the formerly hypoxic cells and render them oxygenated. These oxygenated cells will then be killed by subsequent doses of irradiation. As this process is repeated, the hypoxic cells are gradually converted into oxygenated cells and the tumor cell population becomes depleted through radiation cell killing [5]. It is thought that the persistence of hypoxic cells may be responsible for failure of irradiation to sterilize all tumor cells within the irradiated volume.

d. Redistribution of Normal and Tumor Cells

The radiation sensitivity of mammalian cells varies throughout the cell cycle. Cells are most sensitive to radiation during the mitotic (M) phase of the cycle and most resistant during the DNA synthesis (S) phase. Therefore, in any given group of cells, some of which are in the different phases of the cell cycle, a dose of radiation will selectively kill those cells which happen to be in the sensitive phase of the cell cycle at the time of irradiation. The timing of the subsequent progression of the surviving cells through the next cell cycle to a sensitive phase will have a critical relationship to the interval between radiation fractions if it is desired to maximize cell killing by irradiating the cells when they are likely to be in a sensitive phase. During standard daily radiotherapy fractionation at 24-hour intervals, it is unlikely that any synchronization of the tumor cell population will result from radiation therapy and it is, therefore, unlikely that any significant differential effect between tumor and normal tissue will result in preferential depletion of tumor cells as a result of redistribution in the cell cycle during clinical radiotherapy.

6. THE OXYGEN EFFECT AND THE OXYGEN ENHANCEMENT RATIO (OER)

Oxygen is the most potent radiosensitizing agent. Sensitization of cells can be detected at oxygen concentrations of 100 parts per million (ppm). Maximum sensitization occurs at oxygen concentrations above 25,000 ppm, corresponding to an oxygen tension in the blood of 20 mmg Hg. The sensitization is at half-maximum levels at an oxygen concentration of 4000 ppm or 3 mm Hg. Since the venous blood contains oxygen at a level of 40 mmg Hg, all normal tissues are considered fully oxygenated [4].

The degree to which oxygen improves the cell killing effect of radiation is the same at all levels of survival (i.e. at 10%, 1%, 0.1% survival, etc.) The ratio of the doses of radiation under hypoxic conditions to those under oxygenated conditions to produce the same biologic effect is called the oxygen enhancement ratio (OER) [3]. The OER is measure of the radiosensitizing effect of oxygen. For mammalian cells irradiated with sparsely ionizing (low LET) X and gamma irradiation, the OER is 2.5 – 3.0 [3, 4]. For intermediately ionizing radiation, such as neutrons, the effects of oxygen on survival are less than for sparsely ionizing radiation and the OER is about 1.5 [3]. For densely ionizing radiation, the oxygen effect is negligible and the OER is 1 (Figure 2). Oxygen must be present during the radiation exposure for it to be effective in sensitizing the cell. If oxygen is added as soon as 10 milliseconds after the exposure, it will not sensitize the cell [3, 4]. It is not known precisely how oxygen acts to sensitize cells to radiation, but it is felt that it acts at the level of free radicals to produce organic peroxides that 'fix' the radiation damage to the cell [3].

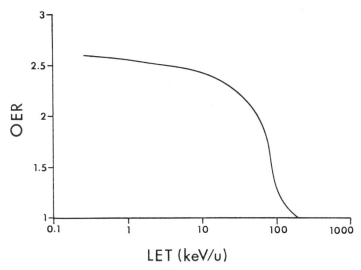

Figure 2. The relationship of the oxygen enhancement ratio (OER) to the linear energy transfer (LET) of radiation. KeV/u = kilo electron volts per micron.

7. RELATIVE BIOLOGICAL EFFECTIVENESS (RBE)

It has been observed that equal doses of different kinds of ionizing radiation, delivered in similar fractionation regimens, do not produce equal biological effects. This observation led to the concept of the relative biological effectiveness (RBE) of radiation. The RBE is the ratio of the doses of a standard (reference) radiation to that of the radiation being evaluated which will produce equivalent damage (i.e. cell killing) in a specific biological system. The standard radiation against which others are measured is 250 KV x-rays. The RBE for ^{60}Co may be expressed, therefore, as

$$\frac{\text{Dose (rad) 250 KV}}{\text{Dose (rad) }^{60}Co}$$

The RBE will vary depending on the particular biological system being used to compare the effects of the two types of radiation, and on the endpoint chosen within the biological system, such as the fraction of surviving cells in culture. The RBE can also vary with the dose rate (number of rads per minute), the number of individual fractions used, total dose, and quality of the radiation. This last factor is particularly important, as the RBE of densely ionizing radiation, such as neutrons, will vary with the dose per fraction, being greater with a course of a few large fractions of radiation than it would be for a course of many smaller fractions. Neutrons and other densely ionizing radiation do not produce effectively reparable sublethal damage (Figure 1a) and there is no 'wastage' of dose with fractionation of neutrons that there is for less densely ionizing x-rays. Therefore, in a course of fractionated irradiation, the neutron RBE will be greater than the x-ray RBE, since all of the neutron dose will go into cell killing (nonreparable lethal damage) while much of the x-ray dose will produce reparable sublethal damage with each fraction.

This relationship of RBE to fraction size is very important in neuro-oncology, as will be discussed further in this chapter, as the tolerance of the CNS to radiation is critically dependent on fraction size. In particular, if neutrons or other densely ionizing radiation are to be more effective clinically than x-rays in the treatment of brain tumors, then the RBE for tumors would have to be larger than the RBE for normal brain tissue. This relationship:

$$\frac{\text{RBE tumor}}{\text{RBE normal tissue}}$$

is called the Therapeutic Gain Factor and should be larger than unity if the radiation being evaluated is to have true clinical value.

RBE is a function of the LET of any particular radiation in that the RBE increases as the LET increases, up to a point. As the density of ionizing events in each cell increases with increasing LET, the radiation becomes more efficient at

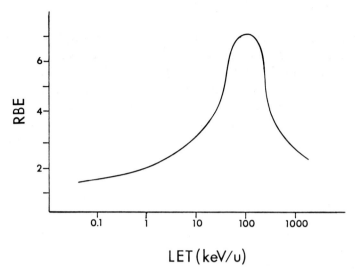

Figure 3. The relationship of the relative biological effectiveness (RBE) to the linear energy transfer (LET) of radiation. KeV/u = kilo electron volts per micron.

cell killing by being more able to deposit enough energy per cell to hit the 'critical' targets. As the LET increases further, above the maximum number of targets hit per cell, the extra ionizations per cell are, in effect, wasted, as they just 'overkill' the cell. There is no benefit to depositing more ionizing events in a cell than the maximum number needed to kill the cell. The extra ionizations are wasted and the higher LET, therefore, is less biologically efficient at cell killing than a lower LET irradiation that would deposit just enough ionizations to kill the cell (Figure 3).

8. NORMAL TISSUE TOLERANCE TO IONIZING RADIATION

The concept of normal tissue tolerance is critical to the understanding of the uses and limitations of radiation therapy in the treatment of malignant disease. Due to the wide variation in total doses, fractionation, and overall treatment times among the different radiation therapy regimens, a problem arises regarding the establishment of a common means of equating these variables to determine the 'equivalency' of the different regimens. The earliest attempts at equating different regimens of total dose and fractionation involved analyzing the results of the treatment of skin cancer and the creation of 'iso-effect curves' (Figure 4). Subsequent studies on pig skin and clinical correlation with results in humans confirmed the validity of 'iso-effect' curves in equating different time-dose-fractionation schemes for the control of skin cancer and late skin necrosis [5]. A

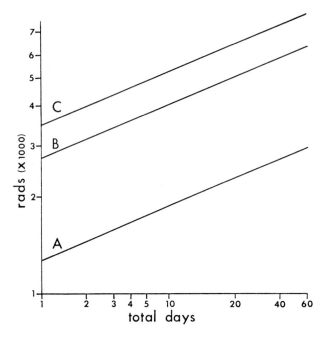

Figure 4. Typical iso-effect curves for the treatment of skin cancer by irradiation. Total days are counted from the first day of treatment to the last day of treatment. Curve A: The iso-effect curve for skin erythema. Curve B: The iso-effect curve for the cure of skin cancer. Curve C: The iso-effect curve for the development of late skin necrosis.

formula was subsequently proposed to relate total dose, time, and number of fractions to normal tissue tolerance. This formula or equation, called the nominal standard dose (NSD) equation, is based on the iso-effect curves for skin and is commonly written:

$$D = NSD \cdot T^{0.11} \cdot N^{0.24}$$

where D = total dose (rads) given in N fractions
       N = number of fractions
       T = overall treatment time in days
       NSD = nominal standard dose

The exponential functions of T and N were determined from the iso-effect curves for skin and represent the relative contributions of overall time and number of fractions to the slope of the iso-effect curve for skin tolerance. The value NSD is the constant of proportionality and its unit is the *ret*, for 'rad equivalent therapeutic', It must be emphasized that the use of the NSD equation in clinical radiotherapy is only valid for comparing the equivalence of two or more time-dose-fractionation schemes that approach full normal connective tissue tolerance. The formula can equate different schemes to the accepted limits of normal tissue tolerance (generally, for connective tissue, equivalent to 6000 rads in 30 fractions of 200 rads each or 1773 ret). It is not valid for determining the tumoricidal equivalence of different schemes nor for comparing time-dose-fractionation schemes which are below the limits of normal tissue tolerance. It is also not appropriate for courses of radiation consisting of less than 5 or more than 30 fractions or for T > 100 days [5]. The NSD formula does not include a factor for field size, as large fields tolerate less radiation and small fields more radiation [3].

To simplify the use of the NSD concept for practical purposes and avoid the necessity for complex calculations of fractional powers of N and T, a simplification was introduced in the form of TDF (time-dose-fractionation) tables [3]. These tables, compiled for treatment schedules of from 1 to 5 fractions per week, consist of factors which are proportional to partial tolerance doses. These factors may be added to compare the equivalence of regimens of less than full tissue tolerance [3]. As mentioned previously, the concept of NSD was based on skin tolerance. The tolerance of the central nervous system has been determined to be different than that for skin [6, 7]. A modification of the NSD formula for brain tolerance has been devised to take this difference in tolerance into consideration. The modified ret value, caled a 'neuret' uses different exponents for fraction number (N) and total time (T), giving more weight to number of fractions and less to time [6, 7]. The modified formula is:

$$\text{Neurets} = D \cdot N^{0.44} \cdot T^{0.06}$$

and its use is based on published spinal cord and brain tolerance data [6, 7].

9. TOLERANCE OF THE CENTRAL NERVOUS SYSTEM TO IONIZING RADIATION

Our knowledge of the tolerance of the human central nervous system to ionizing radiation results from an analysis of the cases of proven radiation damage reported in the literature. Studies have focused on both the pathophysiologic nature of the radiation-induced lesion and the radiation factors of dose, fractionation, time, volume, and quality of radiation which produced the radiation lesions in the central nervous system (CNS).

It has long been believed that neurons are highly resistant to the direct cell killing effects of ionizing radiation due to their lack of mitotic activity [8]. This has led to the search for mechanisms other than a radiation-induced damage to DNA synthesis (which would be manifested as mitotic death) as the causative factor in damage to neural tissue [9]. Studies involving skin damage had focused on the late vascular changes observed after therapeutic doses of radiation. From these studies and reports of vascular changes in specimens of CNS tissue examined after therapeutic radiation, the theory of late vascular damage was proposed to explain the phenomenon of late radiation necrosis of the CNS [8–11].

These late vascular changes have involved the development of progressive degeneration, necrosis, and occlusion of capillaries and arterioles. This resulted in thickening of the vessel wall, with narrowing of the lumen, causing degeneration and necrosis of the neurons and neuroglial cells supplied by the vessels, with resulting demyelination of the white matter in the involved areas of the brain and spinal cord [8]. The necrosis resulting from vascular lesions was then followed by proliferation of glial cells (gliosis) in a manner similar to the connective tissue fibrosis developing in other areas of the body after therapeutic doses of radiation [8]. The vascular etiology of late cerebral necrosis was also felt to explain the observed histopathologic changes outside of the radiation portals [9].

More recently an opposing view has been presented to explain the late changes found after radiation therapy to the CNS [12]. This view is based both on a critique of the vascular origins of late CNS injury and observations of specific lesions that point toward etiologies other than blood vessel injury. The observations include:

a. Lack of a universal 'late effect' sydrome with, instead, diverse manifestations that develop at different rates in different tissues.

b. The wide range in the times at which late injury appears, being consistent with a large variation in the cell cycle kinetics of 'target' cells.

c. The endothelial injury usually appearing much earlier than the parenchymal (neural) injury.

d. The lesion typically found in spinal cord injury being demyelination of long tracts with little or no change in the unmyelinated neurons of gray matter.

e. Some areas of the gray and white matter of the spinal cord being supplied by the same terminal ateriolas of the anterior and posterior spinal arteries, making it unlikely that with a common blood supply, vascular injury would consistently spare gray matter and produce only demyelination of the nerve tracts in white matter.

f. The arteries and arterioles showing no appreciable abnormalities at the time demyelination develops, making capillary thrombosis an unlikely cause.

g. The oligodendrocytes which produce myelin for the nerves of the white matter are slowly dividing cells and radiationinduced loss of reproductive capacity would cause their slow depletion and slow demyelination.

It is felt that the oligodendrocytes, therefore, are the more logical target for radiation myelitis than blood vessels and that Schwann cell depletion, not vacular injury is responsible for radiation neuropathy [12]. This theory also has support in experimental spinal cord damage in animals, but the role of the vascular lesions versus demyelination has not been resolved [13].

In attempting to determine what constitutes 'safe' levels of irradiation to the CNS, it is only by reviewing the published cases of late radiation damage that an approximation can be made of the tolerance levels. In a recent extensive review on the published literature on brain necrosis, over 100 cases were found where radiation either caused or significantly contributed to the necrosis [6, 7]. Excluding patients receiving multiple separate courses of irradiation; those without specified dose, fractions, or overall time; those treated with radon seeds or neutrons; and those with superimposed infection obscuring the radiation effect, 83 evaluable patients were analyzed [6, 7].

Some general observations were made from this published data regarding the tolerance of the brain and the development of late necrosis in the 83 evaluable cases:

a. Most (45 of 83) of the patients received total doses exceeding 7000 R or rads.
b. About one in four patients (22%), however, received doses of 5000 rads or less, but in 14 of these patients the daily fraction sizes ranged from 250 to 3500 rads. Only 8 had daily fractions in the range of 180 to 225 rads.
c. The interval from irradiation to necrosis was 6 months to 3 years in 78% of the patients.
d. Nineteen of 37 operated patients improved (50%), but the follow-up was short in many patients, probably indicating that the final outcome was worse than the available follow-up data suggests.
e. Necrosis was most frequent at ret doses of 1700 to 1800 and neuret doses of 1050 to 1100, corresponding to conventional curative dose schemes of 6000 rads/30 fractions.

The one important piece of information that the published cases of brain necrosis lack is an estimate of the risk of developing necrosis, as the size of the population at risk (all patients irradiated for a particular reason) was not known in most cases. It is, therefore, difficult to state precisely what is the absolute tolerance of the brain to radiation, the data suggesting only that fraction size is more critical for brain tolerance than it is for skin necrosis [6, 7]. The total dose when given in conventional fractionation has a bearing on the incidence of necrosis, with doses of 5000 rads in 25 fractions appearing to be a reasonably 'safe' dose, with an expected incidence of necrosis less than 5%. Reviews of the published data on radiation tolerance of the spinal cord also suggest strongly that fraction size is of critical importance [14 – 16]. Total dose is also important as published data show that the incidence of myelitis increases beyond 5% as doses exceed 5000 rads at conventional fractionation [16]. The length of irradiated cord has also been shown to affect the likelihood of myelitis but no absolute correlation between cord length, total dose, or fraction size can be stated with certainty [16].

10. IMPROVING THE THERAPEUTIC RATIO IN CNS RADIOTHERAPY

The tolerance of the brain and spinal cord to radiation has been a major limiting factor in the ability to control tumors arising in the brain or involving the spinal cord. The limitations of total doses of irradiation that can be delivered to volumes of normal brain in the treatment of primary brain neoplasms, especially astrocytomas, has resulted in a high local recurrence rate and poor survival [17]. Various methods of improving the therapeutic ratio in the radiotherapeutic management of CNS neoplasms have been explored in an attempt to increase the probability of controlling CNS neoplasms with irradiation while not exceeding the normal tissue tolerance of the CNS.

10.1 Improving the Therapeutic Ratio with Chemotherapy

Several methods of interaction may be seen when radiation therapy is administered in conjunction with cytotoxic chemotherapeutic agents.

The two modalities may act independently of one another. An example of this is cranial irradiation in CNS prophylaxis of acute leukemias where irradiation can reach leukemic cells in the brain that are protected from the action of cytotoxic drugs by the blood brain barrier. There may also be independence of the toxicities of the two modalities with no drug effects on the normal tissue within the irradiated fields, although the drug may have cytotoxic effects on the tumor within the irradiated field [18].

Enhancement of effects may occur when the response of tumor or normal tissue to radiation is greater with combined modality therapy than with radiation alone. The cytotoxic effects of combined modality therapy may be less than the simple sum of the expected effect of each modality acting separately, in which case the toxicities are subadditive. If the effects of the combined modalities are greater than the sum of the effects of the separate modalities, the toxicities are supra-additive or synergistic [18].

A cytotoxic drug may enhance the tumoricidal activity of radiotherapy by a number of mechanisms. There may be tumor shrinkage secondary to the drug which permits reoxygenation of hypoxic cells, which then become more sensitive to irradiation. The drug may kill a certain percentage of tumor cells causing progression of the remaining cycling cells into a more radiosensitive phase of the cell cycle, or block progression of the cycle at a more radiosensitive phase. Either modality may interfere with the cells' ability to repair injury caused by the other modality [18]. In the central nervous system, there is evidence that methotrexate enhances radiation damage and produces effects more damaging than those seen in the CNS attributable to methotrexate alone. There is also evidence that vincristine and adriamycin may enhance radiation damage to the spinal cord. There are reports of radiation myelitis occurring at doses below the tolerance of the cord when drug regimens containing adriamycin are used in conjunction with radiation. Damage to the CNS has occurred when the interval between radiation and chemotherapy was varied widely. The sequencing of radiation and chemotherapy has also varied, so no maximum potentiating time or sequence schedule has been established. In the treatment of gliomas, enhancement of the tumoricidal effects of radiation by BCNU was shown by improvement in the mean survival of patients receiving concomitant chemotherapy [19]. There is also evidence that the addition of hydroxyurea to BCNU further improves the mean survival in patients with glioblastoma [18].

Most of the improved clinical results with combined modality therapy have resulted from an additive cell kill by chemotherapy to that obtained with radiation. There has been no clear evidence for a supra-additive effect. The improvements in survival in patients with gliomas has resulted from the cytotoxic effects of chemotherapy added to that of radiation therapy by the use of multicourse chemotherapy given over a period of many months. The response rates for malignant gliomas to chemotherapeutic agents range from 40% for a combination of BCNU and vincristine to 60% for a combination of procarbazine, CCNU, and vincristine, although the responses to chemotherapy alone are short lived, ranging from a median of 4 months for the former combination to 9 months for the latter [20].

More effective chemotherapeutic agents need to be developed before there is significant improvement in the therapeutic ratio in malignant gliomas treated with a combination of radiotherapy and chemotherapy [17, 21].

10.2. Improving the Therapeutic Ratio with Hypoxic Cell Sensitizers

Oxygen is the most potent radiosensitizing agent. The dose of radiation needed to kill cells is 2.5 to 3 times higher in the absence of oxygen than in the presence of oxygen [22]. Since malignant tumors are known to contain poorly oxygenated cells, the decreased radiosensitivity of these hypoxic tumor cells has been implicated as a major cause of tumor recurrence after conventional low LET radiotherapy [22–24]. It was found that electron affinic compounds were capable of sensitizing cells to radiation, as they mimicked the radiosensitizing effects of molecular oxygen [22]. A search for chemical compounds was begun that would meet proposed criteria for an effective clinically useful radiosensitizer [3, 25]. The criteria include:

a. High electron affinity.
b. Selective sensitization of hypoxic cells at concentrations that are not toxic to normal oxygenated cells.
c. Sufficient chemical stability to prevent the sensitizer from being metabolized too rapidly.
d. High solubility in water or lipids, and the capability of diffusing through normal tissue to poorly vascularized tumors.
e. Effectiveness throughout most of the cell cycle.

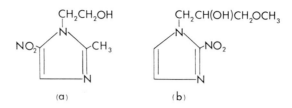

Figure 5. The structures of two hypoxic-cell sensitizers. a) The structure of metronidazole; b) The structure of misonidazole (Ro-07-0582).

f. Effectiveness at the daily dose fractions used in clinical radiotherapy.
g. Lack of whole animal toxicity at effective sensitizing concentrations.

These criteria for a clinically useful radiosensitizer were met by only a few compounds, notably the 2-nitroimidazoles and the 5-nitroimidazoles. The 5-nitroimidazole, metronidazole (Figure 5a), used clinically as an antitrichomonal agent (Flagyl) for many years, was found to have modest radiosensitizing potential and tolerable toxicity [26]. In a randomized study of the treatment of glioblastoma multiforme, it was found that patients treated with moderate doses of radiation (3000 rads in 9 fractions) preceded by high dose (6 g ms/m^2) metronidazole had a significantly greater median survival than did patients treated without metronidazole [27]. Although the overall survival of the sensitizer-treated patients was not different from prior groups of patients treated with high dose (6000 rads) radiation alone, the comparable survival of the metronidazole-treated patients using lower radiation doses was evidence of the presence of a sensitizing effect on the tumor cells achieved by the electron affinic compound.

It was subsequently discovered that the 2-nitroimidazole compounds were significantly more potent radiosensitizers than the 5-nitroimidazoles [9]. The most promising of the 2-nitroimidazoles, Ro-07-0582 (misonidazole), (Figure 5b) was found to be more effective than metronidazole in selectively sensitizing hypoxic cells without enhancing radiation damage to normal tissue [29 – 31]. The effectiveness of misonidazole in selectively sensitizing hypoxic tumor cells lies in its ability to diffuse into the hypoxic cell from the capillaries (distances of more than 130 um) without being metabolized along the way, as is oxygen [32]. This observation suggested that misonidazole might be of value in the treatment of gliomas, which were shown by the metronidazole study to contain viable hypoxic cells that influenced the radiocurability of the tumor [23, 27, 28]. A randomized study compared results in patients with glioblastoma multiforme treated with standard radiotherapy (5656 rad/28 fractions), unconventionally fractionated radiotherapy (4352 rad/12 fractions of 294 or 500 rads), and the unconventional fractionation schedule with misonidazole (3 gm/m^2 4 hours before the 500 rad fractionation) , and showed no difference in median survival [33]. However, the small numbers of patients in the study and the lack of data on the optimum time sequence for giving radiotherapy after misonidazole makes this study inconclusive in establishing the value of hypoxic cell radiosensitization in the treatment of gliomas [33]. There are limits to the potential benefits of electron affinic compounds in improving the results of radiotherapy in the nervous system as in other tumor sites. Not only is there a correlation between the radiation sensitizing ability and electron affinity of hypoxic cell sensitizers, but there is also a correlation between the compound's cytotoxicity to oxygenated cells and its electron affinity [32]. Therefore, compounds which are more electron affinic than misonidazole, hence more potent sensitizers, may also be too toxic to normal cells to be useful. However, neurotoxicity remains the principle factor limiting the clinical use of misonidazole. This is felt to be a function of its serum half-life and lipophilicity [25, 32]. A metabolite of misonidazole, Ro-05-9963, with a shorter half-life and less lipophilicity is being investigated for use as a potentially less toxic sensitizer, as are two other 2-nitroimidazoles, known as SR-2508 and SR-2555 [25].

10.3. Improving the Therapeutic Ratio with Radioprotectors

Selectively increasing the radiosensitivity of tumors without increasing the sensitivity of normal tissue is one method which might improve the therapeutic ratio. An alternative method would be to selectively decrease the sensitivity of normal tissue to the effects of radiation without affecting the sensitivity of the tumor.

The earliest attempts to develop compounds that would protect against the effects of ionizing radiation examined the action of sulfhydryl compounds such as cysteine as radioprotectors [34]. The mechanism of protection by sulfhydryl-containing compounds is not completely understood. It has been proposed that sulfhydryl compounds act as free radical 'scavengers' which compete with molecular oxygen in binding to free radicals created by the interaction of ionizing radiation with water molecules at the cellular level. This mechanism seems plausible because the sulfhydryl protective effect parallels the oxygen effect in protecting oxygenated cells to a much greater degree than hypoxic cells [3, 25]. It also explains why the effect of sulfhydryl protectors is observed with sparsely ionizing (low LET) X and gamma irradiation but not with densely ionizing (high LET) irradiation, where the oxygen effect is minimal or absent. The early sulfhydryl compounds such as cysteine and cysteamine proved too toxic to be of practical use. After screening large numbers of related compounds, it was found that thiophosphate derivatives of cysteamine were sufficiently promising to warrant further study as possible clinical radioprotectors [35]. Of these thiophosphate derivatives of cysteamine, one, S-2- (3-aminopropylamino) ethyl phosphorothioic acid hydrate, known as WR-2721, has been singled out for more intensive clinical studies because of its radioprotective properties [36].

WR-2721 has been found to protect animals against the lethal effects of whole body irradiation with a dose reduction factor (DRF) of 2.7 [37]. The dose reduction factor is the ratio of radiation dose in the presence of radioprotector to radiation dose without protector to achieve the same level of lethality.

The action of WR-2721 in protecting normal tissue, but not tumors, is the result of several possible mechanisms [37, 38]:

a. Most tumors have a poor blood supply and are not well perfused by drugs or oxygen. Therefore, molecules of radioprotectors will not be able to diffuse into most tumors to nearly the same extent that they diffuse into normal tissue cells which are well perfused by blood.

b. There is increased binding of WR-2721 in normal tissue as opposed to tumor cells even though the drug enters both types of cells by the same passive diffusion, producing consistently higher levels of protection in normal tissue than either blood, plasma, or tumor tissue.

c. The sulfhydryl derivative radioprotectors (such as WR-2721) are known to protect primarily against radiation damage to oxygenated cells and only slightly in hypoxic or anoxic cells. In this regard they work in opposite fashion to hypoxic cell sensitizers.

WR-2721 has been shown to protect a wide variety of normal tissue against radiation injury to varying degrees, including bone marrow, skin, small and large intestine, esophagus, kidney, liver, salivary gland, testis, and to a lesser extent, lung [37, 38]. Unfortunately, the drug is not concentrated in the brain or spinal cord because of its low lipophilicity and does not protect these critical organs. Therefore, this radioprotector would not be useful in altering the therapeutic ratio in the irradiation of CNS malignancies or in non-CNS malignancies where brain or spinal cord is the dose-limiting critical organ within the radiation field [25, 37, 38]. Other radioprotectors in the aminothiol category are being investigated for both protective ability and CNS protection specifically, but no drug has been identified at this time that is protective of normal CNS tissue [25, 36, 37].

10.4. Improving the Therapeutic Ratio with Hyperbaric Oxygen

The poor result obtained with conventional radiotherapy in patients with malignant gliomas has been felt to be due to extensive necrosis often observed within these tumors, which implies that a substantial fraction of the cells are anoxic or hypoxic [39, 40]. The radioresistance of the hypoxic or anoxic cells rather than their tissue of origin has been felt to be the major cause of recurrent tumor after definitive radiotherapy [39, 40]. Since normal tissue is already fully oxygenated, an improvement in tumor oxygenation would be expected to selectively increase the radiosensitivity of these hypoxic tumor cells relative to that of normal brain tissue. This would improve the therapeutic ratio in that more efficient killing of malignant cells would then take place with the same amount of radiation which had been observed to be within the tolerance level of the surrounding normal CNS.

A randomized study comparing the effects of radiation therapy and hyperbaric oxygen at 3 atmospheres showed no significant difference in overall survival between the two groups [40]. This lack of demonstrable difference in overall survival may have been due to the limited numbers of patients in the study, as there was a substantial difference in median survival between patients receiving 6000 rads/30 fractions with oxygen (45 weeks) and those receiving the same dose without oxygen (15 weeks) [40]. Although it has been suggested that the usefulness of hyperbaric oxygen may be self-limiting because of cerebral vasoconstriction in the presence of increasing oxygen tension, this has not been proven [4, 40]. The optimal use of hyperbaric oxygen in the improvement of the therapeutic ratio in the management of CNS neoplasms has probably not yet been established due to the limited number of studies in which it has been employed and the small numbers of patients involved. It deserves further investigation, both to perfect the technique and to evaluate its role in conjunction with other modalities of therapy.

10.5. The Use of Interstitial Radiation Therapy

One method of delivering a high dose of radiation to an intracerebral tumor, while avoiding the potential hazards of external beam irradiation of large volumes of normal brain, is to use an interstitial implant of radioactive isotope. This method may offer several potential advantages over external beam irradiation [41, 42]. The first advantage is that the dose of radiation delivered to the normal surrounding brain tissue is lower than that delivered to the neoplasm, in contrast to the reverse situation with external beam irradiation where the dose to surrounding normal brain may be higher than the dose to the tumor. The amount of radiation reaching normal brain tissue beyond the implant varies inversely with the square of the distance from the interstitial source. Another feature of interstitial brachytherapy is that the 'dose-rate effect' is lessened. The dose-rate effect is a main factor in the cell killing efficacy of external beam radiotherapy. As the dose rate (the number of rads per minute delivered to the tumor) is lowered in teletherapy, the effect of a given dose is reduced because of the continuous repair of cells which have suffered sublethal damage. If the dose rate falls low enough, the cells will continue to undergo mitosis while being irradiated. In contrast to this situation in teletherapy, the dose rate used in brachytherapy is in the range of 10 – 100 rads/hour and the radiation is continuous. The dose-rat effect is lessened and cells are reoxygenated while being irradiated, making them more susceptible to radiation injury.

A number of centers have used implants in conjunction with external beam irradiation and/or surgery for primary or recurrent gliomas. Doses of interstitial irradiation ranging from 1000 – 2000 rads in primary treatments (with 3600 – 4000 rads external beam irradiation) and 400 – 10,000 rads in recurrent gliomas after external beam failure have been tried [41, 42]. Responses have been good but the overall cure rates remain poor. The role of interstitial implant has only begun to be evaluated and remains a fertile field for improving the therapeutic results in central nervous system irradiation.

10.6. Improving the Therapeutic Ratio with High LET Irradiation

The problem of the relative lack of radiosensitivity of hypoxic cells in tumors [39, 40] may also be addressed by the use of high LET radiation as a means of improving the tumoricidal effect of a given dose of radiation. High LET (such as alpha particle) and intermediate LET (such as neutron) irradiation has certain radiobiological advantages over conventional low LET irradiation, including:

a. A higher relative biological effectiveness (RBE). The RBE of high LET radiation may reach 3 – 8 for particles with an LET in the range of 100 – 150 KeV/u and 2 – 5 for intermediate LET neutrons [9].

b. A reduction in the OER with high LET irradiation. The effects of oxygen on the response of tumor to irradiation diminishes as the LET increases. With low LET irradiation, the OER is in the range of 2.5 to 3. With an increase in the LET to the intermediate range (i.e. neutrons), the OER falls to the range of 1.5 to 1.7. High LET irradiation (alpha particle) has an OER of 1, that is the presence or absence of oxygen in the cells has no effect on the cell killing ability of the high LET irradiation (Figure 2).

c. Repair of sublethal damage is less with high LET irradiation than with low LET irradiation. This is manifested *in vitro* and *in vivo* as an absence of the shoulder on the cell survival curve after fractions of high LET irradiation (Figure 1a). Neutrons demonstrate a reduced shoulder to the cell survival curve and also are less dependent on fractionation.

A series of 21 patients with glioblastoma multiforme were treated with fast neutrons in doses ranging from 1550 neutron rads/10 fractions to 1850 neutron rads/17 fractions [43]. Of the 17 patients completing therapy, the median survival of 8 months and 65% survival at 6 months were not different from that seen in patients irradiated with conventional photon irradiation. Seven patients underwent autopsy and all were found to have coagulative necrosis at the site of the former tumor, with a surrounding rim of granular tissue containing sparsely scattered tumor cells. Five of these 7 patients demonstrated clinical deterioration without corresponding anatomic lesions that would account for it. It was suggested that a non tumor-related effect of neutrons on normal tissue repair mechanisms accounted for this observation. Similar results were seen in an expanded series of patients from the same institution [44]. The necrotic center at the tumor site was replaced by a transitional zone of demyelination without definite evidence of persistent tumor cells. The previously described abnormal astrocytes which were scattered around the necrotic site were felt on subsequent histopathologic evaluation to be reactive astrocytes which represented a response to neutron irradiation by the normal astrocytes, as they were not seen in patients dying shortly after completing irradiation and were not producing expanding lesions in patients with longer survivals. This diffuse white matter demyelination leading to death was corroborated by the extensive neutron experience of the Hammersmith Hospital [45]. Comparisons were made between patients treated for glioblastomas using photon irradiation of 5000 rads/5 weeks to 5500 rads/6 weeks, with patients receiving one of 3 neutron irradiation schedules to doses of either 1560 neutrons rads, 1400 neutron rads, or 1300 neutron rads. There was no difference in the patient groups for the first 2 – 3 months after irradiation. Subsequently, neutron irradiated patients were less alert, with several developing increasing generalized dementia without focal neurologic signs. In contrast, photon irradiated patients manifested localizing symptoms and signs as evidence of tumor recurrence. Twenty-three patients had either 'second-look' craniotomies (in 6) or autopsies (in 17). In photon treated patients, 6 out of 7 had evidence of moderate to gross tumor recurrence within the treated volume compared to 4 of 17 (25%) of the patients treated with neutrons. In contrast to the photon irradiated patients who had no significant change in areas of brain remote from the tumor, the neutron irradiated patients were found to have diffuse and focal degenerative changes in the white matter.

The findings of the Hammersmith trials showed that:
a. Neutrons produced greater tumor destruction than photon irradiation, but at the price of fatal demyelination of white matter.
b. Reducing the neutron dose from 1560 to 1300 rad did not change the overall survival. Fewer patients died of demyelination at the lower neutron doses, but a correspondingly greater percentage died of recurrent tumor.
c. No therapeutic margin has been demonstrated between tumor destruction and unacceptable normal brain tissue damage.

The value of neutron therapy alone is, therefore, limited by its toxicity to normal brain. Studies are in progress to see whether some combination of photon and neutron irradiation to a restricted tumor volume will overcome the limitations in the use of neutrons alone.

10.7. The Use of Heavy Charged Particles in Improving the Therapeutic Ratio

A beam of photon irradiation will pass completely through body tissue, although the intensity of the beam as it traverses the tissue is diminished by absorption and attenuation. Beams of fast neutrons will display a depth dose distribution similar to that of photon irradiation [46]. Therefore, there is no advantage to the use of neutrons from a purely dosimetric standpoint. Heavy charged particles, however, have a finite range in tissue, which gives them a distinct potential for precise beam localization that photons and neutrons do not have. In addition, these beams, excluding protons, have high LET characteristics as well. The entrance dose of a beam of high energy (megavoltage) photons is low and rapidly rises in the region of 'build-up' until it reaches a maximum at a depth

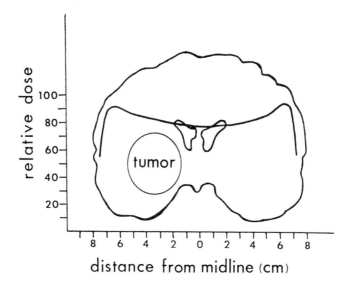

Figure 6. Graphic representation of a typical 'beam profile' across a coronal section of brain, using parallel opposing megavoltage portals. The relative dose to the temporal lobes is 10% higher than the dose to the mid-portion of the brain.

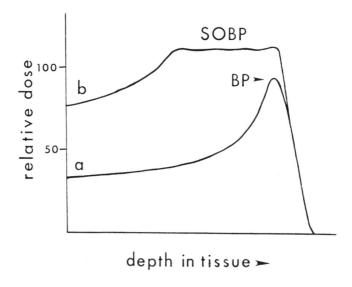

Figure 7. Beam profiles of: a) An unmodified charged particle beam which produces a characteristic 'Bragg peak' (BP) at the end of its path. The ratio of the relative dose at the Bragg peak to the entrance dose is, in this case, approximately 3 to 1; b) A modified or 'modulated' charged particle beam in which the Bragg peak has been spread out (SOBP – spread out Bragg peak) to make the width of the peak suitable for clinical applications. In this beam the ratio of the relative dose at the SOBP to the entrance dose has been reduced to 4 to 3.

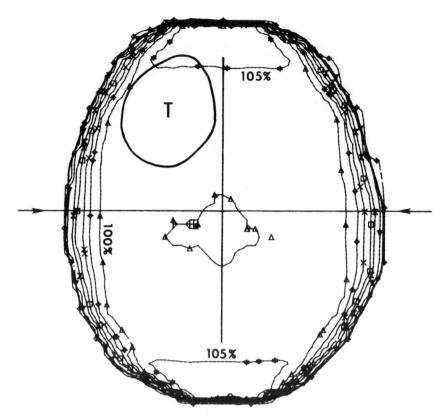

Figure 8. Isodose distribution within the brain produced by parallel opposed 19 × 20 cm portals on a 6 MeV linear accelerator. The dose to the normal frontal and occipital tissue is 5% higher than the dose to the frontal lobe tumor (T).

in tissue dependent on the energy of the photon beam. The higher the energy the deeper is the maximum. The dose then decreases as the beam passes through the tissue and is attenuated. The exit dose is often a substantial proportion of the maximum dose. As a result, a considerable dose is delivered to the normal tissue that surrounds the tumor (Figure 6). A more desirable depth dose distribution, from the standpoint of maximizing the dose to the tumor while keeping the dose to normal tissue to a minimum, would be provided by a beam of irradiation that would have a low entrance dose, rise to a peak dose at a certain depth below the surface, and rapidly fall off beyond the peak, delivering a minimal dose to the normal tissue beyond the tumor. This peak is a characteristic of charged particle beams and is referred to as the 'Bragg peak' (Figure 7a).

For unmodified particle beams, the Bragg peak is too narrow for it to be of clinical use in the treatment of tumors. Beams must be modulated or spread out to broaden the Bragg peak so it will irradiate a thickness of tissue corresponding

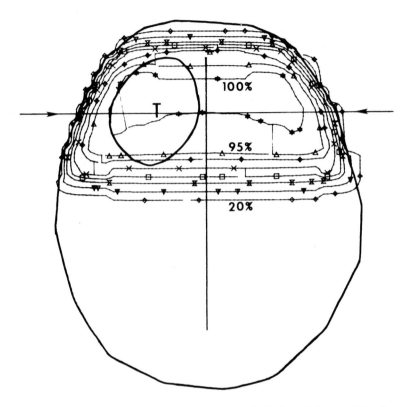

Figure 9. Isodose distribution within the brain produced by limited parallel opposed 6 × 8 cm portals on a 6 MeV linear accelerator. The doses received by the contralateral normal frontal lobe are identical to those received by the frontal lobe tumor (T).

to the tumor diameters seen in clinical practice. This modulation has, as its drawback, the effect of increasing the entrance dose of the beam relative to the peak dose, thereby removing some of the normal tissue-sparing effect (Figure 7b). However, the dose still falls off rapidly beyond the 'spread out' Bragg peak, and modulated charged particle beams, like unmodulated beams, have no exit dose. This spares the normal tissue that lies distal to the tumor (i.e. between the tumor and the exist surface of the patient).

In general, the advantage in clinical radiotherapy of the use of heavier charged particles (protons, helium ions) is in the improvement in dose localization within the tumor volume, positioning the spread out Bragg peak in the tumor, and keeping the entry and exit doses as limited as possible [46–48].

10.8. Improving the Therapeutic Ratio by Photon Beam Arrangement

The standard radiotherapeutic approach to the treatment of most malignancies of the central nervous system is to use conventional photon irradiation beams.

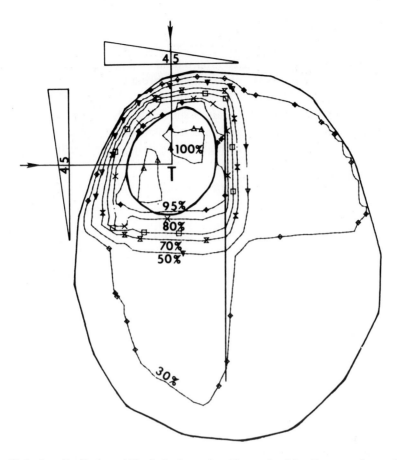

Figure 10. Isodose distribution within the brain produced by a pair of 6 × 7 cm portals on a 6 MeV linear accelerator, each filtered by a 45° wedge filter, and intersecting at 90° angles at the center of the frontal lobe tumor (T). The high dose regions approximate closely the contours of the tumor (T). Only a relatively low dose (less than 50% of maximum) reaches any significant volumes of normal brain tissue remote from the tumor.

Large field irradiation to the entire intracranial contents is the most common approach to the initial radiotherapeutic treatment of gliomas (Figure 8), as extension beyond the portals is known to be a significant cause of treatment failure in patients treated only with localized irradiation portals to gliomas [49]. However, the exclusive use of parallel opposed large field irradiation to the entire intracranial contents for the treatment of tumors which may be eccentrically located within the brain can deliver excessive doses of irradiation to the normal brain tissue (Figure 8). This arrangement of portals can predispose to late brain necrosis in patients who survive more than 6 months after irradiation. A better ratio of tumor dose to normal tissue dose can be obtained with conventional photon irradiation by the use of 'shrinking fields' (Figure 9) to confine the radiation beam to a progressively smaller area encompassing the tumor plus a limited margin of normal brain, for a portion of the total course of irradiation [50, 51]. The reduced portal is usually added after a total dose to the brain of 4000 – 5000 rads is reached with large fields, and is used to bring the tumor dose to a total of 6000 – 7000 rads. Even the use of parallel opposed 'shrinking fields' can deposit high doses of radiation in normal brain tissue (Figure 9). To overcome this problem, a pair of 'wedge filtered' beams intersecting the tumor at a 90° angle can be used to further restrict the area of high dose radiation (Figure 10).

11. CONCLUSION

The basic principles of radiation physics and biology that govern the use of radiotherapy in general also apply to the radiotherapy of central nervous system tumors.

The value and limitations of conventional photon beam irradiation have been established by years of clinical experience. Simple modifications in the dose and fractionation schemes have not been able to overcome the resistance of many central nervous system neoplasms to photon beam irradiation, nor have they been able to circumvent the problem of normal tissue tolerance. The use of older and newer methods of improving the therapeutic ratio has demonstrated the limitations of each of these methods when used alone. The future direction of radiotherapy in central nervous system neoplasms may lie in the use of combination of methods to both improve the tumoricidal effects of irradiation and spare normal tissue.

REFERENCES

1. Johns HE, Cunningham JR: The Physics of Radiology, Springfield, Charles C. Thomas, 1974.
2. Meredith WJ, Massey JB: Fundamental Physics of Radiology, Manchester, John Wright and Sons, 1977.
3. Hall EJ: Radiobiology for the Radiologist, Hagerstown, Harper and Row, 1978.
4. Ritter MA: The radiobiology of mammalian cells, Semin. Oncol. 8:3 – 17, 1981.
5. Duncan W, Nias AHW: Clinical Radiobiology, Edinburgh, Churchill Livingstone, 1977.
6. Sheline GE, Wara WM, Smith V: Therapeutic irradiation and brain injury. Int J Radiat Onco Biol Phys 6:1215 – 1228, 1980.
7. Sheline GE: Irradiation injury of the human brain: a review of clinical experience. In Radiation Damage to the Nervous System: A Delayed Therapeutic Hazard, Gilbert HA, Kagan AR (eds), New York, Raven Press, 1980, pp 39 – 58.
8. Casarett GW: Radiation Histopathology, Boca Raton, CRC Press, 1980.
9. Rubin P, Cooper RA, Phillips TW (eds): Set R.T. 1: Radiation oncology, radiation biology and radiation pathology syllabus, professional self evaluation and continuing education program for radiation therapy, Chicago, American College of Radiology, 1975.

10. Hopewell JW: The importance of vascular damage in the development of late radiation effects in normal tissues. In: Radiation Biology in Cancer Research, Meyn RE, Withers HR (eds), New York, Raven Press, 1980, pp 449–459.
11. Hildebrand J: Lesions of the nervous system in cancer patients, Monograph Series of the European Organization for Research on Treatment of Cancer, Volume 5, New York, Raven Press, 1978.
12. Withers HR, Peters LJ, Kogelnik HD: The pathobiology of late effects of irradiation. In: Radiation biology in cancer research, Meyn RE, Withers HR (eds), New York, Raven Press, 1980, pp 439–448.
13. Van der Kogel AJ: Mechanisms of late radiation injury in the spinal cord. In: Radiation biology in cancer research, Meyn RE, Withers HR (eds), New York, Raven Press, 1980, pp 461–470.
14. Kagan AR, Wollin M, Gilbert HA, Nussbaum H, Hintz BL, Rao A, Chan PYM: Comparison of the tolerance of brain and spinal cord to injury by radiation. In: Radiation damage to the nervous system: a delayed therapeutic hazard, Gilbert HA, Kagan AR (eds), New York, Raven Press, 1980, pp 183–190.
15. Lambert PM: Radiation myelopathy of the thoracic spinal cord in long term survivors treated with radical radiotherapy using conventional fractionation. Cancer 41:1751–1760, 1978.
16. Abbatucci JS, Delozier T, Quint R, Roussel A, Brune D: Radiation myelopathy of the cervical spinal cord: time, dose and volume factors. Int J Radiation Oncol Biol Phys 4:239–248, 1978.
17. Walker MD, Green SB, Byar DP, Alexander E, Batzdorf U, Brooks WH, Hunt WE, MacCarty CS, Mahaley MS, Mealey J, Owens G, Ransohoff J, Robertson JT, Shapiro WR, Smith KR, Wilson CB, Strike TA: Randomized comparisons of radiotherapy and nitrosoureas for the treatment of malignant glioma after surgery. New England J Med 303:1323–1329, 1980.
18. Phillips TL: Clinical and experimental alteration in the radiation therapeutic ratio caused by cytotoxic chemotherapy. In: Radiation biology in cancer research, Meyn RE, Withers HR (eds), New York, Raven Press, 1980, pp 567–588.
19. Walker MD, Strike TA: An evaluation of methyl-CCNU, BCNU, and radiotherapy in the treatment of malignant glioma. Proc Am Assoc Cancer Res 17:163, 1976.
20. Wilson CB, Levin V, Wara W, Sheline G, Edwards M: Adjuvant approaches to the treatment of malignant glial tumors. In: Adjuvant therapy of cancer II, Jones SE, Salmon SE (eds), New York, Grune and Stratton, 1979, pp 447–453.
21. Edwards MS, Levin VA, Wilson CB: Brain tumor chemotherapy: an evaluation of agents in current use for phase II and III trials. Cancer Treat Reports 64:1179–1205, 1980.
22. Chapman JD: Hypoxic sensitizers – implications for radiation therapy. New England J Med 301:1429–1432, 1979.
23. Belli JA, Hellman S: Hypoxic cell radiosensitizers. New England J Med 294:1399–1400, 1976.
24. Chapman JD, Urtasun RC: The application in radiation therapy of substances which modify cellular radiation response. Cancer 40:484–488, 1977.
25. Phillips TL: Sensitizers and protectors in clinical oncology. Semin Oncol 8:65–82, 1981.
26. Urtasun RC, Strumwind J, Rabin H, Band PR, Chapman JD: 'High dose' metronidazole: a preliminary pharmacological study prior to its investigational use in clinical radiotherapy trials. Brit J Radiol 47:297–298, 1974.
27. Urtasun RC, Band P, Chapman JD, Feldstein ML, Mielke B, Fryer C: Radiation and high-dose metronidazole in supratentorial glioblastomas. New England J Med 294:1364–1367, 1976.
28. Carabell SC, Bruno LA, Weinstein AS, Richter MP, Chang CH, Weiler CB, Goodman RL: Misonidazole and radiotherapy to treat malignant glioma: a phase II trial of the radiation therapy oncology group. Int J Radiat Oncol Biol Phys 7:71–77, 1981.
29. Brown JM: Selective radiosensitization of the hypoxic cells of mouse tumors with the nitroimidazoles metronidazole and Ro-07-0582. Radiat Res 64:633–647, 1975.
30. Dennenkamp J, Harris SR: Tests of two electron-affinic radiosensitizers in vivo using regrowth of experimental carcinoma. Radiat Res 61:191–203, 1975.

31. Dische S, Saunders MI: Clinical experience with misonidazole. Brit J Cancer 37: Suppl 3:311–313, 1978.
32. Fowler JF: Hypoxic cell radiosensitizers, present status and future promise. In: Radiation biology and cancer research, Meyn RE, Withers HR (eds), New York, Raven Press, 1980, pp 533–546.
33. Bleehen NM: The Cambridge glioma trial of misonidazole and radiation therapy with associated pharmacokinetic studies. In: Radiation sensitizers: their use in the clinical management of cancer, Brady LW (ed), New York, Masson, 1980, pp 374–380.
34. Patt HM, Tyree EB, Straube RL: Cysteine protection against X-irradiation. Science 110:213–214, 1949.
35. Yuhas JM, Storer JB: Chemoprotection against three modes of radiation death in the mouse. Int J Radiat Biol 15:233, 1969.
36. Davidson DE, Grenan MM, Sweeney TR: Biological characteristics of some improved radioprotectors. In: Radiation sensitizers: their use in the clinical management of cancer, Brady LW (ed), New York, Masson, 1980, pp 309–320.
37. Phillips TL: Rationale for initial clinical trials and future development of radioprotectors. In: Radiation sensitizers: their use in the clinical management of cancer, Brady LW (ed), New York, Masson, 1980, pp 321–329.
38. Yuhas JM, Spellman JM, Culo F: The role of WR-2721 in radiotherapy and/or chemotherapy. In: Radiation Sensitizers: their use in the clinical management of cancer, Brady LW (ed), New York, Masson, 1980, pp 303–308.
39. Kramer S: Radiation therapy in the management of malignant gliomas. In: Seventh national cancer conference proceedings, Philadelphia, Lippincott, 1973, pp 823–826.
40. Chang CH: Hyperbaric oxygen and radiation therapy in the management of glioblastoma. Natl Cancer Inst Monog 46, 1977, pp 163–169.
41. Hosobuchi Y, Phillips TL, Stupar TA, Gutin PH: Interstitial brachytherapy of primary brain tumors: preliminary report. J Neurosurg 53:613–617, 1980.
42. Goldson AL: Past, present, and prospects of intraoperative radiotherapy (IOR). Semin Oncol 8:59–64, 1981.
43. Parker RG, Berry HG, Gerdes AJ, Soronem MD, Shaw CM: Fast neutron beam radiotherapy of glioblastoma nultiforme. Am J Roentgenol 127:331–335, 1976.
44. Laramore GE, Griffin TW, Gerdes AJ, Parker RG: Fast neutron and mixed (neutron/photon) beam teletherapy for grades III and IV astrocytomas. Cancer 42:96–103, 1978.
45. Catterall M, Bewley DK; Fast neutrons in the treatment of cancer, London, Academic Press, 1979.
46. Castro JR: Particle radiation therapy: the first forty years. Semin Oncol 8:103–109, 1981.
47. Munzenrider JE, Shipley WU, Verhey LJ: Future prospects of radiation therapy with protons. Semin Oncol 8:110–124, 1981.
48. Suit HD, Goitein M: Rationale for the use of charged-particle and fast neutron beams in radiation therapy. In: Radiation biology in cancer research, Meyn RE, Withers HR (eds), New York, Raven Press, 1980, pp 547–565.
49. Salazar OM, Rubin P, McDonald JV, Feldstein ML: Patterns of failure in intracranial astrocytomas after irradiation: analysis of dose and field factors. Am J Roentgenol 126:279–292, 1976.
50. Salazar OM, Rubin P, McDonald JV, Feldstein ML: High dose radiation therapy in the treatment of glioblastoma multiforme: a preliminary report. Int J Radiat Oncol Biol Phys 1:717–727, 1976.
51. de Schryver A, Greitz T, Forsby N, Brun A: Localized shaped field radiotherapy of malignant glioblastoma multiforme. Int J Radiat Oncol Biol Phys 1:713–716, 1976.

… continues

9. Radiotherapy of Adult Primary Cerebral Neoplasms

GLENN E. SHELINE

INTRODUCTION

During the last four or five decades radiation therapy has been used with increasing frequency for the treatment of primary cerebral neoplasms, however, prospective, randomized, controlled studies have only been reported in the last ten years. Previously, and for most histologic types of tumor even at present, opinions regarding the efficacy of radiation therapy for the treatment of these neoplasms were based upon retrospective analyses of patients treated in single institutions over extended periods of time. Numerous variables combine to make interpretation of retrospective reviews difficult. Patient selection factors have differed between institutions and within institutions as perceptions as to the value of radiation therapy have changed. Earlier, radiation therapy was given with orthovoltage equipment, but there has been a progressive shift to higher energy radiation. Virtually all patients are now treated with megavoltage irradiation. Opinions regarding the size of the volume to be irradiated, the dose-fractionation pattern and total dose have varied.

The large majority of adult primary cerebral neoplasms remain localized and rarely metastasize either to other parts of the central nervous system (CNS) or outside the nervous system. Thus, they are lesions which, theoretically, should be curable by a local form of therapy. For neoplasms that cannot be totally resected for reasons of excessive morbidity or mortality, radiation therapy would seen potentially beneficial. In practice, the limitations of such therapy are based largely on the radiation sensitivity of critical CNS structures that must of necessity fall within the irradiated volume. While there is a substantial literature on adverse effects of radiation on the adult human CNS, it is largely anecdotal. Data relating radiation dose to effect as a function of treatment time and fractionation are meager. With few exceptions, reports of brain necrosis fail to include the size of the population at risk. Information on the effects of conventional radiotherapy on function of the adult brain are virtually non-existent.

This chapter will summarize selected data on radiation therapy of adult cerebral neoplasms. An attempt will be made to analyze and place into perspective the variables referred to above. The place of radiation therapy in the treatment of adult cerebral astrocytomas, 'malignant gliomas', oligodendrogliomas and ependymomas will be discussed. A consideration of CNS tolerance to irradiation will be included. While there has been much recent research and many publications on the treatment of malignant gliomas and on CNS tolerance to irradiation, there has been a paucity of research or new information relative to the astrocytomas, oligodendrogliomas and ependymomas. Therefore, the sections concerning these three entities will be comparatively brief. By setting the background on what can be accomplished with conventional radiation therapy, this review will, hopefully, provide a basis for the design of more effective therapeutic strategies.

ASTROCYTOMAS

Studies which include patients of all ages with infratentorial and supratentorial lesions provide fairly convincing evidence that for most incompletely resected astrocytomas radiation therapy improves the survival rate over that obtained by surgery alone. For example, Leibel et al. [1] (including patients of all ages) had recurrence-free survival rates at 5, 10 and 20 years of 19%, 11% and 0% with incomplete resection compared with 46%, 35% and 23% when radiation therapy was added. Data relating specifically to the results of treatment of cerebral astrocytomas in the adult, however, are exceedingly sparse. Many reports do not separate results for adults from those for childhood astrocytomas; the latter are known to have a better survival rate. Davidoff [2] reported that of 40 mature adults with cerebral astrocytomas, only one was alive and recurrence-free 13 years after surgical resection. At least four other patients survived 5 years. In the material of Leibel et al. [1] there were 78 patients 10 years of age or over with cerebral astrocytomas; 7 of those were between 10 – 19 years of age and the other 71 were 20 years of age or greater. Thirty-two had partial resection only and 46 had a partial resection plus postoperative radiation therapy. The radiation fields generously encompassed the tumors but did not include whole brain. The usual treatment consisted of 5000 – 5500 rad delivered in daily fractions of about 180 rad with 5 treatments per week. The recurrence-free 5- and 10-year survival rates for surgery alone were 23% and 15% as compared with 35% and 24% for surgery plus postoperative radiation therapy. The number of cases was too small to permit meaningful breakdown by histologic subtypes. Marsa et al. [3] in 1976 reported the Stanford experience with irradiated gliomas. Their cases included 40 unresected or partially resected hemispheric astrocytomas. Three patients were less than 15 years of age at the time of treatment. Concepts regarding treatment volume were similar to those of Leibel et al. but the Stanford group tended to use greater radiation doses. Their dose range was 4900 – 6660 rad with a mean of 5870 rad. The actuarial 5-year survival rate was 41%. These data, from retrospective reviews, suggest that postoperative irradiation improves the 5- and 10-year survival rates for incompletely resected astrocytomas, but to date there has been no prospective randomized study in which surgery alone and surgery plus postoperative radiation therapy have been compared for adult patients with cerebral astrocytoma. Furthermore, 5- and 10-year recurrence-free survival rates on the order of 35 – 40% and 24%, respectively, leave much to be desired.

'MALIGNANT' GLIOMAS

The more aggressive gliomas are commonly referred to as 'malignant' gliomas. As seen in the section on astrocytomas, this should not be taken to imply that the less aggressive adult cerebral astrocytoma is not a lethal neoplasm in most cases. The neoplasms generally classed together under the term 'malignant' glioma are a mixed group of tumors, some of which carry a better prognosis than others. Because of differences in natural behavior and of prognosis, it is important to distinguish the subgroups when reporting results and comparing various forms of therapy [4]. Unfortunately, in addition to problems introduced by sampling and change of histology with time, neuropathologists have not yet agreed upon a single system of classification. Kernohan and Sayre [5] subdivided these lesions into grade III and grade IV astrocytoma with the distinction based upon the apparent degree of malignancy. Others [4, 6] have classified on the basis of histologic type. This system subdivides the malignant gliomas into malignant astrocytoma and glioblastoma multiforme. Unfortunately, many authors do not distinguish between these lesions but lump them all together and report them only as malignant gliomas or, worse yet, as glioblastoma or glioblastoma multiforme. Since there does appear to be a difference in prognosis between these two major subdivisions [4], it is virtually impossible to compare one reported series with that of another without knowing the distribution of neoplasms included. Furthermore, most authors do not clearly identify and separate patients according to extent or type of therapy, location of the neoplasm and age of patient. These factors are of prognostic significance. For example, with presumably similar histologies, younger patients have better survival rates than older ones [7, 8]. The failure to stratify results or treatment protocols according to histology, location, age, extent of resection and radiation dose makes much of the literature useless for the present purpose, namely assessing the place of radiotherapy in treatment of adult cerebral neoplasms.

In spite of problems with interpretation, the results of several retrospective analyses are of interest for specific points. Kramer [9] reported that 5 of 23 patients with 'grade III malignant glioma' survived five years after surgery and radiation therapy. Of 55 patients with grade IV lesions there were no 5-year survivors. Age distribution and number of patients at risk for five years were not stated. Three-fourths of the patients with grade III gliomas returned either to normal performance status or at least to a level of self-care, whereas one-third of those with grade IV gliomas returned to a 'reasonable performance status for varying periods of time'. Marsa et al. [3] had a 20% 5-year actuarial survival for 19 patients (including one < 15 years of age) who received radiation therapy for malignant astrocytoma. There were no survivors among 48 patients (3 < 15 years of age) irradiated for glioblastoma multiforme. Sheline [10] reported the results of treatment for 49 patients (8 < 20 years of age) with malignant astrocytoma and 90 (2 < 20 years of age) with glioblastoma multiforme. All were at risk a minimum of five years. Ten with malignant astrocytoma and 50 with glioblastoma multiforme had surgery only and none survived. Seven (18%) of the 39 with malignant astrocytoma who received postoperative radiation therapy survived at least five years. One of the seven died of recurrence at six years and one of intercurrent disease at eight years. Four of the five who were alive and recurrence free at the time of reporting were leading productive, essentially normal lives. In contrast, there was no 5-year survivor among the 40 patients irradiated for glioblastoma multiforme; the median survival in this group was ten months. Salazar et al. [11] also reported a higher survival rate for grade III compared with grade IV irradiated astrocytomas. Although the 1979 review from the Mayo Clinic [7] failed to find a difference in survival between grade III and grade IV astrocytoma, the majority of the recent retrospective reviews suggests that from the viewpoint of prognosis there are at least two histologic types of malignant glioma. One has a 5-year survival rate in the order of 20% with irradiation but few survivors without, and the other has a 5-year survival rate approaching zero irrespective of therapy given.

Several recent and/or on-going prospective controlled randomized clinical trials have outdated much of the voluminous literature on malignant gliomas. Data from some of these studies are available and will be reviewed. Since results with chemotherapy are the subject of a different chapter, this review will pertain primarily to the radiotherapy aspects. The U.S. Brain Tumor Study Group (BTSG) has conducted a number of clinical trails. One four-armed randomized study compared the best conventional supportive care with radiotherapy and/or 1,3-bis-(2-chloroethyl)-1-nitrosourea (BCNU) [12]. This trial accrued 222 patients who met protocol criteria. All patients underwent definitive surgical resection and all pathology was reviewed by a central pathology review committee. Nine percent were classed as anaplastic astrocytoma, 90% glioblastoma multiforme and 1% other anaplastic gliomas. Patients in the radiotherapy arm received 170 – 200 rad whole brain irradiation per day, 5 days a week, to a total of 5000 – 6000 rad. The median age for patients receiving supportive care only was 57 years (range 10 – 79). The median age for the radiotherapy group was 56 years (range 28 – 78). The median survival time was 14 weeks for the supportive care group and 35 for those receiving radiotherapy. Independent of other variables, younger age and better initial performance status (as judged by Karnofsky rating) correlated with median survival time. This was the first reported controlled trial of radiation therapy demonstrating that conventional radiotherapy with 5000 – 6000 rad whole brain provides a modest, but statistically significant, improvement in median survival time for patients with glioblastoma multiforme.

In 1974 the Radiation Therapy Oncology Group (RTOG) and the Eastern Cooperative Oncology Group (ECOG) initiated a joint study of the treatment of malignant gliomas. This study is now closed to patient accrual and reports are in preparation [8, 13]. The RTOG-ECOG study included biopsy-proven supratentorial malignant gliomas in patients 70 years of age or less. There were four treatment options, two of which involved radiation alone while the other two combined irradiation with chemotherapeutic agents. The control arm called for 6000 rad whole brain irradiation given in 170 – 200 rad daily fractions, five fractions per week. Treatment was given via bilateral coaxial fields with both fields treated daily. Only megavoltage radiation was allowed. The second radiation only arm of the study utilized the same whole brain irradiation followed by a 1000 rad boost to the estimated tumor volume plus a generous margin. The boost dose was given with daily fractions of 150 – 200 rad. A total of 626 patients were accrued during the 5-year study period. Midway through the study, a panel of neuropathologists was formed to review histologic material on all patients. Although the participating institutions had classified the tumors as astrocytoma grade III or IV, the review pathologists used a system in which they were classified as astrocytoma with anaplastic foci (AAF) or glioblastoma multiforme (GB). Distinction between the two groups was based on the presence of foci of coagulation necrosis involving tumor cells in GB. The pathology review diagnosis for the radiation control group (6000 rad) was 10% AAF and 69% GB with 21% not reviewed. The group with 1000 rad boost had 16% and 71% of AAF and GB, respectively (13% not reviewed). In the control group, 20% were less than 40 years of age and 26% greater than 60 years of age. Sixteen percent of the boost group were less than 40 and 26% greater than 60. Thus with respect to age and histology the two radiation only groups were similarly composed.

Age was a very important prognostic factor in the RTOG-ECOG study (Table 1). For all patients under 40 years of age the 18-month survival was 64% compared with 20% for those 40 to 60 and only 8% for the group over 60 years of age. Although age was related to diagnosis and performance (Karnofsky) status, those associations did not explain why age was such a strong prognostic factor. Anaplastic astrocytomas were found in 46% of the under 40 group, 14% of the 40 – 60 and 7% in the over 60 age group. Twenty-five percent of the younger group, 42% of the 40 – 60 group, and 63% of the over 60 age group were nonambulatory. However, when both pathological diagnosis and performance status were held constant (using a stepwise Cox model program) the younger patients still had a greater median survival.

Table 1. RTOG-ECOG study

Age of patients	18 Mo. survival	AAF	Non-ambulatory
< 40	64%	46%	25%
40 – 60	20%	14%	42%
> 60	8%	7%	63%

AAF = Astrocytoma with anaplastic foci.

Twenty-four percent of the RTOG-ECOG patients showed improvement in symptoms three months after treatment. The improvement rate was the same for all treatment arms and the improvement probably can be attributed to radiotherapy. The median survival for all patients in the control group was 9.9 months compared with 8.4 months for those who received the additional 1000 rad boost. The survival curves virtually overlap. This suggests that if the 1000 rad boost succeeded in controlling more tumors, the benefit must have been offset by increased radiation mortality.

Interestingly, in the RTOG-ECOG study, when survival was related to tumor grade as determined at the institution entering the patient, there was no difference between grade III and grade IV astrocytoma. Patients with grade III astrocytoma had a median survival of 296 days compared with 302 days for those with grade IV. However, when the review pathologist panel reclassified using the histologic classification, a distinct difference was evident (Table 2). The median survival for all patients with AAF was 27.4 months versus 8.4 months for those with glioblastoma multiforme. The improved median survival for AAF, compared with GB, was maintained in all age groups. Also the higher the Karnofsky rating, the greater the median survival. It is evident that age, histologic classification and performance status are important variables and must be taken into account in any study in which treatment modalities are to be compared. Further, the histologic diagnoses should be reviewed by a single group or panel of pathologists who have intercompared their classification criteria. Failure to con-

Table 2. RTOG-ECOG study

| | Median survival, months | | | |
	Age ≤ 40	Age 40 – 60	Age > 60	All patients
Anaplastic astrocytoma	39.2	23.9	5.2	27.4
Glioblastoma multiforme	16.7	9.0	6.0	8.4
Karnofsky 70 – 100	31.7	11.2	8.4	
Karnofsky 40 – 60	16.8	7.4	4.7	
Karnofsky 20 – 30	–	3.1	4.1	

trol these variables is likely to result in misleading conclusions.

As noted above, the RTOG-ECOG study failed to find a difference in survival or improvement in symptoms between those patients who received 6000 rad whole brain and those who received 6000 plus a 1000 rad tumor boost. The BTSG, showed an improved median survival for 6000 rad whole brain as compared with 5000. Walker *et al.* [14] reviewed the relation between survival and radiation dose for 621 patients entered into three successive BTSG protocols. The median survival for patients who received approximately 5000 rad was 28 weeks compared with 36 weeks for 5550 rad and 42.0 weeks for 6000 rad. The three radiation dose groups were comparable with regard to radiation therapy parameters other than total dose, distribution of pathology types, use of corticosteroids, age, sex and initial performance status. It was concluded that radiotherapy had a significant influence on patient survival and that there was a clear cut dose-effect relationship. Patients receiving 6000 rad had a 1.3 times increase in median life span as compared with those who received 5000 rad ($p = 0.004$).

Salazar *et al.* [15] explored the use of radiation doses as high as 8000 rad. With astrocytoma grade IV they found that increasing the maximum dose from 5200 to 6000 and then to 7500 rad resulted in a progressive increase in median survival but produced no cures. The difference in median survival between the group receiving a median maximum dose of 7500 and the group who received 5200 rad was significant. However, the differences between the intermediate 6000 rad group and the other two groups were not statistically significant. For astrocytoma grade III, as the median maximum dose increased from 5200 to 5800 to 7450 rad the median survival increased from 43 to 82 to 204 weeks, respectively. The differences between median survival was statistically significant for all three groups. Again, the higher doses did not improve the cure rate; only one patient was recurrence-free at 200 weeks. The data of Salazar *et al.* must be interpreted with caution. It was not a randomized trial and the different dose groups were treated at different periods of time. Earlier patients had received the lowest doses with the highest dose group being the most recent. Between 1958 and 1973, 70 patients were treated with doses in the range of 5000–6000 rad; 51% were classed as astrocytoma grade III [11]. Subsequently, 28 patients were treated with 4000–6000 rad whole brain plus a local tumor boost that increased the maximum to 6100–8100 rad. In this latter group, only 21% were classed as astrocytoma grade III. This marked change in the fraction assigned to grade III as compared with grade IV suggests there was a change either in referral pattern or in the criteria used for grading. If the latter is correct, it could easily account for the difference in median survival found for the various dosage groups of astrocytoma grade III. In the necropsy material, Salazar *et al.* [15] found radiation effects in normal tissue at the periphery of recurrent tumors but not in more distant normal brain. This is not surprising since even in their very high dose group the more distant brain received total doses in the 4000–6000 rad range; it was only in the boost area, i.e., that of the tumor and the immediately adjacent tissue, that the high doses were delivered [11]. These data should not be used to imply that large volumes of normal brain can be treated with 7000–8000 rads, conventional fractionation, without incurring a substantial risk of necrosis.

There have been numerous reports and much debate relative to the volume that should be irradiated for malignant glioma but interpretation of the data are confused by patient selection factors. In 1960, Concannon et al. [16] compared postmortem tumor volumes with previously planned radiation therapy treatment volumes. These patients died shortly after the treatment plans had been formulated. Premortem estimates of tumor volume had been based on contrast roentgenologic examinations. In only two of the 21 patients examined was the gross tumor plus a 1 cm surrounding zone included in the planned treatment volume. There was a definite miss in 8 patients and in 11 others, tumor coverage was questionable. If the high dose treatment volume had been increased from a cylinder of approximately $8 \times 8 \times 10$ cm to $9 \times 10 \times 12$ cm the number of clear misses would have been reduced to one-fifth and those of questionable coverage to one-tenth of the patients. It should be kept in mind that this was a highly select group of patients, all of whom had such aggressive disease that they had died shortly after admission to the hospital. Concannon et al. recommended that large fields should be used. Subsequently, Kramer [17] recommended that the whole of the intracranial contents should be irradiated for glioblastoma (astrocytic glioma grade III and IV). Salazar et al. [18] concluded that patients treated to the whole brain survived longer than those who had limited fields. They went so far as to suggest that perhaps grade II astrocytoma should be treated with whole brain irradiation; apparently the basis for this recommendation was the fact that only 54% survived five years. Todd [19] suggested, in 1963, that for supratentorial glioma the irradiated volume should include 2–3 cm around the evident limits of the tumor. For patients with glioblastoma multiforme, Ramsey and Brand [20] reported an average survival of 20.4 months with limited volume irradiation compared with 8.5 months for whole brain irradiation. Scanlon and Taylor [7] failed to find a statistically significant difference in survival between limited and large volume treatment. Other authors have joined the fray regarding the volume to irradiate. Onoyama et al. [21] concluded that for glioblastoma it is necessary to deliver a high dose to a volume limited by the extent of the tumor. Fossati et al. [22] found that during the first year radiation field size did not affect survival rate but the 2-year survival rate was only 4% for whole brain irradiation and 27% for 'focal irradiation'. Hochberg et al. [23, 24] believed that decreasing the field size would diminish the incidence of radiation necrosis and of intellectual deterioration in long term survivors. With the aid of high quality serial CT scans, it has become evident that in the vast majority of the patients with malignant glioma recurrent tumor first appears at the initial tumor site. Hochberg and

Pruitt [23] found that CT scans defined, within 2 cm, the gross and microscopic tumor volume in 29 of 35 patients scanned within two months of postmortem examination. Multicentricity occurred in only 4% of untreated patients and in each case was identified by CT scan. Ninety percent of 42 patients with glioblastoma studied by serial CT scans showed tumor recurrence within a 2 cm margin of the primary site; recurrence outside the 2 cm margin was delineated by CT scans in all instances.

Unfortunately, there has been no controlled randomized trial of generous field versus whole brain irradiation and most of the available data are from patients treated prior to the availability of modern neurologic diagnostic techniques. At present the common dogma is that for a malignant glioma the entire intracranial contents should be treated, however, there is no convincing evidence that this increases the survival rate over that obtained by generous field irradiation. There is increased morbidity with whole brain irradiation, particularly if the extension brings the ears and auditory canals into the irradiated volume. Obviously there are data and advocates pro and con whole brain irradiation. It is the present reviewer's opinion that with the use of modern diagnostic methods, including good quality CT scanning, it is often unnecessary to include the entire intracranial contents. This is particularly important if limiting the volume, e.g., with a frontal lobe lesion, permits exclusion of tissues that may lead to troublesome morbidity or loss of function.

Kristiansen et al. [25] have reported a prospective, randomized study in which one arm included partial brain radiation therapy. The supratentorial area was treated with bilateral opposed fields, 5 days a week for 5 weeks with a total absorbed dose of 4500 rad. Another group of patients received conventional care without radiotherapy or chemotherapy. The median survival rates were 10.8 months for those irradiated and 5.2 months for those receiving conventional care only. Approximately 30% of the irradiated patients returned to full or partial working capacity and in 29% this was maintained for at least one year. Non-irradiated patients achieved working capacity in less than 10% at six months and none was able to work one year after resection. These results are similar to those of the BTSG and the RTOG-ECOG studies in which whole brain irradiation was used. However, lack of information regarding the percentage of grade III and grade IV gliomas in the Kristiansen study, makes direct comparison with the two U.S. studies impossible.

Methods to improve upon results obtained by postoperative conventional photon radiotherapy include the use of unconventional fractionation schemes, irradiation with hyperbaric oxygenation, photon irradiation in conjunction with hypoxic cell radiosensitizing agents and the use of high linear energy transfer (high LET) radiations such as fast neutrons. Chemotherapy and immune therapy are subjects of other chapters and will not be considered here.

Hypoxic cells are less sensitive to photon irradiation than are those irradiated under aerated conditions. The degree to which the presence of oxygen increases sensitivity to irradiation is known as the oxygen enhancement ratio (OER). OER's as high as 3 have been obtained for cells in culture given photon irradiation under standard aerobic conditions compared to those radiated under degrees of hypoxia consistent with cell survival. Malignant gliomas are believed to have hypoxic areas and it is thought that radioprotection conferred by hypoxia may be one reason that conventional photon radiotherapy has not been more successful. Chang [26] irradiated 38 patients in a chamber pressurized to three atmospheres with pure oxygen and 42 in air at one atmosphere. After 18 months the survival rate for the hyperbaric oxygen group was 28% and for the air-control group was 10%, but by 27 months the two survival curves converged. This was not a randomized study and the temporary difference in the two survival curves is of questionable significance.

Another approach to circumvention of the protective effects of hypoxia is the use of high LET radiations, the effects of which are less dependent on oxygen concentration. Neutrons, pions and several heavier charged particles are under investigation, but the only results reported to date are for Cyclotron produced neutrons. In 1976 Parker et al. [27] treated glioblastoma multiforme (grade III and IV astrocytoma) patients with neutrons at the University of Washington in Seattle. It was assumed that the relative biological effectiveness (RBE) of neutrons compared with megavoltage photons was approximately 3 and most patients received 1850 ± 50 rad given in approximately 13 fractions over 43 days. The six month survival rate, 62%, was not significantly different from that for a historical control group treated by conventional photon therapy. The average post-treatment survival actually appeared to be shortened. Autopsy findings in seven patients showed tumor replacement by coagulative necrosis. Although the entire brain was irradiated, the appearance of brain distant from the tumor was thought to be within normal limits. Because of these poor results, namely the unexplained CNS deterioration, the Seattle group switched to a mixed beam treatment plan [28]. This involved 60 rad fractions of neutrons on Mondays and Fridays with 180 rad fractions from ^{60}Co irradiation on Tuesdays, Wednesdays and Thursdays. Under this treatment plan the whole brain received 600 to 660 neutron rad and 2800 to 3200 photon rad followed by a boost to the tumor region of 120 to 180 neutron rad and 500 to 1000 photon rad. Results for the Seattle neutron patients were updated by Laramore et al. [28] in 1978. Again, historical controls were used but this time results were given according to grade of astrocytoma. The average survival in months for astrocytoma grade III treated by photons, neutrons or mixed beam was 26.0, 12.6 and 6.0, respectively. For astrocytoma grade IV the average survival was 9.9, 7.0 and 8.7 months, respectively. As judged by survival, patients with astrocytoma grade III did worse when treated by neutrons either alone or mixed with photons than did patients treated by conventional photon therapy. With grade IV lesions, it made little or no difference which form of therapy was used. By the time of the second report, postmortem data were available for 15 neutron patients. All neutron patients, whether treated by neutrons alone or mixed beam therapy, showed replacement of the bulk of the tumor by a localized mass of coagulative necrosis. This was surrounded by a zone of demyelinization before intact white matter was reached. In contrast to the initial report [27], diffuse degeneration and demyelination of white matter were described in areas far from the original tumor volume. These changes occurred in cerebellum, brain stem and cerebrum. A diffuse gliosis was present. A similar lack of improvement in survival and of diffuse damage to normal brain was reported by Catterall *et al.* [29] from the Hammersmith Hospital in London. A recent report by Hornsey *et al.* [30] provides a possible explanation for the unexpectedly great damage to the brain noted in the two clinical studies. The Hornsey group investigated the effects of neutrons on rat brain and spinal cord. White matter necrosis and paralysis of the hind limbs, respectively, were used as endpoints. Compared with 200 rad per fraction gamma rays, the RBE for the Hammersmith neutron beam was 5.2. If the RBE for the human brain is similar, this would explain the unexpected injury observed in the human trials. Laramore *et al.* concluded that either the whole brain neutron dose should be reduced or that conventional whole brain photon irradiation should be followed by a neutron boost to the primary tumor volume; this approach has been adopted for trial in a presently on-going RTOG protocol. Perhaps these variations will retain the beneficial effect of neutrons on the glioma and yet avoid the deleterious effects on normal brain.

Electron affinic chemicals which can substitute for oxygen in fixing DNA damage by radiation but, unlike oxygen, are not metabolically utilized as they diffuse through tissue are under investigation. A large number of such compounds have been studied in tissue culture systems and with animal tumors. They selectively radiosensitize hypoxic cells with the degree of radiosensitization dependent upon oxidation-reduction potential and drug concentration [31]. Because of relatively favorable oxidation potentials, pharmacologic properties and biological half-lives in humans, a 5-nitroimidazole, metronidazole, and 2-nitroimidazole, misonidazole have reached clinical trials. The main dose limiting factors for metronidazole and misonidazole are gastrointestinal and neurotoxicity. Neurotoxicity usually appears as a peripheral sensory polyneuropathy. The incidence of neuropathy is closely related to the total dose of the drug rather than size of the individual doses [32–34]. Numerous dose fractionation schemes designed to take advantage of the pharmacologic properties and total dose limitations of these compounds are being tried in various clinical centers.

Urtasun *et al.* [35] used radiation and high dose metronidazole for treatment of supratentorial glioblastomas. Thirty-six patients, stratified according to functional level, were randomly allocated to receive either radiation alone or radiation with metronidazole. The total tumor dose was 3000 rad of megavoltage irradiation given in 9 fractions over 18 days. Metronidazole was administered orally four hours before each radiation fraction. While once failures began to occur the failure rate was about the same in both groups of patients, there was a delay of about 4½ months before the process of dying began in the group of patients receiving the radiation sensitizer. The difference in survival was statistically significant; p = 0.02. The median survival for the irradiated only group was 15 weeks, a survival comparable to that obtained with operation alone in the BTSG study. Metronidazole improved the poor results obtained with radiation alone but only up to the level obtained with irradiation in the BTSG study. The BTSG gave 6000 rad in 30 – 35 fractions which according to the Ellis formula represents a nominal dose of approximately 1700 ret as compared to the approximately 1300 used by Urtasun *et al*. Would metronidazole or another imidazole similarly improve results with higher dose irradiation?

Following completion of the metronidazole study described above, Urtasun *et al*. conducted a prospective randomized study comparing higher dose irradiation alone with radiation plus either metronidazole or misonidazole. Twenty-five percent of their patients had grade III and 75% had grade IV astrocytoma. The Karnofsky ratings were below 50% in 25% of their patients. The radiation alone arm (21 patients) of the study involved three-quarter brain radiation to a total dose of 5800 rad in 30 fractions over a period of six weeks. The metronidazole patients [12] received 9 fractions of 435 rad each for a total of 3915 rad in three weeks. Metronidazole, 6 gm/m^2, was administered prior to each radiation exposure. The misonidazole group, (24 patients) received radiation as per the metronidazole group. Misonidazole, 1.25 gm/m^2, was given prior to each irradiation. CCNU, 80 mg/m^2, was administered at the time of relapse for all patients. Kaplan-Meier survival plots were prepared and the data subjected to the Wilcoxon-Gehan test. There was no statistical difference among the three treatment groups.

The RTOG recently completed a phase II non-randomized study of misonidazole and radiotherapy for treatment of grade III and IV tumors [37]. Four hundred rad was given each Monday for six weeks. Irradiation was preceded four hours by 2.5 gm/m^2 of misonidazole. Each Tuesday, Thursday and Friday for six weeks a 150 rad dose of irradiation was administered. During the seventh week, 150 rad were given per day for five fractions. The total radiation dose was thus 6000 rad in 29 fractions over 7 weeks. For the 35 evaluable patients the median survival was 37 weeks, a result similar to that obtained by the BTSG with radiation alone. The median survival for patients with grade IV tumors treated with radiation and misonidazole was 30 weeks compared with 36 weeks for irradiation alone in the BTSG study in which 90% had glioblastoma multiforme. Encouraged by the modest toxicity encountered, and in spite of the negative survival results in the phase II misonidazole study, the RTOG has launched a randomized prospective study comparing radiotherapy plus chemotherapy against radiotherapy with misonidazole plus chemotherapy [37].

OLIGODENDROGLIOMAS

The combination of low incidence and variable, often long, natural history of patients with oligodendroglioma make evaluation of therapy difficult. Table 3 presents data from the literature [3, 38–43]. With surgical resection alone, Bailey and Bucy [38], Earnest et al. [39] and Sheline et al. [42] reported 5-year survival rates, after correcting for surgical deaths, of 23–37%. A more recent report by Chin et al. [43] reported an 82% 5-year survival rate with surgical resection, but only one patient was free of recurrence at five years. Exclusive of the Chin material, the 5-year survival rate for surgery plus postoperative radiation therapy ranged from 53–85%. In 24 patients, Chin et al. had a 100% 5-year survival rate with postoperative radiotherapy and 80% were free of evidence of recurrence.

The patients treated at UCSF [42] when reviewed in 1964 showed an 85% 5-year survival with irradiation versus 31% without. The difference was significant; $p = 0.02$. At 10 years the survival rates were 55% and 25%, respectively. By coincidence there were 13 patients in each group. These patients were not randomized but, insofar as could be ascertained, the two treatment groups were similar. The records of the six irradiated patients who had survived ten years by 1964 were reviewed again in 1975. One of the six had subsequently died of recurrent tumor but the other five were alive and without known recurrence at last observation (10, 13, 14, 21 and 22 years post-therapy). Most of the UCSF patients maintained an active, useful life for the duration of their observation or until near time of death for those who had a recurrence of the oligodendroglioma.

None of these series of cases was randomized, but taken together they suggest

Table 3. Oligodendrogliomas*

	Year reported	No. of patients	Therapy	Survival rate 5 years	10 years
Bailey and Bucy [38]	1929	8	S	37%	–
Earnest, Kernohan and Craig [39]	1950	112	S	23%	–
Richmond [40]	1959	22	S + RT	53%	–
Bouchard and Peirce [41]	1960	9	S + RT	56% (5/9)	33% (1/3)
Sheline et al. [42]	1964	13	S	31% (4/13)	25% (2/8)
	1964	13	S + RT	85% (11/13)	55% (6/11)
Marsa et al. [3]	1975	14	S + RT	74%**	36%**
Chin et al. [43]	1980	11	S	82% ((9%))	–
	1980	24	S + RT	100% ((80%))	–

*Based only on patients surviving operation; **Actuarial survival; RT = Radiation Therapy; S = Surgical Resection; (( )) = No Evidence of Recurrence.

that postoperative radiation therapy significantly improves the survival rate up to at least 5 years and probably to 10 years or longer.

EPENDYMOMAS

Ringertz and Reymond [44] reported their experience with ependymomas treated by surgery alone (Table 4). They had 11 supratentorial ependymomas in patients of 15 years of age or older who had been observed at least five years. Only two of the 11 patients survived 5 years and one of these expired 2 years later of recurrence. Including patients of all ages they had 21 supratentorial ependymomas. Seven of the 13 which were located in the cerebral hemispheres were cystic, often separated from the lumen of the ventricle only by a very thin membrane. With the 8 more medially situated tumors the walls of the ventricles exhibited polypoid growths, occasionally with small nodular seeding. The ependymomas arising in the region of the septum pellucidum had a tendency to invade both lateral ventricles and the third ventricle. These findings undoubtedly explain the tendency for ependymomas to spread within the ventricular system and occasionally to the subarachnoid space of the spinal cord.

Including all age groups, Ringertz and Reymond had 14 patients with supratentorial ependymomas who survived both the diagnostic and the surgical procedures. At the time of reporting, 3 of the 4 with histologically malignant lesions and 4 of the 10 with histologically benign lesions had recurred locally. In the case of the malignant ependymomas, recurrences appeared between 8 and 12 months, whereas with the benign variety one recurred at 9 months and the others at 3, 3½ and 7 years. Bouchard and Peirce [41] reported 58% 5-year and 50% 10-year survival rates for a group of 12 patients given postoperative radiation; this material included an unspecified mixture of ages and intracranial sites. Bouchard and

Table 4. Supratentorial ependymomas

	RT	5-year survival	Patient's age	Reported spinal seeding
Ringertz and Reymond [44]	No	2/11	$\geq$ 15 Yrs	None
Bouchard and Peirce [41]	Yes	7/12*	All ages	None
Kricheff et al. [45]	Yes	2/9	All ages	None
Phillips et al. [46]	<3500r	0/7*	All ages	?
	>4500r	13/15*	All ages	?
Kim and Fayos [49]	Yes	5/11	All ages	1

*Includes infratentorial ependymomas; RT = Radiation Therapy.

Peirce did not make a practice of treating the spinal cord. Kricheff et al. [45] reported 65 cases treated by surgery plus irradiation between 1943 and 1960. There were 9 patients, of all ages, with supratentorial ependymomas who survived surgery and were followed a minimum of 5 years. The survival rate was only 22% (2/9). Their radiation fields tended to be rather small (minimum 6 × 6 and maximum 8 × 10 cm) for lesions with a propensity to spread locally. Furthermore, from 1943 through 1953 low dose repeated courses of roentgentherapy were used. Since their paper was presented in April 1963, most of the patients at risk five years must have been treated by the now outmoded small field, low dose technique. Phillips et al. [46] presented results for 42 patients with supra- and infratentorial intracranial ependymomas. Thirty-one patients survived the postoperative period. Adults with supratentorial tumors were not separated from other patients. This paper is mentioned here because of the dose effect noted. The 5-year survival rate for doses less than 3500r was 0% (4 supratentorial and 3 infratentorial lesions). With doses greater than 4500r 87% (5 supratentorial and 10 infratentorial lesions) survived 5 years and 8 of 13 or 62% survived 10 years. The use of low doses and small fields probably explains the low survival experience of Kricheff et al.

In spite of the high local failure rate, none of Kricheff's patients developed spinal cord metastasis from a supratentorial tumor. Phillips et al had two instances of spinal cord seeding. One of these was in a 14-year-old with an unbiopsied brain stem lesion treated with only 5 × 5 cm opposed fields. Barone and Elvidge [47] reported no instance of spinal seeding in 47 patients (including all ages) with intracranial ependymomas; the extent to which irradiation was used is uncertain. Fokes and Earle [48] presented data on 180 cases of ependymoma, 167 were from the Armed Forces Institute of Pathology (AFIP) files. Nineteen of 32 supratentorial tumors were noted to be in proximity to the ependymal lining of the ventricle. There were five instances of seeding from supratentorial primary sites found in autopsied cases. However, the manuscript does not state how these 5 patients were selected for entry into the AFIP files, whether they had radiation therapy, the degree of malignancy or whether there was a recurrence of tumor at the primary site at the time of autopsy. Surprisingly only 8 of their supratentorial tumors, all histologically malignant, received radiation therapy. As the authors point our, supratentorial ependymomas may metastasize. However, it would be incorrect to use these data to infer the frequency with which seeding occurs at the time of initial diagnosis or when the primary tumor has been controlled; this is especially important for the lower grade ependymoma. Kim and Fayos [49] had 11 patients with supratentorial tumors, all of whom were irradiated using local fields. The 5-year actuarial survival rate was 46.3%. Including the 11 supratentorial and 21 infratentorial lesions the 5-year survival rate for those receiving more than 4500 rad was 46% compared with 20% for a dose less than 4500 rad. One patient with a poorly differentiated supratentorial primary developed spinal subarachnoid implants. In fact, of the 11 ependymomas with subsequent spinal implants, 9 had had poorly differentiated tumor and the 2 with well differentiated tumors were subtentorial in origin. It appears that the Kim and Fayos material was highly biased toward poorly differentiated (21 of 32 cases) ependymomas. In a group of 28 patients irradiated for intracranial ependymomas, Salazar et al. [50] found that 10% of patients with doses below 1350 ret were alive as compared with 56% for patients given higher doses. Only 21% survived after partial brain irradiation versus 57% for those treated with whole brain irradiation. Glanzmann et al. [51] reported 24 patients with intracranial ependymomas. As with most other papers it was not possible to separate cases by age, site, tumor grade and therapy. Glanzmann's highest survival rate was in patients who had irradiation with tumor doses above 4500 rad with inclusion of the whole brain up to at least 3000 rad. It was concluded that prophylactic irradiation of the spinal axis is not indicated for a well differentiated ependymoma. They did recommend whole cerebrospinal axis irradiation for high grade malignant ependymomas but questioned whether this would improve cure rates.

The reports discussed above include small numbers of patients often selected for some unknown reason to receive a particular form of therapy. Problems related to patient selection plus the many other significant variables such as tumor site, histology, type of treatment, radiation dose and irradiation volume make interpretation uncertain and it is thus difficult to defend any particular position regarding treatment. Based on the evidence that is available, it is the reviewer's opinion that postoperative radiation therapy should be given whenever the tumor is incompletely resected. Currently, the entire intracranial content is irradiated to 4000–4500 rad and the primary tumor site boosted to a total tumor dose in the order of 5500 rad. Treatment is given via bilateral coaxial fields with both fields being treated daily and the daily increment being about 180 rad. For malignant ependymomas whole axis irradiation is being utilized although there is question whether even this will improve the long-term survival rate for these very aggressive tumors.

RADIATION TOLERANCE OF THE BRAIN

The limiting factor in irradiation of cerebral neoplasms is the tolerance of the surrounding normal brain tissue that must be included in the irradiated volume. Any localized neoplasm could be destroyed. Since it is brain tolerance that determines whether localized CNS neoplasms can be cured with radiation therapy, a discussion of tolerance levels is in order. Adverse CNS reactions to radiation can be considered in three groups according to time of appearance, namely acute, early delayed and late delayed. Acute reactions, thought due to edema, occur during the course of radiotherapy. With conventional fractionation, i.e., in the

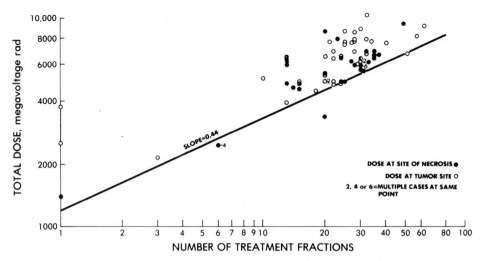

Figure 1. Total dose in megavoltage rad equivalents plotted against number of treatment fractions. The line is plotted so that most reported instances of necrosis are above it. Reproduced with permission from Sheline et al. [56].

order of 180–200 rad per day, these reactions are uncommon, generally not severe, and amenable to corticosteroid therapy for symptomatic relief. Recovery is generally spontaneous.

A few weeks to two or three months following a course of conventional photon irradiation, some patients will develop CNS signs and symptoms. Often these take the form of an accentuation of prior symptoms [52, 53]. This syndrome probably occurs in 20–25% of patients and is generally mild requiring no specific therapy. Occasionally corticosteroid therapy may be useful. Lampert et al. [54] and Lampert and Davis [55] reported three patients who expired approximately three months after radiation therapy. Postmortem examination showed disseminated patches of demyelination with central necrosis and petechial hemorrhages. Lampert and Davis emphasized that although the early delayed effects of irradiation on the CNS are usually transient and go unrecognized, they may be severe and in extreme cases lethal.

Late delayed reactions which lead to permanent radiation necrosis constitute the major hazard of high dose CNS irradiation. Little is known about the effects of radiation therapy on function of the adult human brain when the injury is short of necrosis. Sheline et al. [56] recently reviewed the literature and found 80 cases of documented brain necrosis in patients treated with a single course of radiation therapy and in which data regarding radiation dose and fractionation pattern were stated. Whenever the doses were not given in terms of megavoltage rad, appropriate conversion factors were applied. Total dose in megavoltage rad was then plotted against number of treatment fractions on a log-log graph (Figure 1). The line in Figure 1, drawn just below most of the points in the scattergram, has a slope of 0.44. This implies that for brain necrosis there is fairly strong dependence not only upon total dose but on the number of treatment fractions. Unfortunately, lack of denominators prevents construction of a true isoeffect curve. If the usual Ellis formula, NSD (RET) = $D \times N^{-0.24} \times T^{-0.11}$*, is modified according to the slope 0.44 and it is assumed that the T exponent is -0.03, the modified formula will be: NSD (NEURET) = $D \times N^{-0.41} \times T^{-0.03}$. As will be seen, there is considerable support for the concept that a formula with approximately these exponents is applicable to the entire CNS. Van der Kogel [57] reviewed the literature on the tolerance of the human spinal cord; a log-log plot of tolerance dose versus number of fractions gave a slope of 0.4. He concluded that for treatment periods up to six weeks the time factor is negligible. Pezner and Archambeau [58] concluded that for human brain tolerance the N exponent is about 0.45 and T 0.03. Wara et al. [59] derived exponents of 0.377 and 0.058 for the human spinal cord. Hornsey et al. [60] determined the isoeffect curves for x-ray induction of brain damage in the rat. Death within one year was used as the endpoint. Based on their data the modified Ellis formula would read NSD = $D \times N^{-.38} \times T^{-.02}$. Van der Kogel found that for the rat spinal cord the slope of the isoeffect curve varied from 0.42 to 0.46, depending upon the endpoint used. He concluded that the effect of overall treatment time was very small. The above data suggest that tolerance of the CNS is highly dependent on number of fractions with a minimal dependence, at least up to six weeks, on overall treatment time. Within limits then, the best way to protect the CNS is to extend fractionation.

In addition to knowing the slope of the line relating threshold dose to number of fractions, it would be exceedingly useful to know the incidence of necrosis attendent upon a given combination of dose and fraction number. Some inferences can be made, but there are few hard data for the human brain. It is evident from Figure 1 that a total dose of 6000 rad given in 30 fractions over a total of six weeks is on the threshold where necrosis may occur. Considering the large number of patients who have received such a dose and the small number of reported instances of necrosis, even in well controlled clinical trials where patients have been carefully followed with serial CT scans, it would seem that the incidence of necrosis with this dose-fraction combination must be rather low.

Boden [61] had a 25% incidence of brain stem necrosis in 24 patients treated for extracranial lesions. His doses when coverted to megavoltage rad equivalents ranged from 5300–6540. Treatment was given in 17 days with approximately 13 fractions. These cases fall well above the threshold dose line plotted in Figure 1. Salazar et al. [11, 15] used maximum doses up to 8000 rad, conventional fractionation, and reported absence of brain necrosis in areas distant from the tumor site. In evaluating normal brain tolerance levels, these data must be interpreted

---

*D = total dose in rad, N = number of daily fractions, and T = overall treatment time in days.

carefully. Radiation therapy was administered with a boost technique such that the high dose volume was limited to the region of the tumor site. Distant normal brain received only 4000–6000 rad. In the final report [15], these authors describe histologic changes suggestive of marked radiation effect occurring in normal tissue at the periphery of recurrent tumors, i.e., the high dose volumes showed radiation changes. Marks et al. [62] recently reported on 139 patients who had received 4500 rad or more and in whom there were 7 cases of proven brain necrosis (Table 5). Marks used parallel opposed fields, treated one field per day and gave a calculated midplane dose of 180–200 rad per day. To reduce their various dose fractionation schemes to a common denominator, they used the modification of the Ellis formula developed by Wara et al. [59] for human spinal cord. This formula was: equivalent dose (ED) = $D \times N^{-0.377} \times T^{-0.058}$. With this formula 6000 rad in 35 fractions over 7 weeks gives an ED of 1250. Six thousand rad in 29 fractions would be approximately ED 1330. With an ED less than 1250, Marks found no necrosis in 51 patients. When the ED was $\geq$ 1250 to < 1330 there were two instances among 60 patients. With an ED of 1330 to 1460 there were five cases among 28 patients. These data are consistent with a relatively steep dose response curve. Had Marks treated both fields daily, it is like that the ED associated with a given incidence of necrosis would be higher.

The above discussion regarding radiation tolerance of human brain assumes that one treatment is given per day without concomitant other therapy that might alter the radiosensitivity of normal CNS structures. The addition of chemotherapy (concomitant or otherwise), hyperthermia or multiple daily radiation exposures might influence the radiosensitivity. Addition of irradiation to a chemotherapeutic regimen also may increase brain sensitivity to the chemotherapeutic agent. As noted by Shapiro et al. [63], Bleyer and Griffin [64] and others,

Table 5. Incidence of human cerebral radionecrosis modified from Marks et al. [62].

Number of irradiated patients: 139
Radiation therapy
  Total dose: $\geq$ 4500 rad
  Daily dose: 180–200 rad
  Fields treated per day: 1
  Equivalent dose (ED)* = total dose $\times N^{-0.0377} \times T^{-0.058}$
Proven necrosis: 7 patients
  White matter: 7 patients
  Grey matter: 1 patient
  ED vs incidence necrosis
    < 1250**   0/51
    > 1250 to < 1330*** 2/60
    > 1330 to 1460   5/28
---

*Formula of Wara et al. developed for human spinal cord radiation tolerance; **ED 1250 equivalent to ~ 6000 rad/35 fractions/7 weeks; ***ED 1330 equivalent to ~ 6000 rad/29 fractions/6 weeks.

radiation exposure increases the likelihood of permanent brain damage from methotrexate.

SUMMARY

Postoperative radiation therapy increases the length of survival and median survival times for adults with incompletely resected cerebral astrocytoma, glioblastoma multiforme, oligodendroglioma and ependymoma. Effectiveness is limited by the tolerance of the CNS to irradiation. With adequate conventional radiotherapy, 5-year recurrence free survival rates range from zero for glioblastoma multiforme to 35 – 40% for astrocytoma and to 60 – 80% for well differentiated ependymoma. Efforts to improve the results of conventional photon radiotherapy have met with little success to date. It is hoped that a better combination of factors, perhaps using high LET radiation, a radiosensitizer, chemotherapy and/or immunotherapy will improve these results.

REFERENCES

1. Leibel SA, Sheline GE, Wara WM, Boldrey EB, Nielsen SL: The role of radiation therapy in the treatment of astrocytomas. Cancer 35:1551 – 1557, 1975.
2. Davidoff LM: A thirteen year follow-up study of a series of cases of verified tumors of the brain. Arch Neurol Psychiat 44:1256 – 1261, 1940.
3. Marsa GW, Goffinet DR, Rubinstein LJ, Bagshaw MA: Megavoltage irradiation in the treatment of gliomas of the brain and spinal cord.
4. Sheline GE: The importance of distinguishing tumor grade in malignant gliomas: treatment and prognosis. Int J Rad Oncol Biol Phys 1:781 – 786, 1976.
5. Kernohan JW, Sayre GP: Tumors of the central nervous system. Armed Forces Institute of Pathology, Fascicle 35, Washington, D.C., 1952, p 17 – 42.
6. Rubinstein LJ: Tumors of the central nervous system. Armed Forces Institute of Pathology, Fascicle 6, Washington D.C., 1972, pp 42 – 55.
7. Scanlon PW, Taylor WR: Radiotherapy of intracranial astrocytomas: analysis of 417 cases treated from 1960 through 1969. Neurosurg 5:301 – 308, 1979.
8. Chang CH, Schoenfeld D, Horton J, Salazar O, Perez-Tamayo R, Kramer S: Comparison of postoperative radiotherapy and combined postoperative radiotherapy and chemotherapy in the multidisciplinary management of gliomas. Cancer. In Press.
9. Kramer S: Radiation therapy in the management of malignant gliomas. Seventh National Cancer Conference Proceedings, Lippincott & Co, 1973.
10. Sheline GE: Conventional radiation therapy of gliomas. In: Recent Results in Cancer Research, Gliomas – Current Concepts in Biology, Diagnosis and Therapy. Hekmatpanah J (ed), Springer-Verlag, 1975, pp 125 – 134.
11. Salazar OM, Rubin P, McDonald JV, Feldstein ML: High dose radiation therapy in the treatment of glioblastoma multiforme: a preliminary report. Int J Rad Oncol Biol Phys 1:717 – 727, 1976.
12. Walker MD, Alexander E, Hunt WE, MacCarty CS, Mahaley MS, Mealey J, Norrell HA, Owens G, Ransohoff J, Wilson CB, Gehan EA, Strike TA: Evaluation of BCNU and/or radiotherapy in the treatment of anaplastic gliomas. J Neurosurg 49:333 – 343, 1978.

13. Nelson JS, Schoenfeld D, Tsukada Y, Fulling K: Histologic criteria with prognostic significance for malignant glioma. In: Tumors of the Central Nervous System, Modern Radiotherapy in Multidisciplinary Management. Chang CH, Housepian E (eds), Masson Pub., In Press.
14. Walker MD, Strike TA, Sheline GE: An analysis of dose-effect relationship in the radiotherapy of malignant gliomas. Int J Rad Oncol Biol Phys 5:1725–1731, 1979.
15. Salazar OM, Rubin P, Feldstein ML, Pizzutiello R: High dose radiation therapy in the treatment of malignant gliomas: final report. Int J Rad Oncol Biol Phys 5:1733–1740, 1979.
16. Concannon JP, Kramer S, Berry R: The extent of intracranial gliomata at autopsy and its relationship to techniques used in radiation therapy of brain tumors. Am J Roentgenol 84:99–107, 1960.
17. Kramer S: Tumor extent as a determining factor in radiotherapy of glioblastomas. Acta Radiol 8:111–117, 1969.
18. Salazar OM, Rubin P, McDonald JV, Feldstein ML: Patterns of failure in intracranial astrocytomas after irradiation: analysis of dose and field factors. Am J Roentgenol 126:279–292, 1976.
19. Todd IDH: Choice of volume in the x-ray treatment of supratentorial gliomas. Brit J Radiol 36:645–649, 1963.
20. Ramsey RG, Brand WN: Radiotherapy of glioblastoma multiforme. J Neurosurg 39:197–202, 1973.
21. Onoyama Y, Abe M, Yabumoto E, Sakamoto T, Nishidai T, Suyama S: Radiation therapy in the treatment of glioblastoma. Am J Roentgenol 126:481–492, 1976.
22. Fossati F, Jucker C, Tosi G: Effect of field size and unconventional dose fractionation in the radiotherapy of malignant cerebral gliomas. International Symposium on Multidisciplinary Aspects of Brain Tumor Therapy, June 1979, p 106.
23. Hochberg FH, Pruitt A: Assumptions in the radiotherapy of glioblastoma. Neurology 30:907–911, 1980.
24. Hochberg FH, Slotnick B: Neuropsychologic impairment in astrocytoma survivors. Neurology 30:172–177, 1980.
25. Kristiansen K, Hagen S, Kollevoid T, Torbik A, Holme I, Nesbakken R, Hatlevoll R, Lindgren M, Brun A, Lindgren S, Notter G, Anderson AP, Elgen K: Combined modality therapy of operated astrocytomas grade III and IV. Confirmation of the value of postoperative irradiation and lack of potentiation of Bleomycin on survival time: A prospective trial of the Scandanavian Glioblastoma Study Group. Cancer 47:649–652, 1981.
26. Chang CH: Some new radiotherapeutic approaches and combined protocol trials in the management of malignant gliomas. In: Recent Results in Cancer Research, Gliomas – Current Concepts in Biology, Diagnosis and Therapy. Hekmatpanah J (ed), Springer-Verlag, 1975, pp 135–150.
27. Parker RG, Berry HC, Gerdes AJ, Soronen MD, Shaw CM: Fast neutron beam radiotherapy of glioblastoma multiforme. Am J Roentgenol 127:331–335, 1976.
28. Laramore GE, Griffin TW, Gerdes AJ, Parker RG: Fast neutron and mixed (neutron/photon) beam teletherapy for grades III and IV astrocytomas. Cancer 42:96–103, 1978.
29. Catterall M, Bloom HJG, Ash DV, Walsh L, Richardson A, Uttley D, Gowing NFC, Lewis P, Chaucer B: Fast neutrons compared with megavoltage x-rays in the treatment of patients with supratentorial glioblastoma: a controlled pilot study. Int J Rad Oncol Biol Phys 6:261–266, 1980.
30. Hornsey S, Morris CC, Meyers R, White A: Relative biological effectiveness for damage to the central nervous system by neutrons. Int J Rad Oncol Biol Phys 7:185–189, 1981.
31. Sheline G, Wasserman T, Wara W, Phillips T: The role of radiosensitizers in therapy of malignant brain tumors. In: Multidisciplinary Aspects of Brain Tumor Therapy, Paoletti P, Walker MD, Butti G, Knerich R (eds), Elsevier-North Holland Biomedical Press, 1979, pp 197–208.
32. Dische S: Hypoxic cell sensitizers in radiotherapy. Int J Rad Oncol Biol Phys 4:157–160, 1978.

33. Urtasun RC, Chapman JD, Feldstein ML, Band RP, Rabin HR, Wilson AF, Marynowski B, Starreveld E, Shnitka T: Peripheral neuropathy related to misonidazole: Incidence and pathology: Brit J Cancer 37:271–275, 1978.
34. Wasserman TH, Phillips TL, Johnson RJ, Gomer CJ, Lawrence GA, Sadee W, Marques RA, Levin VA, VanRaalte G: Initial United States clinical and pharmacologic evaluation of misonidazole (Ro-07-0582) an hypoxic cell radiosensitizer. Int J Rad Oncol Biol Phys 5:775–786, 1979.
35. Urtasun R, Band P, Chapman JD, Feldstein ML, Mielke B, Fryer C: Radiation and high-dose metronidazole in supratentorial glioblastomas. New Engl J Med 294–1365–1367, 1976.
36. Urtasun RC, Miller JD, Feldstein ML: A comparison of misonidazole and metronidazole in the treatment of glioblastomas. (In preparation).
37. Carabell SC, Bruno LA, Weinstein AS, Richter MP, Chang CH, Weiler CB, Goodman RL: Misonidazole and radiotherapy to treat malignant glioma: a phase II trial of the Radiation Therapy Oncology Group. Int J Rad Oncol Biol Phys 7:71–77, 1981.
38. Bailey P, Bucy PC: Oligodendrogliomas of brain. J Path Bact 32:735–751, 1929.
39. Earnest F, Kernohan JW, Craig WM: Oligodendrogliomas: review of 200 cases. Arch Neurol Psychiat 63:964–976, 1950.
40. Richmond JJ: Malignant tumors of the central nervous system. In: Cancer, Vol 5, Raven R (ed), Butterworth & Co Ltd, 1959, pp 375–389.
41. Bouchard J, Peirce CB: Radiation therapy in the management of neoplasms of the central nervous system with a special note in regard to children: twenty years' experience, 1939–1958. Am J Roentgenol 84:610–628, 1960.
42. Sheline GE, Boldrey E, Karlsberg P, Phillips TL: Therapeutic considerations in tumors affecting the central nervous system. Radiology 82:84–89, 1964.
43. Chin HW, Hazel JJ, Kim TH, Webster JH: Oligodendrogliomas. I. A clinical study of cerebral oligodendrogliomas. Cancer 45:1458–1466, 1980.
44. Ringertz N, Reymond A: Ependymomas and choroid plexus papillomas. J Neuropath & Exper Neurol 8:355–380, 1949.
45. Kricheff II, Becker M, Schneck SA, Taveras JM: Intracranial ependymomas. A study of survival in 65 cases treated by surgery and irradiation. Am J Roentgenol 1:167–175, 1964.
46. Phillips TL, Sheline GE, Boldrey E: Therapeutic considerations in tumors affecting the central nervous system: ependymomas. Radiology 83:98–105, 1964.
47. Barone BM, Elvidge AR: Ependymomas: a clinical survey. J Neurosurg 33:428–438, 1970.
48. Fokes EC Jr, Earle KM: Ependymomas: Clinical and pathologic aspects. J Neurosurg 30:585–594, 1969.
49. Kim YH, Fayos JV: Intracranial ependymomas. Radiology 124:805–808, 1977.
50. Salazar OM, Rubin P, Bassano D, Marcial VA: Improved survival of patients with intracranial ependymomas by irradiation: dose selection and field extension Cancer 35:1563–1573, 1975.
51. Glanzmann C, Horst W, Schiess K, Friede R: Considerations in the radiation treatment of intracranial ependymoma. Strahlentherapie 156:97–101, 1980.
52. Boldrey E, Sheline G: Delayed transitory clinical manifestations after radiation treatment of intracranial tumors. Acta Radiol 5:5–10, 1967.
53. Hoffman WF, Levin VA, Wilson CB: Evaluation of malignant glioma patients during the post-irradiation period. J Neurosurg 50:624–628, 1979.
54. Lampert P, Tom MI, Rider WD: Disseminated demyelination of the brain following Co-60 (gamma) radiation. Arch Pathol 68:322–330, 1959.
55. Lampert PW, Davis RL: Delayed effects of radiation on the human contral nervous system. 'Early' and 'late' delayed reactions. Neurology 14:912–917, 1964.
56. Sheline GE, Wara WM, Smith V: Therapeutic irradiation and brain injury. Int J Rad Oncol Biol Phys 6:1215–1228, 1980.

57. Van der Kogel AJ: Late effects of radiation on the spinal cord. Publication of The Radiobiological Institute, TNO Rijswijk, The Netherlands, 1979, pp 118–121.
58. Pezner RD, Archambeau JP: Brain tolerance unit: a method to estimate risk of radiation brain injury for various dose schedules. Int J Rad Oncol Biol Phys 7:397–402, 1981.
59. Wara WM, Phillips TL, Sheline GE, Schwade JG: Radiation tolerance of the spinal cord. Cancer 35:1558–1562, 1975.
60. Hornsey S, Morris CC, Myers R: The relationship between fractional and total dose for x-ray induced brain damage. Int J Rad Oncol Biol Phys 7:393–396, 1981.
61. Boden G: Radiation myelitis of the brain-stem. J Fac Radiol 2:79–94, 1950.
62. Marks JE, Baglan RJ, Prassad SC, Blank WF: Cerebral radionecrosis: incidence and risk in relation to dose, time, fractionation and volume. Int J Rad Oncol Biol Phys 7:243–252, 1981.
63. Shapiro WR, Allen JC, Horten BC: Chronic methotrexate toxicity to the central nervous system. Clin Bulletin 10:49–52, 1980.
64. Bleyer WA, Griffin TW: White matter necrosis, mineralizing microangiopathy, and intellectual abilities in survivors of childhood leukemia: Associations with central nervous system irradiation and methotexate therapy. In: Radiation Damage to the Nervous System, Gilbert HA, Kagan AR (eds), New York, Raven Press, 1980, pp 155–174.

10. Tumor Cell and Host Response Parameters in Designing Brain Tumor Therapy

PAUL L. KORNBLITH, BARRY H. SMITH and MAURICE K. GATELY

1. INTRODUCTION

In the study of patients with malignant brain tumors, there are certain cell biological and immunological parameters which may prove useful as adjuncts in the diagnostic and therapeutic approaches to these patients. This chapter deals with certain selected biological tumor cell characteristics and host cell response factors which bear upon understanding how such parameters may influence clinical care. Important cell biological factors include tumor cell growth characteristics, malignant potential, and the cellular basis of sensitivity or resistance to chemotherapeutic agents. Host response factors are comprised of the humoral and cellular responses of brain tumor patients to their own tumors and the clinical significance and/or effectiveness of these defense mechanisms.

In vitro assays to characterize the cell biological properties of individual patient glioma cells, to determine their sensitivity to specific chemotherapeutic agents and to determine the patient's own immunological responses to his or her tumor are appealing theoretically because they may provide a basis for improved therapy and prediction of clinical course without adding any patient risk. Various *in vitro* models have been utilized in recent years for both chemotherapy and immunology testing in glioma patients [1 – 9]. The practicality and reliability of such *in vitro* colony formation and microcytotoxicity assays have been amply shown. In our laboratory, significant variation in responsiveness of cultured individual patient glioma cell lines to 1,3-bis (2-chloroethyl)-1-nitrosourea (BCNU) is clear [4] although the basis of the difference has not been shown. These data have enabled separation of *in vitro* glioma BCNU-sensitive and insensitive populations. In addition, tissue culture analysis of cerebrospinal fluid for the presence and nature of tumor cells has been utilized as an adjunct to the diagnosis and monitoring of therapy [10].

With regard to immunological responses, allogeneic and autologous microcytotoxicity testing and an immune adherence assay have indicated that a significant proportion of brain tumor patients have a humoral immune response to their tumors. This humoral immune response appears to involve IgG and IgM for the observed cytotoxicity and primarily IgM for the immune adherence [11].

Of course, despite the practicality of such *in vitro* testing, the question of its predictive relevance for the clinical situation must be asked. The fact is that the data in hand to date indicate that the *in vitro* characterization, chemotherapeutic sensitivity and immune evaluatory techniques do have diagnostic and predictive value [10, 12, 13].

In this chapter we will describe these various *in vitro* approaches, emphasizing the kinds of information that can be gleaned from them and their presently known clinical relevance.

Clearly, the future will see more advanced and sophisticated techniques become available to screen, monitor and guide therapy in brain tumor patients. It is also apparent that approaches of using cellular and host response data serve best as supplements to the standard screening, diagnostic and neuropathological techniques.

2. MATERIALS AND METHODS

2.1. Tissue Culture

Initially, tissue obtained during surgery is minced into 1 mm³ pieces and planted in plastic flasks (Falcon, Oxnard, CA.) using sterile technique. Once a monolayer of cells has established itself on the flask surface, the cell line is subcultured. The cells are washed twice using Hanks Balanced Salt Solution (Ca^{++} − and Mg^{++} − free, GIBCO) and incubated at 37°C for 10 minutes with 1.0 ml trypsin-EDTA solution (0.5% trypsin, 0.02% EDTA). The resulting cell suspension is then diluted and divided between two flasks. When a cell line has successfully been subcultured a minimum of two times, it is considered ready for use in the chemotherapy and immunologic microtiter assay [4, 8].

All lines were characterized by morphological, ultrastructural, biophysical and biochemical techniques [14–17]. The cultures are routinely examined for mycoplasma in the Mycoplasma Laboratory, Dept. of Bacteriology, Massachusetts General Hospital, using the stained agar technique of Dienes [18] or by Dr. Richard DelGuidice at the Fredrick Research Center by fluorochrome technique [19].

2.2. CSF Tissue Culture

CSF specimens obtained from patients are placed in tissue culture medium and examined for growth [10]. These specimens normally reach our laboratory within 24 hours of their collection. For culture, 0.7 cc aliquots of human CSF are placed in 0.7 cc of F-10 nutrient medium supplemented by 10% fetal calf serum. These are incubated at 34°C without added CO_2 in three sterile Leighton tubes, each with a well fitting coverslip. After 3 days (occasionally after 1 of 7 days) the coverslips are removed, fixed with Zenker's fluid, stained with hematoxylin and

Giemsa, and mounted in Canada balsam for microscopic examination.

For millipore cytological examination at our hospital, cells are examined after being washed with saline under no applied pressure through a $5\text{-}\mu$ millipore filter. The filter is then fixed with ethanol and examined for cells after Papanicolau staining [20]. Where results of millipore filtration and CSF tissue culture are compared, they are taken from the same CSF specimen.

For analysis of these results, the records of 122 patients were reviewed with special attention to site of tumor, operative procedure, biopsy or autopsy diagnosis, and findings on routine cytological examination. In 13 records, data were incomplete and these patients were excluded. The slides of the 109 patients from whom information was complete were reviewed and photographs were made of any cellular growth. These photographs were then examined to assess abundance and character of growth, nuclear pattern, and cellular morphology. At the time of review, the investigator did not know the diagnosis of the patient under consideration.

To be classified as unequivocally malignant, growth has to fulfill at least four of the following seven criteria adapted from Barker and Sanford [21]. These criteria have been widely applied in the categorization of transformed and non-transformed cells and serve as a firm and stringent basis for classification in this study. In these cases, either quantitative (criteria 1 and 2) or morphological observations (criteria 3 through 7) are used for classification.

1. More than five cells per high-power field
2. Clusters of cells (more than three contiguous cells)
3. More than one nucleus per cell
4. Prominent nucleoli or more than three distinct nucleoli
5. Nuclear to cytoplasmic ratio greater than 0.5 (nuclear area greater than one-half cell area)
6. Pleomorphic growth (widely variant cellular morphology in a given specimen)
7. Very basophilic cytoplasm.

Cells that do no clearly meet these criteria are classified as 'doubtful' (less than four categories) and cells with none of them are called 'benign'. These terms are chosen to correspond with those used in cytological reports to aid in comparing our results with cytological studies.

2.3. Biological Characterization

A variety of techniques, based on tissue culture, are utilized to study the characteristics of the tumor cells. Parameters of interest include simple determinations of cell morphology and growth characteristics [17], cell kinetics, response to dibutyrl cyclic AMP [22] and dimethylformamide [23] karyotyping and DNA flow cytometry [24], cytoplasmic and secreted proteins [25], glucose and glycogen metabolism [26], S-100 [27], glial fibrillary acidic protein [28], electrical properties [14, 16] and growth through filters [29]. Details of the methodology for each parameter are given in above references.

Scanning and transmission electron microscopy have also been of interest. For transmission microscopy small chunks of tumor removed at surgery are immersed in a fixative consisting of 4% paraformaldehyde and 5% 0.1 M cacodylate buffer (at pH 7.4). The tissue is subsequently divided into 1 mm cubes and allowed to fix for an additional 24 hours. After rinsing, the tissue is postfixed in 1% OsO_4 in 0.1 M cacodylate for 1 hour and rerinsed. Dehydration is achieved in graded methanol and propylene oxide and embedded in Epon 812. Thin sections cut on a diamond knife with an LKB Ultratome are counterstained with uranyl acetate and lead citrate on uncoated copper grids. Sections are examined on either a Phillips 200 EM or a JEOL 100 CX at 60–80 kv.

Preparation of tissue culture cells for electron microscopy is achieved by pouring warmed fixative (37°C) into the Falcon culture dish and allowing it to stand for 1 hour followed by rinse, postfixation in 1% OsO_4 in 0.1 cacodylate, rinse, rapid dehydration in methanol and Epon 812 embedding. Thin sections cut parallel to the surface of the monolayer are placed onto grids, counterstained and examined as above.

For scanning electron microscopy tissue fixed in 4% paraformaldehyde, 5% glutaraldehyde in 1.0 M cacodylate buffer at pH 7.4 is critical point dried in a Tousimas critical point dryer and sputtercoated with gold. Tissue or cells are examined on a JEOL JSM-35C microscope. When possible, the scanning images are quantified using a Bausch and Lomb FAS II image analysis system interfaced to the scanning microscope through an EM-2 module.

2.4. Chemotherapy

The BCNU stock solution is initially prepared by dissolving the drug in absolute ethanol and diluting with sterile distilled water to a concentration of 3.3 mg/ml and stored at $-80°C$. The serial dilutions are freshly prepared for each use from the stock solution, using warmed (37°C) Hams F-10 medium (GIBCO) as the diluent [12, 30].

The spontaneous decay of BCNU in aqueous solution is an important consideration in the determination of the dosage of this agent actually delivered to the cells in culture. Concentrations in the first reports of our technique (165 µg/ml, 66.0 µg/ml, 43.5 µg/ml, 33.0 µg/ml, and 22.0 µg/ml) did not take this into account. Actual decay is, of course, dependent on many factors in storage and preparation as well as actual test conditions [31, 32]. Based on the decay curve of Levin *et al.* [31] as well as DeVita [32] describing such decomposition prior to and during exposure, the concentration delivered to the cells in the microcytotoxicity plates can be estimated at 33 µg/ml, 13.2 µg/ml, 8.7 µg/ml, 6.6 µg/ml and

---

†NSC-79037; 1-(2-chloroethyl)-3-cyclohexyl-1-nitrosourea. Manufactured by Philips Roxane Laboratories, Inc., Columbus, Ohio for Bristol Laboratories, Division of Bristol Myers, Syracuse, NY.

4.4 µg/ml. Spectroscopic and chromatographic studies are being utilized to ascertain precisely the drug decay under our conditions. In control experiments the diluent is not cytotoxic at the concentrations used.

Comparison of these *in vitro* concentrations with those actually achievable physiologically in patients is difficult. By the method of Dickson and Suzanger, [33] and calculated with pharmacokinetic data, our estimate of the maximum concentration of BCNU obtainable in plasma and solid tumor tissue after intravenous administration of 100 mg/ml/m^2 BCNU was 33 µg/ml. More direct measurement using selected ion monitoring chemical ionization mass spectroscopy by Levin *et al.* [31] to study patient sera indicate that the highest levels of BCNU achievable are of the order of 1 – 3 µg/ml given a dosage of 80 mg/m^2 infused over 30 – 45 minutes. The data of Levin appear to be the more accurate of the two. Why the cytotoxic *in vitro* dose levels are higher than those achieved *in vivo* is not clear, although the sensitivity *in vitro* appears to be predictive of the clinical response [12].

For the testing of aqueous insoluble drugs, a new method of surface plating of agents in a volatile solvent has been developed. Such drugs as CCNU or AZQ may then be more easily and reliably tested. For details, see Smith, Cooke, Leonard and Kornblith, 1983 [34].

Plates are incubated for one hour in a 5% CO_2, 37°C environment; then blotted with a sterile gauze sponge to remove medium and the remaining drug in solution. The plates are washed twice with Hanks Balanced Salt Solution (HBSS, GIBCO) and flooded with medium. Approximately 20 to 24 hours later, plates are rinsed three times with HBSS, fixed with methanol for 10 minutes and stained with Giemsa. The intact cells remaining in each well can be counted in one of two ways. The first involves projecting the light microscopic image of the well onto a semi-opaque screen. The screen is marked at the position of each intact cell and the marks are automatically tabulated. Cell counting is then done by observers who are unaware of the test solutions used on the plates. In the other method, the plates are counted automatically using a Bausch and Lomb FAS-II image analysis system. Full details of this methodology are given in Smith *et al.* [30]. The great advantage of this methodology is that counting of a full plate (48 wells) is accomplished within five minutes (including all statistics, standard deviation and t values) compared to more than 1 hour per plate by hand counting. The cytotoxic index (C.I.) of each dilution is calculated by the following formula:

$$C.I. = 1 - \frac{\text{average number of cells in test wells}}{\text{average number of cells in control wells}}$$

For each dilution testing, the mean and variance for that particular set of wells and for the control sets of wells are calculated. Any observed differences in cell counts between test dilutions and the control are tested by the student t-test with the requirements of a 99% confidence level. An *in vitro* response of C.I. ≥ 0.25 at 6.6 μg/ml BCNU concentration defines a responding cell line while a C.I. < 0.25 at the same BCNU concentration defines a 'non-responder'. The dose level of 6.6 μg/ml represents an approximation to physiologically achievable concentrations.

More than seventy-five brain tumor patients' cell lines have undergone microcytotoxicity testing either at the Massachusetts General Hospital between 1974 and 1979 or at the National Institutes of Health (1979–1980). The subpopulation of 58 glioma patients reported on here is comprised of astrocytomas Grade I, II and mixed (7), astrocytomas Grade III (15), Grade III-IV (11) and Grade IV (25). The mean age for patients having an astrocytoma was 48 and the male: female ratio was nearly 1:1. The patients chosen for the correlation of *in vitro* and clinical responses were drawn from the 58 MGH glioma cases tested *in vitro* as described by Kornblith, Smith and Leonard [12]. Initial qualification for this phase of the study, as for the first phase, was growth of the patient's tumor in tissue culture through a minimum of two subcultures. Approximately 95% of the 865 glioma specimens received in our laboratory have been successfully grown in tissue culture. In addition the following criteria also had to be met: 1) Grade III, III-IV, or IV or glioblastoma multiforme, only. Low grade gliomas and other tumor types were not studied as these less malignant tumors are only rarely treated with chemotherapy and therefore clinical correlations are difficult; 2) patient survival to allow for a) postoperative radiation therapy (4500–5000 rads); b) more than two doses of nitrosourea (BCNU or CCNU) chemotherapy (doses spaced approximately six weeks apart) and c) two or more postoperative computerized tomographic (C.T.) scans.

There were 14 patients who met these criteria and were selected for comparison of the clinical and *in vitro* data. They were comprised of astrocytomas Grade III (3), Grade III-IV (5), and Grade IV (6). The average patient age in each grade was close to the overall mean in Grade III (42 vs 45) and in Grade IV (41 vs 50). The means were slightly farther apart in Grade III-IV (36 vs 54). The overall male: female ratio for the fourteen patients was 5:2. The clinical history of these patients is fully described in Kornblith *et al.* [12].

Patient data were collected from hospital records as well as individual senior physician data. Correlations of C.I. data with various parameters of patient clinical condition and therapeutic response included age, sex, tumor location, duration of symptoms prior to surgery and pretreatment condition, extent of surgery, 'Karnofsky scale' functional rating [35], and use of steroids and anticonvulsants. Also correlated were type and amount of radiation therapy and chemotherapy, change in tumor size, postoperative period prior to evidence of recurrence and survival times.

C.T. scans were examined for each patient in the study. To assess the action of chemotherapy in comparable situations all patients evaluated received radiation therapy of the same approximate dosage (average 5000 rads) and had two or more postoperative C.T. scans. Of the 14 patients used for the *in vitro*-clinical correlation, none had less than three scans and the average was four postoperative scans generally performed at three-month intervals. To determine tumor volume, measurements of all three coordinates were made at each level of the scan in which the mass was visible and the volume calculated from the best geometric fit whenever possible [12]. The independent assessment of tumor size and/or change by the neuroradiologists was compared to the laboratory estimate.

2.5. Humoral Immunology

2.5.1. Immunologic Microcytotoxicity Assay. A well-characterized tumor cell line originating from a patient with a Grade III-IV astrocytoma can be used as the allogeneic target line for the microcytotoxicity assay [5, 6, 8]. This line and the other lines used for autologous study [36] are derived from tumor tissue obtained during surgery. The tissue is minced into 1 mm^3 bits, and placed into a 75 cm^2 flask using Ham's F-10 nutrient medium supplemented with 10% Fetal Calf Serum (FCS). Thus explanted, cultures are maintained at 34°C. The medium is changed weekly until confluency is reached; thereafter the line is subcultured as needed. These lines are used for studying humoral responses by two primary methods – microcytotoxicity and immune adherence.

The microcytotoxicity assay is performed in Falcon 3034 microtest plates. Blood is drawn in a sterile manner from tumor patients and from normal blood bank donors. After each blood sample has retracted and clotted, it is centrifuged at 1000 × g, and the serum aliquoted and stored at −70°C. Additive-free pooled human serum (PHS) from fasting adults (Miles Laboratories) is used as the control serum. A flask containing the desired target cells is washed twice with Hank's Balanced Salt Solution (HBSS), Ca^{++} and Mg^{++} free and incubated with 1 ml of trypsin-ethylenediamine tetraacetic acid (EDTA) to remove the cells from the flask. After 10 minutes of incubation, cells are counted in a hemocytometer and the suspension brought to the desired dilution by addition of F-10 with FCS. A Terasaki syringe is used to deliver 10 µl aliquots to each well of the plate. Plates are ready for experimentation after 18 hours of incubation in a 5% CO_2 humidified atmosphere. They are then washed twice with Hanks BSS and blotted with sterile gauze sponges pressed into the plates with a sterile nylon roller. A Hamilton microtiter syringe is used to deliver 10 µl to each well: two rows of six wells each receive PHS as a control and one row of six wells receives test serum. Serum is blotted after an hour's incubation in the CO_2 chamber and then 10 µl of complement preparation is added to each well. The complement source is from sera drawn from six-week old New Zealand rabbits by cardiac puncture. This is mixed with the appropriate amount of serum obtained from the human umbilical cord at delivery. After two hours of incubation with complement, plates are flooded with 8 ml of F-10 with FCS (with penicillin and streptomycin) and replaced in the incubator for an additional 18–20 hours. Plates are then rinsed with Hank's, fixed with methanol, and stained with Geimsa. The remaining cells are microscopically counted and the amount of killing (Cytotoxic Index) is expressed as for the chemotherapy assay (See 2.4 above). A positive result is possible when cytotoxicity in the experimental wells differs significantly from control wells $p \leq .05$, as determined by the student t-test.

2.5.2. Immune Adherence Assay. The Immune Adherence Assay has been adapted from Tachibana *et al.* [37] for use in the same Falcon 3034 microtiter plates. Cells are trypsinized into plates 18 hours prior to the experiment as described above [38].

Serum titrations are performed on ice with Veronal Buffered Medium with 5% Fetal Calf Serum (VBM-FCS). Nutrient medium is decanted from the plates which are then washed twice with HBSS. The residual solution is removed from the plates by blotting with an absorbent gauze sponge. Each well receives 10 μl aliquots of titrated serum, is incubated at 37°C for one hour and then washed twice more in HBSS and blotted. Pooled human serum and VBM-FCS are used as controls. Erythrocyte indicator solution with 1/30 fresh guinea pig serum in VBM is delivered in 10 μl volumes to each well using a Terasaki syringe. After a further 30 minutes at 37°C, in a humidified 5% CO_2 atmosphere, plates are inverted and gently tapped for 30 minutes. They are then washed with HBSS until all nonspecifically-adherent red blood cells (RBC) have been removed, as judged by the appearance of control wells, and evaluated by phase-contrast microscopy using a counting grid. A target cell is considered positive if at least 3 RBC's are attached. The percentage of positive cells per well is calculated only when control wells are less than 5% positive. Duplicate titration series are performed whenever test serum volume permits and duplicate or triplicate controls are performed. For the correlation of *in vitro* and clinical data for both the immune microcytotoxicity and adherence assays, postoperative survival data are taken from patient records and survival plots are constructed.

2.6. Cellular Immune Assays

Blood from tumor patients and from normal blood bank donors is drawn into syringes containing preservative-free heparin, 5 units/ml. Mononuclear cells are isolated by centrifugation on discontinuous Ficoll-Hypaque gradients [39] followed by centrifugation on discontinuous sucrose gradients [40] to remove contaminating platelets. The final cell pellets are resuspended in culture medium consisting of a 1:1 mixture of RPMI-1640 and Dulbecco's modified Eagle's medium supplemented with 5% human AB serum, 5×10^{-5}M 2-mercaptoethanol, nonessential amino acids, 2 mM L-glutamine, 100 units/ml penicillin, and 100 µg/ml streptomycin. Peripheral blood mononuclear cells prepared in this manner are assayed for: 1) their ability to lyse autologous and allogeneic glioma cells, 2) their ability to make a cytolytic lymphocyte response against allogeneic lymphocytes in mixed lymphocyte cultures (MLC) and against autologous or allogeneic gliomas in mixed lymphocyte-tumor cultures (MLTC), 3) their ability to respond to the mitogens concanavalin A (Con A), phytohemagglutinin (PHS), and pokeweed mitogen (PWM) and 4) natural killer (NK) activity.

Glioma lines are derived and maintained as described above. Freshly trypsinized glioma cells which are to be used as targets in cytolytic assays are incubated with 100 µCi sterile ^{51}Cr-sodium chromate in 0.5 ml culture medium for 1 hour at 37°C. The cells are then washed three times and resuspended at a final concentration of 5×10^4 cells/ml. One-tenth ml aliquots of this suspension are added to the wells of Costar 3696 'half-area' microtest plates. The plates are incubated at 37°C for 18 hours in a humidified atmosphere of 5% CO_2 in air to permit the cells to adhere and spread on the floors of the wells. The culture medium is then aspirated from each well and replaced with 0.1 ml of lymphocyte suspension. Lymphocytes are incubated with glioma cells for 24 hr at 37°C. At the end of this time, 0.05 ml of culture medium is withdrawn from each well, and the amount of ^{51}Cr in each sample is measured in a gamma counter. The total amount of ^{51}Cr in the glioma cells is determined in controls in which the glioma cells are lysed by the addition of 1% sodium dodecyl sulfate. The percent specific ^{51}Cr release from glioma cells incubated with lymphocytes is calculated as $[(e - c)/(100 - c)] \times 100$ where e is the percentage of ^{51}Cr released from glioma cells incubated with lymphocytes and c is the percentage of ^{51}Cr released spontaneously from glioma cells alone. Spontaneous ^{51}Cr release usually varies from 20 to 35% of the total releasable ^{51}Cr in 24 hours.

Lymphocytes cytotoxic for allogeneic lymphocytes or glioma cells are generated by incubating responder lymphocytes with stimulator lymphocytes and/or glioma cells in 2 ml culture medium in the wells of Costar 3524 tissue culture plates. Stimulator lymphocytes receive 2000 R of gamma irradiation prior to culture and glioma cells receive 10,000 R of gamma irradiation. Responder and stimulator lymphocytes are each cultured at a density of 2×10^6 cells/well. Optimal glioma cell density has been found to be $0.5 - 1 \times 10^5$ cells/well for the cell lines studied thus far. Responder lymphocytes are harvested from the wells after 7 – 12 days and assayed for their ability to lyse ^{51}Cr-labeled glioma or lymphocyte targets. Lymphocytes to be used as targets in cytolytic assays are incubated with PHA-L (E-Y Laboratories) for 3 days and with ^{51}Cr-sodium chromate for 18 hr prior to the assay.

To assay for mitogen responsiveness, 2×10^5 mononuclear, cells are mixed with Con A, 40 µg/ml (Miles Laboratories), PHA-L, 1 µg/ml (E-Y Laboratories), or PWM, 1% (Grand Island Biological) in 0.15 ml culture medium in the wells of Falcon microtest II plates. Control wells receive mononuclear cells without mitogen. The plates are incubated for 48 hr at 37°C, and then 0.25 μCi ^{3}H-thymidine is added to each well. The plates are incubated for an additional 18 hr, after which the contents of the wells are harvested onto glass-fiber filters by means of a Titertek multiple sample harvester. The filters are dried, and the amount of ^{3}H-thymidine which has been incorporated into DNA and thus retained on the filters is measured by means of a liquid scintillation counter.

NK activity is assessed by measuring the ability of peripheral blood mononuclear cells to lyse cells of the K562 line, which was originally derived from a patient with chronic myelogenous leukemia in blast crisis [41]. This cell line has been shown to be highly susceptible to the cytolytic action of human NK cells [42]. K562 cells are labeled with ^{51}Cr-sodium chromate in a manner similar to that described above for glioma cells, and 0.1 ml aliquots of cell suspension containing 2×10^4 labeled K562/ml are added to the wells of Linbro 76−311−05 round-bottom microplates. To these are added 0.1 ml aliquots of mononuclear cell suspensions from which adherent cells have been removed by incubation on plastic tissue culture dishes for 30 minutes at 37°C. The microplates are centrifuged at $500 \times G$ for 5 minutes at room temperature, and then incubated for 4 hr at 37°C. At the end of this time supernatants containing released ^{51}Cr are harvested using a Titertek supernatant collection system, and radioactivity is measured with a gamma counter. Percent specific ^{51}Cr release is calculated as above.

3. RESULTS

3.1. Biological Characterization

The biological characterization of human glioma-derived cells in tissue culture is, of necessity, a multifactorial process of which space here permits only a few general statements. Gliomas, even of the same pathological grade, present a rather heterogeneous picture of cell properties, paralleling the heterogeneity of their clinical behavior. In general, however, the more malignant the tumor, the more quickly it establishes itself in tissue culture. The selective process, determining what cells of the original mix come to predominate is, of course, not understood and it is not clear the extent to which all of the original cell types or variants of the intact tumor are represented. In addition, the more malignant as compared to more benign CNS tumors (such as glioblastoma vs. meningioma) show a more erratic pattern of cell doubling times. The more malignant lines tend to show less well-differentiated cell morphology, multinucleate-giant cell patterns, variable DNA content (from normal to tetraploid and aneuploid) and ability to be maintained in culture more or less indefinitely (well beyond the 15th passage). Drift in chromosomal pattern also has been reported [43] and may be an important feature of malignancy. It may also, however, result from diverse pressures in tissue culture. Finally, ability to grow in soft agar and to grow through the pores of nucleopore filters may be useful additions to the malignancy characterization [29].

At the level of electron microscopy, the major features indicative of malignancy coincide with those of the light microscopic studies. In addition, surface membrane features, in particular microvilli, indicate neoplastic growth properties, although not invariably so. Other features which may be indicative of malignancy include nuclear shape, irregularity and chromatin patterns; relative paucity of actin cables; and relative frequency of surface bleb formation. Most other features are quite variable and, at least to this point, have not been found to be very helpful.

From the histochemical and biochemical point of view a great many features have been studied both in this and other laboratories. Glial fibrillary acidic protein appears to be one of the most useful markers for glial cells at present. This can be applied to both the solid tumor mass as well as to the tissue culture explants and cells [28]. The major difficulty is that there is not a clear correlation between degree of malignancy and presence or absence of GFAP, let alone the meaning of its presence or absence in culture [44]. Certainly, when positive, GFAP assays are useful in identifying glial derived cells. S-100 is another marker of interest. In our hands, S-100 has been most readily identified in the benign acoustic schwannomas [45]. In gliomas, correlation with malignancy is currently not possible with any degree of certainty.

For DNA analysis we have used flow cytometric techniques which, of course, can indicate relative proportions of the population in cell cycle stages leading to mitosis and so provide a measure of malignancy. For finer resolution more study is required.

Morphological responses to the administration of dibutyrl cAMP and dimethylformamide are also of some interest. In our experience well over 50% of glioma-derived cell lines respond to db-cAMP by undergoing a morphological change involving shrinkage of the cell soma and the formation of long, branched processes [22, 23]. In addition there is a decreased rate of cell growth that is maximally of the order of 30%. Both malignant and benign glioma-derived cell lines show this response. Frequently, however, the lower grade, glioma-derived lines show a more differentiated astrocytic morphology prior to db-cAMP treatment. With DMF, a different morphological change takes place – transformation to a bipolar cell shape. This occurs in 75% of tumors in our experience. More will have to be done to determine whether there is any significant correlation with the degree of malignancy. Of interest is the fact that the morphological responses of the db-cAMP- and DMF- treated glioma-derived cells are not seen in control fibroblasts from the same or other patients. With regard to electrical properties, and most specifically membrane properties, microelectrode techniques have been applied to single glioma cells in culture [14, 16]. The resistance and capacitance of the glioma-derived cells clearly mark them as of glial origin. Correlation with tumor grade is not apparent.

Finally, a new area being developed in conjunction with the positron emission computed tomographic scanning of gliomas in patients [46] is that of glioma cell glucose/glycogen metabolism [26]. The hypothesis here has been that despite the tissue culture environment, some critical aspects of glioma cell metabolism should remain constant and qualitatively like those in the host. This, in fact, appears to be true for the overall rate of glucose consumption. The average glucose uptake rate for Grade III tumors is 2.50 and for Grade IV tumors, 2.42 m moles/min/mg protein, so that there is no significant difference between these grades. Work is proceeding to attempt to differentiate Grade I-II from III-IV by this means.

3.2. CSF Tissue Culture

There were 109 cerebrospinal fluid (CSF) specimens from patients with varying neurological disorders which were incubated in tissue culture medium for 1, 3, and sometimes 7 days. Strict criteria for malignancy were applied to cells found at these intervals. In 35 patients with verified central nervous system neoplasms, eight cases had malignant cells and 11 others had 'doubtful' cells by tissue-culture analysis (Table 1). Thirty-three of these cases were also examined with standard millipore cytological techniques: six had malignant cells and four had 'doubtful' cells. Of 50 cases with inflammatory or other non-neoplastic conditions, cells were cultured in 13. None was considered malignant by our criteria. Tissue culture of CSF has several potential benefits. Even with stringent criteria, it is possible to demonstrate the presence of unequivocally malignant cells in CSF by tissue culture. The systematic application of such criteria may eventually increase the positive identification of malignancies. Further, since these cells are growing, the degree of malignancy may be more accurately determined by a study of growth in culture. Such a study cannot be done by conventional methods. Finally, tissue culture can help to guide therapy in certain instances in which a surgical biopsy cannot be obtained.

3.3. Chemotherapeutic Microcytotoxicity Assays

For the 42 glioma cell lines reported here, 35 (70%) had a significant cytotoxic index (CI) in response to BCNU at 6.6 μg/ml in the aqueous microcytotoxicity assay. The majority of the glioma lines tested (78%) were Grade III, III-IV, and IV (glioblastoma multiforme) and most of these (80%) were 'responders' to BCNU. Pathological grading does not have a strict relationship to *in vitro* microcytotoxicity C.I. with a wide variation in sensitivity noted at each tumor grade.

For the 14 patients meeting all the criteria outlined in 'Materials and Methods'

Table 1. A comparison of CSF tissue culture and millipore filtration cytology in 35 neoplasms with confirmed diagnoses

Tumor type	No. of specimens	Millipore filtration			CSF tissue culture		
		Neg.	Doubtful	Pos.	Neg.	Doubtful	Pos.
Astrocytomas							
solid mass	6	4*	1	0	3	1	2
with gliomatosis	2	0	1	1	0	2	0
Medulloblastoma	6	4*	0	1	3	2	1
Reticulum cell sarcoma	6	4	2	0	2	2	2
Pineal region neoplasm	4	3	0	1	3	1	0
Metastatic neoplasms							
solid mass	2	1	0	1	0	1	1
with meningeal carcinomatosis	2	0	0	2	0	0	2
Miscellaneous (meningioma, craniopharyngioma, acoustic neuroma, chordoma)	7	7	0	0	5	2	0
Total	35	23	4	6	16	11	8

*One specimen was unsatisfactory for millipore cytological examination.
Reproduced with the permission of J. Neurosurg. [10].

above, there were 3 Grade III astrocytomas, 5 Grade III-IV astrocytomas, and 6 Grade IV astrocytomas. In Table 2 the correlation of C.I. and its corollary, *in vitro* 'response' (C.I. $\geq$ 0.25) or 'non-response' with change in size of tumor on C.T. scan is given [12].

All three astrocytoma Grade III cell lines responded *in vitro*. Serial C.T. examination showed two of the three tumors to clearly decrease in size (approximately 20 – 50%), while in the remaining patient (No. 9), no evidence of tumor could be found. Three postoperative C.T. scans over a period of one year failed to show tumor presence in this patient, although during the second year, there was CT evidence of increasing mass of tumor. The patient died approximately at the end of the second year. At the time of review patients No. 4 and 38 were alive.

In the astrocytoma Grade III-IV group there were two responders. Patient No. 2 (C.I. = 0.54) was asymptomatic more than five years after surgery. However, recent C.T. scans show recurrence of tumor. Prior to this recurrence his tumor had shown a decrease on C.T. scan that continued for over 45 months.

Among the three non-responders, there was a good correlation of *in vitro* C.I. with clinical progress. Two patients showed a clear increase while one showed no change. Patient No. 51, at 14 months from surgery, had residual but unchanging tumor on his C.T. scan. The other two had low *in vitro* C.I.'s (<0.20) and exhibited a clear increase in tumor size on C.T. scan. Patient No. 22 had evidence of significant recurrence two months after surgery and required reoperation. Patient No. 37 remained asymptomatic for 12 months following surgery. Tumor recurred within this period, however, as shown by C.T. scan taken seven months after the completion of radiotherapy.

Table 2. Chemotherapy responses: *in vivo – in vitro* correlations

Astrocytoma Grade	Case number*	C.I.	Resp/non-resp.	C.T. scan
Astrocytoma Grade III	4	.67	R	D
	9	.51	R	no evid. tumor
	38	.30	R	D
Astrocytoma Grade III-IV	2	.77	R	I
	22	.05	NR	I
	37	.09	NR	I
	44	.54	R	D
	51	.19	NR	no change
Astrocytoma Grade IV	8	.52	R	D
	12	.29	R	I
	20	.24	NR	I
	25	.12	NR	I
	35	.80	R	I
	40	.53	R	D

* = Refer to Table 1; D = Decrease in tumor size on C.T. scan; I = Increase in tumor size on C.T. scan; R = Response *in vitro*, C.I. > 0.25; NR = No response *in vitro*, C.T. $\leq$ 0.25.
Reproduced with the permission of *Cancer* [12].

Within the Grade IV category, two of the four responders showed a decrease in tumor mass on serial C.T. scans and two showed an increase. The two patients (No. 8, 40) whose tumors decreased clinically had similar *in vitro* C.I.'s (approximately 0.50). Patient No. 8 died 22 months following surgery. Patient No. 40, currently 19 months out of surgery, does not show clear evidence of recurrence on C.T. scan, but is suspected of developing recurrent tumor. *In vitro* C.I. varied considerably for the two responders whose tumors increased clinically. Patient No. 12 (C.I. = 0.29) died 18 months after surgery while patient No. 35 (C.I. = 0.80) died 12 months after surgery of recurrent tumor.

The remaining two patients with Grade IV astrocytomas did not respond to chemotherapy either *in vitro* or *in vivo*. Patient No. 25 had evidence of recurrence six months out of surgery and died 11 months later. Patient No. 20 had a very slowly growing tumor. Eight years after presenting with a seizure disorder his first operation was performed. During the next seven years he had four further operations. Following his second reoperation, he received radiation and chemotherapy. He died 87 months after initial surgery, having functioned at a 90% Karnofsky level for most of this period.

Overall, as indicated by Table 2, six of nine 'responses' detected *in vitro* could be correlated with C.T. scan documented decrease in tumor mass. All five non-responding cell lines were correlated with increasing tumor mass or no change despite nitrosourea therapy. The addition of a new assay designed to enable the screening of aqueous insoluble drugs has substantially broadened the capability of our chemotherapy testing program [34]. Comparison of the BCNU and AZQ (aziridynlbenzoquinone) data achieved with the assay is of some interest. BCNU decays rather quickly under conditions of surface plating, whereas AZQ does not (or at least it has toxic metabolites). In addition, if the surface assay microtiter plates are followed for periods of 72 hours and 168 hours after plating the cells on the drug to be tested, there is, at least for some lines, increased killing of doses of AZQ not toxic at 24 hours, whereas there is no such shift for BCNU. A third drug of interest, spirohydantoin, shows no significant cytotoxicity except at long periods (72 − 168 hrs) of drug exposure. Thus, several drug characteristics as well as variations in sensitivity of the glioma-derived cell lines can be tested in this assay [34]. Twenty glioma lines have now been tested against AZQ and correlations are being made in conjunction with a pilot clinical trial of AZQ.

3.4. Immunologic Microcytotoxicity Assay

In a series of allogeneically studied sera, the majority of astrocytoma patients (81%) reacted with the common target cell line derived from a Grade III-IV astrocytoma. The sera from the lower grade astrocytomas reacted at a higher rate than those from the higher grade tumors (Table 3A) [47]. This variability in response rate with tumor grade is more notable in autologous testing where 67% of Grade I-III tumor patients (15) showed a positive response compared to 10%

Table 3. Incidence of humoral immunity

A. Allogeneic testing	No. positive	No. tested	% Positive	
Gliomas	103	127	81	
Grade I	13	15	87	
Grade II	12	14	86	
Grade III	33	42	79	
Grade IV	45	56	80	
Normal blood bank donors	6	65	9	
Non-neoplastic CNS disease	1	10	10	
Other neuroectodermal tumors	24	44	55	
B. Autologous testing	No. positive	No. tested	% Positive	Average age
Gliomas	11	25	44	48
Astrocytoma (Grades I-III)	10	15	67	40
Glioblastoma	1	10	10	61

for glioblastoma (Grade IV) patients (10) (See Table 3B). In the autologous microcytotoxicity studies (Table 3B) 44% of all glioma patients (11 of 25) tested had positive cytotoxic responses to their own tumor cells in culture [9, 13]. Ten autologous fibroblast lines tested from brain tumor patients showed no positive responses. Cytotoxicity of tumor cells was present most often in preoperative or intraoperative sera.

In allogeneic testing absorption with platelets did not remove reactivity. Testing against normal adult glia showed less than a 10% rate of response and no response was observed versus fetal glial cells. Sera from normal patients were significantly cytotoxic in only 9% of the cases and sera from patients with non-neoplastic CNS disease were correspondingly low (10%). Cross-reactivity with other neuroectodermal tumors against the target line is seen in various sera. By immune adherence technique, 50% of autologous lines tested had positive responses [38]. These responses remained even after absorption with normal brain or autologous fibroblasts and did not correlate directly with survival or with the corresponding autologous cytotoxic responses.

When positive serological microcytotoxicity responses were compared with patient survival, there was a positive correlation, although its statistical significance is unclear [13]. Age of the patient also correlated with positive response in that older patients were less likely to have a positive response to their tumors. Positive immune adherence testing was not predictive of survival.

3.5. Cellular Immune Assays

Studies on the cellular immune status of glioma patients are still at an early stage, and the following represents a summary of our initial observations. Freshly isolated peripheral blood lymphocytes from each of 6 glioma patients failed to lyse autologous cultured glioma cells to an extent significantly greater than normal control lymphocytes when assayed at lymphocyte: glioma ratios as high as 100:1. Likewise, lymphocytes from 6 of 7 glioma patients failed to lyse allogeneic gliomas. Peripheral blood lymphocytes from one patient with a glioblastoma were found to cause substantial lysis of 3 of 3 allogeneic cultured glioma lines and 1 of 2 allogeneic colonic adenocarcinomas or an allogeneic erythroleukemia. Lymphocytes from the glioma patient lysed neither autologous fibroblasts nor autologous cultured glioma. The lytic activity against allogeneic glioma cells was shown to be mediated by T lymphocytes, not by natural killers, and was maintained at essentially constant levels during serial assays over 10 months. The nature of the antigen recognized by this patient's lymphocytes and whether or not this reactivity is related to the patient's disease are unknown. However, similar reactivity was not observed for lymphocytes from any of the 34 normal donors tested thus far. Natural killer activity was normal or slightly elevated in each of 3 patients in which this cellular immune function was examined.

All patients who were studied early in the course of their disease showed only slight or no depression of lymphocyte function as measured by the ability of their lymphocytes to respond *in vitro* to mitogens (8 patients) or to allogeneic lymphocytes (7 patients). Lymphocytes from 6 of 7 patients demonstrated mildly depressed responsiveness to PHA (responses 20 to 50% decreased as compared to normal controls assayed simultaneously). In contrast, lymphocyte responsiveness to Con A and to PWM was normal for 5 of 8 and 4 of 5 patients, respectively. Two patients who were studied serially to the time of their deaths showed declining, and in one case virtually absent, mitogen responsiveness as their disease progressed. Other investigators [48 – 51] have demonstrated the presence of a factor in the serum of many glioblastoma patients which nonspecifically inhibits lymphocyte responses to mitogens and to allogeneic lymphocytes *in vitro*. The level of inhibitory activity was reported to be correlated with the clinical status of the patients [49, 51]. We are initiating studies to examine our patients for the presence of this inhibitory factor.

The above observations indicate that despite the normal or near-normal capability of our patients' lymphocytes to respond to mitogens and to generate cytolytic lymphocyte responses to allogeneic lymphocytes *in vitro*, these patients did not make strong cytolytic lymphocyte responses to their own tumors *in vitro*. Several authors have reported that lymphocytes from glioma patients are weakly cytotoxic toward autologous tumor *in vitro* in assays using higher lymphocyte: glioma ratios and longer incubation times than we have used [52]. However the specificity and meaning of such long term cytotoxicity assays have come under serious question [53, 54]. To further examine possible mechanisms by which gliomas escape cell-mediated immune attack, we have studied the abilities of several glioma lines to elicit allogeneic cytotoxic lymphocyte responses *in vitro*. Results of these experiments are summarized below.

Five of 8 glioma lines were unable to stimulate allogeneic cytolytic T lymphocyte (CTL) responses in mixed lymphocyte-tumor cultures *in vitro*. Analysis of the reasons for the failure of these lines to induce CTL generation revealed three separate mechanisms of importance. In the case of two of the non-stimulatory glioma lines, CTL specific for the gliomas could be generated if responding lymphocytes were cultured with glioma cells in the presence of irradiated lymphocytes from a third individual. Hence, these two lines possessed a defect in immunogenicity which could be overcome by 'help' from an allogeneic mixed lymphocyte reaction [55]. The nature of this 'help' is under further study. One glioma, when added to a mixture of lymphocytes from two allogeneic individuals inhibited the mixed lymphocyte reaction which otherwise would have occurred. Inhibition was shown to be due to the secretion of a macromolecular, nonspecific immunosuppressive substance(s). Two of the non-stimulatory gliomas were shown to secrete a thick coat of mucopolysaccharide which impeded contact between lymphocytes and glioma cells. In the case of one glioma line which was studied extensively, removing this coat with the enzyme hyaluronidase permitted increased generation of CTL specific for the glioma. However, such CTL generation required the presence of both responder lymphocytes and irradiated lymphocytes from a third individual. Hence, the inability of this glioma line to elicit cytolytic lymphocyte responses represented the combined effects of both the presence of a protective mucopolysaccharide coat and a separate defect in immunogenicity.

In summary, the five glioma lines which failed to stimulate allogeneic CTL responses each possessed one or more of the following escape mechanisms (Table 4): 1) an intrinsic defect in immunogenicity, the nature of which is as yet poorly defined, but which may involve a defect in the ability to stimulate helper T cells, 2) the secretion of nonspecific immunosuppressive substance(s) and 3) the production of a protection mucopolysaccharide coat. Three patients who received specific active immunotherapy consisting of one or more injections of irradiated autologous glioma cells failed to make any detectable cytolytic lymphocyte response to these tumors *in vivo*. One of the glioma lines possessed the defect in immunogenicity while a second secreted large amounts of immunosuppressive substance. The third glioma could not be grown in sufficient quantity to permit detailed characterization *in vitro* in addition to its use in immunotherapy *in vivo*. It is thus apparent that further studies of how these escape mechanisms operate and by what means they may be overcome are likely to be of central importance to the successful use of immunotherapy in glioma patients. In addition, characterization of the ability of individual glioma lines to interact with the

Table 4. Properties of cultured glioma lines affecting their abilities to stimulate cytolytic T lymphocyte (CTL) responses *in vitro*

Glioma line	Grade	Directly stimulates CTL response	Stimulates CTL response with 'Help' from allogeneic MLR[1]	Production of immunosuppressive substance	Presence of mucopolysaccharide coat
JM	IV	Yes		N.T.[1]	No
AB	III	Yes		N.T.	Small
LH	II-III	Yes		N.T.	Small; not on all cells
NN	III	No	Yes	N.T.	No
LM	IV	No	Yes	N.T.	Large
RB	IV	No	Yes	Yes	No
GB	IV	No	N.T.	Yes	No
FG	II	No	N.T.	N.T.	Large

[1]MLR = mixed lymphocyte reaction; N.T. = not tested.

cellular immune system *in vitro* may contribute to the planning of individualized immunotherapeutic regimens.

4. DISCUSSION

The potential significance of *in vitro* biological characterization and chemotherapeutic or immunologic *in vitro* assays capable of predicting the biological behavior and response or non-response of a given patient's glioma to a particular chemotherapy agent or the state of the patient's own immune response to that tumor seems clear. Studies in our laboratory of cell biological, chemotherapy and immunologic assays have demonstrated that the data generated can be related to the *in vivo* patient situation and prognosis. Although much more correlation data need to be collected, the results to date support the use of *in vitro* methods to supplement standard screening, diagnostic and neuropathological techniques.

The demonstration of the practicality of such assays is in itself important. In 1978 Kornblith and Szypko [4] reviewed the history of *in vitro* chemotherapy sensitivity testing and factors limiting *in vitro* testing of chemotherapeutic agents on human solid tumors. The latter included 1) the difficulty of establishing lines from many human solid tumors; 2) difficulties in cellular characterization; 3) slow rate of growth even in successful cases so that observations are too late in the clinical course to be of any value and 4) problems inherent in the extrapolation of *in vitro* data to the clinical situation. Each of these problems now has one or more answers. For human gliomas we have had a greater than 95% success rate in establishing monolayer cultures in the more than 850 such tumors received to date. It is furthermore possible to characterize these cultures as to their cellular origin using, with some confidence, a panel of morphological, ultrastructural, biophysical and biochemical parameters as documented in this and other laboratories [14 – 17, 44]. Improvements in these methods will, of course, continue to come. Finally, data now have begun to show that such *in vitro* studies can be useful in the prediction of clinical response and in modification of therapy [3, 12, 56].

The cellular and humoral immunological assays have also been made possible by many of the same factors. Here it has been important to demonstrate that the response is specific (indeed, the autologous positive response is restricted to the tumor cells only and not fibroblasts) and that the humoral or cellular cytotoxic response is present preoperatively as well as postoperatively. In addition, the frequency of response in a normal blood bank population is very low.

The potential usefulness of *in vitro* assays that are predictive of clinical responsiveness goes well beyond the simple testing of available agents. With the better understanding of the biology of glioma cells that is possible *in vitro* (i.e., membrane properties, cytoplasmic and nuclear metabolic defects and proliferation control), it should be possible to design more rational therapeutic agents that are based on combinations of cytotoxicity, differentiation control and improved detectability by the immune system or target cell sensitivity to immune lysis through surface antigenic expression. The cellular immunological studies have begun to elucidate the mechanisms by which glioma cells can escape immune detection and destruction. Such studies should suggest possible ways to defeat these escape mechanisms and so improve patient immune response. If the molecular basis of response or non-response to BCNU or other chemotherapy agents can be elucidated, then ways to avoid the defenses of non-responding cell populations may be devised.

The fact that previous clinical studies of nitrosourea effectiveness have had mixed results is not too surprising in light of the knowledge that the responding and non-responding populations were not separated [57 – 62]. Even without such separation, the most recent data from the EORTC Brain Tumor Cooperative Group (which found a significantly increased survival with BCNU and CCNU [63]) support a role of the nitrosoureas in glioma therapy. The Brain Tumor Study Group data [64], however, underscore the need for new, perhaps multiagent therapy.

On the immunological side, the fact that 44% of glioma patients can mount cytotoxic humoral immune responses to their own tumor cells and that such immunological responses can be correlated with survival, suggests that attempts to enhance the humoral immune response in patients may be worthwhile and such efforts at immunotherapy based on the *in vitro* are now underway in at least two centers in the U.S. and one in Great Britain. Patients with highly malignant tumors may have fixed or circulating antigen [65], or possibly their immune responses are less competent. The observations that glioma patients do not make strong cytolytic lymphocyte responses to their tumors *in vivo* and that the majority of gliomas are unable to elicit allogeneic cytolytic lymphocyte response *in vitro* suggest that efforts to modify these tumors so as to increase their immunogenicity while blocking the production of immunosuppressive factors and of mucopolysaccharide coats may be important if immunotherapy is to be successful. The cellular immune mechanisms beginning to be elucidated hold substantial promise of providing new immunotherapeutic modalities.

The present *in vitro* chemotherapy and immunological assay systems (based on monolayer tissue culture techniques) are not necessarily the optimal or the only predictive assays. Rosenblum *et al.* [2, 66] have developed *in vitro* colony formation assay that is thought to study the cells most likely to proliferate in the *in vivo* tumor, i.e. those with ability to form clones. They have found greater clonogenic capacity for malignant gliomas than for Grade I, II astrocytomas. Salmon *et al.* [3] are also using a colony formation assay to determine *in vitro* sensitivity to chemotherapeutic drugs (and have attempted to correlate their *in vitro* data with clinical response). Tumor cells from individual patients are prepared as suspensions, treated with each drug and plated in agar. Drug effect is measured as reduction of colonies relative to control colonies. These investigators and their collaborators have now studied the chemotherapeutic sensitivities of a wide range of human solid tumors. Clinical correlation studies have indicated that patients who showed resistance *in vitro* do not show *in vivo* tumor responses to the clinical agent. Our limited data parallel these data quite well with all five of our non-responders showing no sign of reduction of tumor size and two-thirds of the *in vitro* responders showing a clinically apparent reduction in tumor size.

The data summarized in this chapter support the idea that *in vitro* biological studies, sensitivity testing to various chemotherapeutic agents and *in vitro* assays of patient humoral and cellular antiglioma responses may be useful in the study, management, and improvement of therapy for glioma patients. Ultimately, they offer the promise of assisting in the development of highly individualized, optimal chemotherapeutic and immunotherapeutic treatment programs for brain tumor patients.

ACKNOWLEDGMENTS

The work described in this report was carried out in the Surgical Neurology Branch, National Institute of Neurological and Communicative Disorders and Stroke, National Institutes of Health, Bethesda, Maryland and the Massachusetts General Hospital, Harvard Medical School, Boston, Massachusetts. Partial support for these investigations was provided by Grants CA 07368 and CA

22613, National Cancer Institute, DHHS, the June Rockwell Levy Foundation, Inc., Dreyfus Medical Foundation, Sherman Fairchild Foundation, Inc. and by gifts from the friends and family of Albert H. Bartlett. We wish to acknowledge the invaluable assistance of Sandra Crum, Eleanor Frishman, and Sally Morris in the preparation of this manuscript.

REFERENCES

1. Holmes HL, Little JM: Tissue culture microtest for predicting response of human cancer to chemotherapy. Lancet 2:985 – 987, 1974.
2. Rosenblum ML, Knebel KD, Vasquez DA, Wilson CB: Brain tumor therapy: Quantitative analysis using a model system. J Neurosurg 46:145 – 154, 1977.
3. Salmon SE, Hamburger AW, Sochnlen B, Durie BGM, Alberts DS, Moon TE: Quantitation of differential sensitivity of human-tumor stem cells to anticancer drugs. N Engl J Med 298:1321 – 1327, 1978.
4. Kornblith PL, Szypko PE: Variations in response of human brain tumors to BCNU in vitro. J Neurosurg 48:580 – 586, 1978.
5. Kornblith PL, Dohan FC Jr, Wood WC, Whitman BO: Human astrocytoma: Serum-mediated immunologic response. Cancer 33:1512 – 1519, 1974.
6. Kornblith PL: Complement-mediated microcytotoxicity of adherent tissue-cultured cells. In: Serologic analysis of human cancer antigens, Rosenberg S. (ed), Academic Press, Inc., New York, 1980, pp 591 – 594.
7. Pfreundschuh M, Shiku H, Takahashi T, Veda R, Ransohoff J, Oettgen HF, Old LJ: Serological analysis of cell surface antigens of malignant human brain tumors. Proc Natl Acad Sci 75:5122 – 5126, 1978.
8. Wood WC, Kornblith PL, Quindlen EA, Pollock LA: Detection of humoral immune response to human brain tumors. Cancer 43:86 – 90, 1979.
9. Coakham HB, Kornblith PL: Serologic analysis of solid tumor antigens: Immune adherence studies on human gliomas. In: Serologic analysis of human cancer antigens, Rosenberg S. (ed), Academic Press, Inc., New York, 1980, pp 65 – 90.
10. Black PM, Callahan LV and Kornblith PL: Tissue cultures from cerebrospinal fluid specimens in the study of human brain tumors. J Neurosurg 49:697 – 704, 1978.
11. Kornblith PL, Pollock LA, Coakham HE, Quindlen EA, Wood WC: Cytotoxic antibody responses in astrocytoma patients: an improved allogeneic assay. J Neurosurg 51:47 – 52, 1979.
12. Kornblith PL, Smith BH, Leonard LA: Response of cultured human brain tumor to nitrosoureas: Correlation with clinical data. Cancer 47:255 – 265, 1980.
13. Coakham HB, Kornblith PL: The humoral immune response of patients to their gliomas. Acta Neurochir, Suppl 28:475 – 479, 1979.
14. Black PMcL, Kornblith PL: Biophysical properties of human astrocytic brain tumor cells in cell culture. J Cell Physiol 105:565 – 570, 1980.
15. Lightbody J, Pfeiffer SE, Kornblith PL, Herschman H: Biochemically differentiated clonal human glial cells in tissue culture. J Neurobiol 1:411 – 417, 1970.
16. Trachtenburg MC, Kornblith PL, Hauptli J: Biophysical properties of cultured human glial cells. Brain Res 38:279 – 298, 1972.
17. Kornblith PL: Role of tissue culture in prediction of malignancy. Clin Neurosurg 25:346 – 376, 1978.
18. Madoff S, Paches WN: Mycoplasma and the L forms of bacteria. In: Rapid diagnostic methods in medical microbiology, Graber C (ed), Baltimore, Williams and Wilkins, 1970.

19. DelGuidice RA, Hopps HE: Microbiological methods and fluorescent microscopy for the direct demonstration of mycoplasma infection of cell cultures. Proceedings of the Institute for Medical Research Workshop on Mycoplasma Infection of Cell Cultures (In press).
20. Kline TS: Cytological examination of the cerebrospinal fluid. Cancer 15:591 – 597, 1962.
21. Barker BE, Sanford KK: Cytologic manifestations of neoplastic transformation in vitro. J Natl Cancer Inst 44:39 – 63, 1970.
22. Smith BH, Liszczak T, Pleasants ER, Kornblith PL: Modulation of cAMP-induced process formation and con-A induced cap formation in a human glioma line. Neurosci Soc Abstr IV:596, 1978.
23. Gumerlock MK, Smith BH, Pollock LA, Kornblith PL: Chemical differentiation of cultured human glioma cells: Morphologic and immunologic effects. Surg Forum XXXII:445 – 477, 1981.
24. Shitara N, McKeever P, Whang Peng J, Smith BH, Schmidt S, Kornblith PL: Cytofluorometric DNA determination and cytogenetic analysis in human cultured cell lines derived from brain tumors. (Submitted).
25. McKeever PE, Quindlen E, Banks MA, Williams U, Kornblith PL, Laverson S, Greenwood MA, Smith BH: Biosynthesized products of cultured neuroglial cells: I. Selective release of proteins by cells from human astrocytomas. Neurology 31:1445 – 52, 1981.
26. Cummins CJ, Galarraga J, DeLaPaz R, Smith BH, Passonneau JV, DiChiro G, Kornblith PL: Some aspects of the aerobic and anaerobic metabolism of human astrocytoma III and IV in culture (In preparation).
27. Dohan FC Jr, Kornblith PL, Wellum GR, Pfeiffer SE, Levine L: S-100 protein and 2', 3'-cyclic nucleotide 3'-phosphohydrolase in human brain tumors. Acta Neuropathol (Berl) 40:123 – 128, 1977.
28. Laverson S, McKeever PE, Kornblith PL, Quindlen E, Howard R: Diagnosis of glioma on frozen section by immunofluorescence for glial fibrillary acidic protein. Lancet 1:674, 1981.
29. Scott RM, Liszczak TM, Kornblith PL: 'Invasiveness' in tissue culture: A technique for the study of gliomas. Surg Forum 29:531 – 533, 1978.
30. Smith BH, Ellis J, Pacheco M, Brown K, Kornblith PL: Automated image analysis of microtiter plate chemotherapy and immunological assays. (In preparation).
31. Levin VA, Hoffman W, Weinkam J: Pharmacokinetics of BCNU in man: A preliminary study of 20 patients. Cancer Treat Rep 62:1305 – 1312, 1978.
32. DeVita VA, Denham C, Davidson JD, Oliverio VT: The physiological disposition of the carcinostatic 1,3-bis (2-chloroethyl)-1-nitrosourea (BCNU) in man and animals. Clin Pharm Ther 8:566 – 577, 1967.
33. Dickson JA, Suzangar M: In vitro sensitivity testing of human tumour slices to chemotherapeutic agents – its place in cancer therapy. In: Human tumours in short term culture, Dendy PP (ed), Academic Press, Inc., New York, pp 107 – 138, 1976.
34. Smith BH, Cooke C, Leonard L, Kornblith PL: Solid phase testing of chemotherapy agents (In preparation).
35. Hochberg FH, Linggood R, Wolfson L, Baker WH, Kornblith PL: Quality and duration of survival in glioblastoma multiforme. JAMA 24:1016 – 1018, 1979.
36. Kornblith PL, Coakhan HB, Pollock LA, Ward W, Green S, Smith BH: Autologous serological responses in glioma: Correlation with tumor grade and survival (In press). Cancer, 1983.
37. Tachibana T, Klein E: Detection of cell surface antigens on monolayer cells. I. The application of immune adherence in microscale. Immunology 19:771 – 782, 1970.
38. Coakham HB, Kornblith PL, Quindlen EA, Pollock LA, Wood WC, Hartnett LC: Autologous humoral response to human gliomas and analysis of certain cell surface antigens: In vitro study with the use of microcytotoxicity and immune adherence assays. J Natl Cancer Inst 64:223 – 233, 1980.
39. Böyum A: Isolation of mononuclear cells and granulocytes from peripheral blood. Scand J Clin Lab Invest 21, Suppl 97:77 – 89, 1968.

40. Perper RJ, Zee TW, Mickelson MM: Purification of lymphocytes and platelets by gradient centrifugation. J Lab Clin Med 72:842–848, 1968.
41. Lozzio CB, Lozzio BB: Cytotoxicity of a factor isolated from human spleen. J Natl Cancer Inst 50:535–538, 1973.
42. Herberman RB, Holden HT: Natural cell-mediated immunity. Adv Cancer Res 27:305–377, 1978.
43. Rankin JK, Shapiro WR, Posner JB: Cellular stability and chromosomal evolution of early passage cells from human gliomas. Proceedings, American Association for Cancer Research. Abst. 21:55, 1980.
44. Bigner DD, Bigner SH, Pontèn J, Westermark B, Mahaley MS, Rouslahti E, Herschman H, Eng LF, Wikstrand CJ: Heterogeneity of genotypic and phenotypic characteristics of fifteen permanent cell lines derived from human gliomas. J Neuropath and Exp Neurol 40:201–229, 1981.
45. Pfeiffer SE, Sundarra N, Dawson G, Kornblith PL: Human acoustic neurinomas: nervous system specific biochemical parameters. Acta Neuropathol (Berl.) 47:27–31, 1979.
46. DeLaPaz R, DiChiro G, Smith BH, Kornblith PL, Quindlen EA, Sokoloff L, Brooks RA, Kessler RM, Johnston GS, Manning RG, Flynn RM, Wolf AP, Fowler JS, Brill B, Blasberg RG, London WT, Sever JL, Kufta CV, Rieth KG, Goble JC, Cummins C: ^{18}F-2-fluoro-2-deoxyglucose positron emission tomography of human cerebral gliomas. Scientific Manuscripts, AANS, 29–30, 1981.
47. Kornblith PL, Quindlen EA, Pollock LA, Coakham HB: Humoral Immunology of Brain Tumors. In: Multidisciplinary aspects of brain tumors therapy, Paoletti P, Walker MD, Butti G, Knerich R (eds), Elsevier/North-Holland, New York, 113–122, 1979.
48. Brooks WH, Netsky MG, Normansell DE, Horwitz DA: Depressed cell-mediated immunity in patients with primary intracranial tumors. J Exp Med 136:1631–1647, 1972.
49. Brooks WH, Caldwell HD, Mortara RH: Immune responses in patients with gliomas. Surg Neurol 2:419–423, 1974.
50. Young HF, Sakalas R, Kaplan AM: Inhibition of cell-mediated immunity in patients with brain tumors. Surg Neurol 5:19–23, 1976.
51. Brooks WH, Latta RB, Mahaley MS, Roszman TL, Dudka L, Skaggs C: Immunobiology of primary intracranial tumors. Part 5: Correlation of a lymphocyte index and clinical status. J Neurosurg 54:331–337, 1981.
52. Levy NL: Specificity of lymphocyte-mediated cytotoxicity in patients with primary intracranial tumors. J Immunol 121:903–915, 1978.
53. Takasugi M, Mickey MR, Terasaki PI: Studies on specificity of cell-mediated immunity to human tumors. J Natl Cancer Inst 53:1527–1538, 1974.
54. Herberman RB, Oldham RK: Problems associated with study of cell-mediated immunity to human tumors by microcytotoxicity assays. J Natl Cancer Inst 55:749–753, 1975.
55. Gately MK, Glaser M, Mettetal RW Jr., Dick SJ, Kornblith PL: The generation in vitro of lymphocytes cytotoxic for allogeneic human glioma cells: Requirement for third-party stimulator lymphocytes. Fed Proc 40:1041, 1981 (Abs).
56. Rosenblum ML, Hoshino T, Levin VA, Wilson CB: Planning of brain tumor therapy based on laboratory investigations – sequential BCNU-5-FU treatment of malignant gliomas. Scientific Manuscripts AANS, 31–32, 1981.
57. Wilson CB, Boldrey EB, Enot KJ: 1,3-bis (2-chloroethyl)-1-nitrosourea, (NSC-409962) in the treatment of brain tumors. Cancer Chemother Rep 54:273–281, 1970.
58. Young RC, Walker MD, Canellos GP, Schein PS, Chabner BA, DeVita VT: Initial clinical trials with methyl-CCNU 1-(2-chloroethyl)-3-(4-methyl cyclohexyl) 1-nitrosourea (MeCCNU). Cancer 31:1164–1169, 1973.
59. Hoogstraten B, Gottlieb JA, Caoili E, Tucker WG, Talley RW, Haut A: CCNU (1-[2-chloroethyl]-3-cyclohexyl-1-nitrosourea, (NSC-79037) in the treatment of cancer. Phase II study. Cancer 32:38–43, 1973.

60. Wasserman TH, Slavik M, Carter SK: Review of CCNU in clinical cancer chemotherapy. Cancer Treat Rev 1:131–151, 1974.
61. Walsh JM, Cassidy JR, Frie E III, Kornblith PL, Welch K: Recent advances in the treatment of primary brain tumors. A Seminar. Arch Surg 110:696–702, 1975.
62. Reagan TJ, Bisel HF, Childs DS, Layton DD, Rhoton AL, Taylor WF: Controlled study of CCNU and radiation therapy in malignant astrocytoma. J Neurosurg 44:186–190, 1976.
63. EORTC Brain Tumor Group: Effect of BCNU on survival, rate of objective remission and duration of free interval in patients with malignant brain glioma – first evaluation. Europ J Cancer 12:41–45, 1976.
64. Walker MD, Green SB, Byar DP, Alexander E, Batzdorf U, Brooks WH, Hunt WE, MacCarty CS, Mahaley MS, Mealey J, Owens G, Ransohoff J, Robertson JT, Shapiro WR, Smith KR, Wilson CB, Strike TA: Randomized comparisons of radiotherapy and nitrosoureas for the treatment of malignant glioma after surgery. N Engl J Med 303:1323–1329, 1980.
65. Sjogren HO, Hellstrom I, Bansal SC, Hellstrom KE: Suggestive evidence that blocking antibodies of tumor-bearing individuals may be antigen-antibody complexes. Proc Natl Acad Sci USA. 68:1372, 1971.
66. Rosenblum ML, Vasquez DA, Hoshino T, Wilson CB: Development of a clonogenic cell assay for human brain tumors. Cancer 41:2305–2314, 1978.

11. Metabolic Therapy of Malignant Gliomas

JAMES T. ROBERTSON, E. STANFIELD ROGERS, W.L. BANKS and HAROLD F. YOUNG

INTRODUCTION

Malignant gliomas, in particularly glioblastoma multiforme, comprise 50% of all primary intracranial tumors affecting the cerebral hemispheres in adults [1]. These dreadful neoplasms have resisted all attempts at permanent cure by combinations of surgery, radiation therapy and chemotherapy. Of the malignant gliomas, clearly, the anaplastic astrocytoma Grade 3 and 4 (glioblastoma multiforme) have universally fatal outcomes. Surgical resection is limited to gross tumor removal to insure adequate decompression of the brain and, although some studies suggest extensive decompressions are followed by longer term survivals, tumor location within the brain prevents radical surgery as employed for cancer elsewhere.

For several years, the Brain Tumor Study Group has systematically combined old and new forms of surgery, radiation and chemotherapy [2]. They have shown that patients with malignant gliomas receiving 5,500 to 6,000 rads of radiation to the brain live significantly longer than those receiving 5,000 rads or less. The median survivals increased from 17 weeks to 38 weeks when this form of radiation was added to extensive resection of the tumor. These investigators have emphasized the close correlation of age, Karnofsky rating, degree of malignancy and length of symptomatology as determinants in long-term survival.

The nitrosoureas are the most commonly used chemotherapeutic agents employed in the treatment of malignant glioma. The best results have been obtained with the use of BCNU (1, 3, -Bis-(2-chloroethyl)-1-nitrosourea) [2]. This agent has increased median survivals to 43 weeks when used in combination with radiation therapy following conventional neurosurgical resection. Of the patients who tolerated the initial courses of BCNU therapy, 25% have survivals which have extended beyond 18 months [2, 6]. Unfortunately, in spite of the Brain Tumor Study Group's efforts, no better chemotherapeutic agent has appeared. Concomitant with longer survival and the higher doses of radiation, delayed radiation damage to the brain has appeared and correlates with the experimental damage to normal brain tissue when 6,000 rads of therapy are administered. Radiation necrosis of significant degree occurred in a relatively low incidence. In addition to the usual toxicity associated with the nitrosoureas, a severe form of pulmonary toxicity with BCNU occurs when a total dose of 1.5 gram/M^2 has been administered [7, 8].

Currently, further progress in the treatment of malignant glioma appears static. It is clear that some new advance in treatment is mandatory. The possibilities include intra-arterial administration of chemotherapeutic agents, continued use and development of new agents, different forms of radiation therapy including intracavitory radiation and efforts to determine specific cell type sensitivity to drugs.

One of the most intriguing therapies, which has received preliminary trial, has been the effort to metabolically classify a given malignant tumor by determining its essential amino acid uptake and then placing a patient on a specific diet which is relatively deficient in the amino acid selectively desired by the malignant tumor [9]. Manipulated nutritional therapy of cancer is not new. Tannenbaum [10] (1940) studied the effects of caloric restriction upon the growth of a variety of tumors. Although the growth rate was reduced and the life spans of the animals with a restricted diet increased, none were cured. Other studies in animal tumor models have emphasized that caloric and/or protein restriction could control the growth of the tumors, but, generally, this was associated with a detrimental effect on the animal's protein reserves [11 – 14]. Diets devoid in specific essential amino acid have been shown to impair the growth of animal tumors, but again significant host weight losses occurred [15, 16]. Skipper and Thompson inhibited growth in sarcomas in mice by feeding diets totally devoid of certain single essential amino acids [15]. Sugimara had similar findings with the walker tumor in rats who were force-fed diets totally devoid of methionine, isoleucine or valine [16]. Phenylalaine restriction resulted in tumor growth inhibition in BW7756 hepatoma, C3HBA mammary adenocarcinoma and S91 melanoma [17, 18]. Tumor growth with restricted phenylalaine was unaffected in sarcoma 180 or sarcoma 37 models. Pine [19] reported increased survivals for L1210 and P388 leukemic mice after feeding diets restricted in phenylalaine. No alterations and survivals for TA3 mammary carcinoma and E14 leukemic models occurred. Pine, Jose and Good [19, 20] have suggested that the tumor regression occurring with reduction of dietary intake of essential amio acids probably resulted from a depression of blocking antibody activity in the serum thus enhancing the cellular immunity of the host. The effect on tumor regression or growth failure was not believed to be secondary to a direct effect from the amino acid depletion.

Of all the amino acid dependent tumors, the rabbit papilloma induced by the Shope virus has been the most exciting [21]. The virus induces the enzyme arginase in the infected cells. This virus coated enzyme greatly lowers the arginine level of the cellular free amino acid pool [22, 23]. This tumor growth can be markedly slowed by supplying sufficient parenteral arginine in the presence of an arginase inhibitor to bypass the effect of low cellular arginine. At the same time, if arginine is deprived, the induced papilloma may markedly regress or disappear. Dialysis of the blood against arginase can be used effectively to lower the circulating arginine level irrespective of the diet [24]. This dialysis therapy in the rabbit papilloma system has a dramatic inhibitory influence upon the papilloma carried by rabbits causing the tumor to nearly vanish in as little as five days. This example of amino acid therapy would appear to be entirely metabolic.

Theuer [25] determined the effects of feeding diets containing various levels of each of the essential amino acids on tumor and host weights in C57BL female mice bearing BW10232 implanted adenocarcinomas. Low dietary levels of phenylalaine, valine or isoleucine decreased tumor weight with no significant detrimental effects on host weight loss. Low dietary levels of tyrosine, threonine, leucine or methionine also produced decreased tumor weights, but the host animal lost weight. Induced low levels of lysine did not result in alteration of either tumor or host weight changes. Therefore, in this encompassing study, tumor weight reduction with no significant reduction in host weight was achieved with diets low in phenylalaine, valine and isoleucine. The study emphasizes that dietary regulation of certain amino acids in these animals was useful in reducing tumor growth without compromising the host. Worthington *et al.* [26] reported that synthetic diets containing 25% of the optimal levels in isoleucine, leucine or phenylalaine-tyrosine were of no benefit in reducing tumor incidence or improving long-term survivals in female Balb/Can mice when methylcholanthrene was implanted to produce tumors. Recently, Lowery *et al.* [27] have indicated that the survivals of methylcholanthrene tumor bearing mice were reduced when the host nutritional status was compromised. It is clear, therefore, that in animals a dietary regimen can effect tumor growth, but a fine balance must be achieved to avoid significant catabolism which may be detrimental.

Specific dietary therapy has been used effectively for several years to control blood levels of certain essential amino acids in children, e.g. phenylalaine in patients with phenyketonuria [28, 29]. Manipulation of nutritional factors in tumor patients has possibly been effective in the therapy of a few specific neoplasms [30, 31]. In these examples, however, the amino acid selected to be deficient in the diet was approached in random fashion and studied in some instances in conjunction with antimetabolites. Turney and Rogers [32] have described no signs of tumor involvement sixteen years following a surgical, radiological and dietary program applied to a brain stem astrocytoma in a child.

METABOLIC CLASSIFICATION OF TUMORS

In 1953, Buzzati-Traverso [33] discovered that he could separate single gene mutants of drosophilia by simply squeezing the flies into chromatographic paper, spraying with ninhydrum after development with butanolacetic acid water, and comparing the pattern of amino acids and peptides. Rogers and Burton [34] applied this system to some of the general problems of pathology. This led to a method of rapidly characterizing the metabolism of individual tumors [35]. Consistent differences were evident between malignant gliomas (Table 1). Generally, a pattern of amino acid uptake occurred in tumors of a specific diagnostic type, although it was clear that individual differences were evident from glioma to glioma. The concept of metabolically characterizing a tumor in addition to microscopic identification of the tumor thus began. The method involves obtaining a specimen of fresh malignant glioma from the operating room, placing it in a sponge moistened with isotonic saline in a sterile dish and, within fifteen to thirty minutes, selecting a sample of the tumor tissue with the use of a dissecting microscope. The tumor tissue is minced, and a sample is examined microscopically after staining a preparation with polychrome methylene blue. This confirms the tissue type and diagnosis. Approximately 50 milligrams of the minced tissue is then suspended in 3 cc's of a synthetic tissue culture medium, TC199 (Difco), and incubated for three hours in a Dubnoff shaker (40 shakes per minute). The cells are then spun out for fifteen minutes at 3,000 rpm, and the supernatant diluted as necessary for amino acid analysis on a Beckman model 120C analyzer. The essential amino acid concentrations of the media before and after the incubation were determined by elution chromatography using the automated amino acid analysis. The tumor dependent essential amino acids were identified as those essential amino acids whose concentrations were decreased in the media as a result of incubation with the minced brain tumor tissue. The essential amino acid that was decreased to the greatest extent by this method was the one that could be considered to be

Table 1. Frequency and rank order of amino acids for anaplastic astrocytoma patients by the assay procedure of Rogers and Woodall [35]

Amino acid	Rank order First**	Second**	Third**
HIS	11	3	1
ILE	7	5	4
MET	5	9	5
VAL	3	3	5
LYS	3	1	3
PHE	2	1	3
THR	1	4	3
LEU	0	5	5
ARG	0	1	1
Number of patients	32	32	30

*The patients that died before beginning radiation are excluded; this analysis is selected but representative.
**These terms refer to the ranking of the amino acids that were lost from the media.

reduced in a dietary regimen applied to patients. Rogers and Robertson [36] divided glioblastoma multiforme into at least four sub-groups relating to the essential amino acids most taken up by the cells and attempted to correlate this uptake with survival. Five groups of tumors were encountered: a group utilizing mainly histidine and methionine; one using mainly phenylalanine and tyrosine; one using largely the branched chain amino acids; and the last using methionine. Another group utilizing histidine-arginine appeared likely. The tumors utilizing mainly methionine took up little or no histidine. This preliminary study indicated that the histidine-methionine group may live longer than the other three principal groups. Rogers [37] has studied the comparative amino acid uptakes with a variety of human tumors in addition to the malignant gliomas. These include adenocarcinoma of the breast, adenocarcinoma of the ovary, lung carcinoma, carcinoma of the cervix and carcinoma of the stomach. There appears to be a consistent different uptake of the tumors. He recommends this *in vitro* metabolic test system be applied as a screening test on each tumor to properly select the best agent for tumor therapy. Banks [38] applied this method of metabolic classification of gliomas in 20 patient specimens and confirmed the reproducibility of metabolic classification. In a combined study of 43 patients from the University of Tennessee and the Medical College of Virginia, Banks considered the apparent tumor dependent amino acids [38]. Sixty-seven percent of the tumors were glioblastomas, 26% were anaplastic astrocytomas, and the remaining three tumors were classified as other malignant gliomas. In ten of the 43 patients studied, histidine showed the greatest loss from the media in comparsion to the other esential amino acids. In five patients, histidine was decreased in the assay by the second greatest percentage; and, in one patient, histidine was diminished in the media by the third greatest percentage. In almost every case, there is at least one essential amino acid, often different for different patients, that gave an increased concentration in the media following incubation with the minced tumor tissue. Histidine, methionine and isoleucine combined ranked as the first essential amino acid for more than 63% of the patients in this study. There was no correlation between the specific neuropathology and the ranking of the various essential amino acids by this assay.

CLINICAL TRIALS OF DIETARY THERAPY FOR MALIGNANT GLIOMAS

In 1975, using a neurosurgical, biochemical and dietary team and the Rogers' method of determining the most essential amino acid taken up by the malignant gliomas of three patients, we initiated a preliminary trial of dietary manipulation in combination with surgery and x-ray treatment [9] (Table 2). We received support in our dietary manipulation to a superb degree from Mead-Johnson & Co. Two of our brain tumor patients were initially managed with diets identical to the therapy

Table 2. Patient profiles upon entry into study*

Group	Sex M	F	Age (years)	Initial neurological status*** (%)
Diet therapy	4	4	39 ± 5** (8)	76 ± 5 (8)
Diet therapy & BCNU	3	6	45 ± 4 (9)	77 ± 2 (9)
BCNU only	1	8	44 ± 5 (9)	73 ± 3 (9)
Control	0	6	45 ± 3 (6)	57 ± 10 (6)

*The patients that died before beginning radiation therapy or did not start diet therapy are excluded but were randomized and included in the statistical analysis.
** $\bar{x}$ ± SEM
***Karnofsky scale.

Table 3. Patient survivals*

Time (days)	Diet therapy	Diet therapy & BCNU	BCNU	Control
180	7/8	9/9	9/9	5/6
360	6/8	7/9	5/9	3/6
540	6/8	4/9	5/9	3/6
720	6/8	4/9	4/9	2/6
900	2/7	2/8	1/6	2/6
Median survival (days)	804	487	398	466
Mean survivals ($\bar{x}$ ± SEM)	696 ± 121	606 ± 97	534 ± 102	536 ± 159
Number of patients	8	9	9	6

*The patients that died before beginning radiation or did not start diet therapy are excluded but were randomized and included in statistical analysis.

of maple syrup urine disease. The dietary preparations were constructed to fit each patient's needs. This was enhanced by specific dietary planning and the availability of product 80056 supplied by Mead-Johnson. Fortunately, action taken in 1971 by the Federal Drug Administration changed the designation of amino acid formulas from drugs to specific dietary foods. In addition, we had the background of the pediatric metabolic manipulation experience. In this preliminary study, the diet was well-tolerated by the patients; the specific amino acid was lowered i the serum; and no serious untoward effects appeared. Two of the three patients appeared to be significantly benefited by this therapy. Subsequently, the Diet, Nutrition and Cancer program of the National Cancer Institute initiated an exploratory study to determine the feasibility of limiting one essential amino acid in the diet of a limited number of patients with malignant astrocytomas as a potential new form of adjuvant therapy (Table 3). This study was performed at the University of Tennessee Center for the Health Sciences and the Medical College of Virginia (Virginia Commonwealth University). Neurosurgical, neuroradiological, biochemical, pathological and dietary teams operated at each institution and, ultimately, 43 patients were included in the preliminary protocol. After appropriate diagnostic studies, the patient underwent surgical resection of the tumor. The tumor was determined an anaplastic astrocytoma. A fresh specimen was taken from the operating theatre for identification of the most dependent essential amino acid by the method of Rogers [39]. The most essential amino acid for a given tumor was that essential amino acid was utilized during the period of incubation in the tissue medium. The patients selected for study were limited to either sex between 21 and 65 years of age. The patients were randomly assigned to four groups of therapy: diet therapy only, diet therapy plus BCNU, BCNU only, and surgery plus x-ray only. Following surgery, all patients received a total dose of 6,000 rads of x-ray therapy administered five days per week for six to seven weeks. For those patients that were randomized into the BCNU only or diet therapy plus BCNU group, chemotherapy was administered intravenously over three successive days at a total dose of 80 milligrams/M^2 body surface area every eight weeks, unless serious signs of toxicity developed. The protocol regarding BCNU was identical to that used by the Brain Tumor Study Group. During the radiation therapy phase, the planning for the diet therapy was completed for those patients in the diet therapy groups. The dietary regimen met the minimum for all of the essential amino acids as determined by Rose [40] for men and Leverton et al. [41] for women. The remaining contents of the diet met the 1974 National Research Council recommended dietary allowances [42]. The dietary program contained a synthetic formula which was devoid in the restricted essential amino acids; a variety of natural foods in prescribed quantities to meet the minimum requirements for the restricted essential amino acid, vitamin and mineral supplements as needed; sufficient fluids, free foods including oils, fats and sugars desired and needed to provide adequate calories and proprietary products to provide additional calories if needed [43]. The formula was prepared using Mead-Johnson product 80056 as the vitamin, mineral and non-protein caloric portion of the formula to which was added all the essential amino acids except the one which was restricted. The dietitian instructed each patient to keep daily food records as a reminder to the patient of the importance of compliance [44]. Extensive dietary instruction to the patient and the family was effected. The food records for the three days prior to the periodic visits were assessed for their adequacy by the dietitian and nutritional deficiencies in the diets consumed were brought to the patient's attention to insure correction. The diet therapy program began approximately two weeks following the completion of the radiation therapy. If necessary, the patients were admitted to the hospital for a one week period to adjust to the regimen and for intensive dietary instruction. The patients were then discharged and followed carefully. At the return visits, appropriate laboratory, neurological, and nutritional evaluations were effected. The status of remaining cerebral tumor was determined every two months by computerized axial tomographic scans. Anthropometric measurements of height, body weight, mid-triceps skin fold and mid-upper arm circumference measurements were performed by the nurse coordinator. At each return, a plasma sample was withdrawn from all patients; a protein free filtrate prepared; and the plasma amino acid profile determined by elution chromatography using an amino acid analysis.

Twenty-three of the 43 patients were randomized to some form of diet adjuvant therapy. In many cases, skin testing for immune responses were determined using monilia and mump skin responses. White blood cell count, particularly lymphocyte counts, were routinely monitored. A review of these tests at one institution indicated a delayed loss of skin sensitivity and suppression of lymphocyte counts that is so characteristic of malignant glioma in the patients with dietary therapy. However, the immune studies were not done in sufficient depth to confirm this apparent trend, but they were apparently enhanced in the diet versus non diet group of patients.

Independent neuropathological study of the 43 patients revealed that 67% had glioblastomas, 26% anaplastic astrocytomas, and three patients were classified as two oligodendrogliomas and one lymphoma. The patient population was homogenous with respect to age and initial performance status as measured by the Karnofsky scale within the four randomized arms of the study. The dietary manipulation was well-tolerated by the patients and proven safe. The total number of patients entered into the study was too small to adequately determine the relative values of the four treatment arms. There was no statistical differences between the mean survival times of the four treatment groups when adjusted for the prognostic variables. There were no statistical differences in the sequence of performance ratings as determined by the Karnofsky [6] ratings for the four groups. On the other hand, the dietary restriction appeared to be without added risk to the limited number of patients studied. The dietary restriction therapy either alone or in combination with chemotherapy was not a curative modality for the groups of patients that were studied who had a poor prognosis by current therapies. There was a trend indicating that the dietary restrictive therapy for non-glioblastoma patients in terms of survival and performance status was beneficial, but the small sample size precluded conclusive support.

In this study, the amounts of the restricted amino acid in the diet were always kept at least at the level of the required amounts determined by Rose and Leverton, but one to two standard deviations below the patient's control serum.

Perhaps more pronounced effects on survival and/or performance status would have been noted had we restricted the diet to a greater extent without producing severe host catabolic effects.

The project concluded with a number of questions yet to be answered, but indicate that this is a feasible adjunctive therapy that deserves further study in a significant number of patients to reach specific conclusions. The patients on some form of dietary therapy clinically appeared much less ill than those on the chemotherapy or surgery plus x-ray arms. Detailed exploration of the immune response factors is clearly indicated in future projects.

With the advent of positron emission computerized tomography, a preliminary collaborative effort has been initiated to study patients with malignant gliomas using C-11 labeled amino acids at the Medical and Health Sciences division in Oak Ridge [45]. Since our preliminary studies showed the selective uptake of certain essential amino acids by malignant gliomas, it was logical to demonstrate that amino acid utilization by brain tumors can be measured and possibly be used to monitor the response of such tumors to treatment by visualizing the metabolic activity using available C-11 labeled amino acids DL-valine and DL-tryptophan. Eleven patients with malignant gliomas have been studied. Selective visualization of the tumor does occur with these labeled amino acids. Uptake of L-valine and L-tryptophan in tumors in normal brain occurs. Uptake of valine and tryptophan appears to be higher in the tumors than in normal brain tissue. This aproach using C-11 labeled amino acids is a new modality that may provide more specific diagnostic information about tumor type and metabolism and may replace the Rogers' method of metabolic classification with the availability of additional labeled amino acids. For example, if it can be shown that the most tumor dependent essential amino acid using the Rogers' method for a given tumor is also the amino acid most concentrated by positron emission computerized tomography in a series of cases, then the *in vitro* method is confirmed *in vivo* and subsequent histological diagnosis may be confirmed. There are numerous exciting possibilities about exploring metabolism with this new adjunct.

SUMMARY

Considerable data is available in animal tumor models to indicate that amino acid manipulation retards tumor growth and can be done without producing serious catabolism. The possibility of considering metabolic classification of tumors as an adjunct in deciding definitive therapy is intriguing. The Rogers' method of classifying malignant gliomas by virtue of their essential amino acid uptake is reproducible and may prove to be a rapid screening method for therapy determinations. Application of this method utilizing positron emission computerized tomography is preliminary but does tend to indicate that malignant gliomas may selectively enhance the C-11 labeled amino acids. Preliminary human trials of dietary manipulation appear to be feasible, safe and worthy of consideration for future adjuvant therapy in malignant glioma.

ACKNOWLEDGMENT

Appreciation is given to the University of Tennessee General Clinical Research Center, Grant RR 211, for nursing care and assistance in dietary instruction.

REFERENCES

1. Wilson CB, Boldery EB, Enot KJ: Bis (2-chloroethyl)-1-nitrosourea (NSC-409962) in the treatment of brain tumors. Cancer Chemotherapy Reports 54:p 273 – 281, 1970.
2. Walker MD, Hunt WE, Mahaley MS, Norrell HA, Ransohoff J, Gehan EA: Evaluation of BCNU and/or radiotherapy in the treatment of anaplastic gliomas. J Neurosurg 49:pp 333 – 343, 1978.
3. Kusske JA, Williams JP, Garcia JH, Pribram HW: Radiation necrosis of the brain following radiotherapy of extracerebral neoplasms. Surg Neurol 6:pp 15 – 20, 1976.
4. Sogg RL, Donaldson SS, Yorke CH: Malignant astrocytoma following radiotherapy of a craniopharyngioma. J Neurosurg 48:pp 622 – 627, 1978.
5. Caveness WF: Pathology of radiation damage to the normal brain of the Monkey. National Cancer Institute Monograph 46:pp 57 – 76, 1976.
6. Shapiro WR: Management of Primary Malignant Brain Tumors. Neurol Neurosurg Update, Biomedia Inc. Princeton, NJ, 1978.
7. Hologe PY, Jenkins EE, Greenberg SD: Pulmonary Toxicity in Long Term Administration of BCNU. Cancer Treatment Reports 60:pp 1691 – 1694, 1976.
8. Jones MPH, Marsden HB, Bailey CC: Fatal pulmonary fibrosis following 1,3-Bis (2-chloroethyl)-Nitrosourea (BCNU) therapy. Cancer 42:pp 74 – 76, 1978.
9. Greer CR, Rogers S, Robertson JT, Molinary S, Smith MA, Ramey DR: Specific therapy of patients with malignant brain tumors by using surgery, irradiation, and metabolic control. Clin Cong of the Am Col of Surgeons, abst 843, p 179, San Francisco (October 1975).
10. Tannenbaum A: Relationship of body weight to cancer incidence. Amer J Cancer 38:335 (1970).
11. White ER, Belkin M: Source of tumor protein. 1. Effect of a low-nitrogen diet on the establishment and growth of a transplanted tumor. J Natl Cancer Inst 5:pp 261 – 263, 1945.
12. White ER: Source of tumor protein. II. Nitrogen-balance studies of tumor bearing mice fed a low nitrogen diet. J Natl Cancer Inst 5:pp 265 – 270, 1945.
13. Sherman CD Jr, Morton JJ, Mider GB: Potential sources of tumor nitrogen. Cancer Res 10:pp 374 – 378, 1950.
14. Babson AL: Some host-tumor relationships with respect to nitrogen. Cancer Res 14:pp 89 – 93, 1954.
15. Skipper HE, Thomson JR: Amino Acids and Peptides with Antimetabolic Activity. Little Brown and Co, pp 38 – 53, 1958.
16. Sugimura T, Birnbaum SM, Winitz M, Greenstein JP: Quantitative Nutritional Studies with Water-soluble, Chemically Defined Diets. VII. The forced feeding of diets lacking one essential amino acid. Arch Biochem Biophys 81:pp 448 – 455, 1959.
17. Lorincz AB, Kuttner RE, Brandt MB: Tumor response to PHE-TYR limited diets. J Amer Dietetic Assn 54:pp 198 – 205, 1969.

18. Demopoulos HB: Effects of low phenylalanine-tyrosine diets on S91 Mouse melanomas. J Natl Cancer Inst 37:pp 185–190, 1966.
19. Pine MJ: Effect of low phenylaline diet on murine leukemia L1210. J Natl Cancer Inst 60:pp 633–641, 1978.
20. Jose DJ, Good RA: Quantitative effects of nutritional essential amino acid deficiency upon immune response to tumors in mice. J Exp Med 137:pp 1–9, 1972.
21. Rogers S: Induction of Arginase in Rabbit Epithelium by the Shope rabbit papilloma Virus, Nature 183, 1815, 1959.
22. Roger S, Moore M: Studies of the mechanism of action of the Shope rabbit papilloma virus. J Exp Med 117:521, 1963.
23. Rogers S: Shope papilloma virus: a passenger in man and its significance to the potential control of the host genome. Nature 212:1220, 1966.
24. Rogers S: Significance of dialysis against enzymes to the specific therapy of cancer and genetic deficiency diseases. Nature 220:1321, 1968.
25. Theur RC: Effect of essential amino acid restriction on the growth of female C57BL mice and their implanted BW 10232 adenocarcinomas, J Nutrition 101:pp 223–232, 1971.
26. Worthington BS, Syrotock JA, Ahmed SI: Effects of essential amino acid deficiencies on syngeneic tumor immunity and carcinogenesis in mice. J Nutrition 108:pp 1402–1411, 1978.
27. Lowery SF, Goodgame T, Norton JA, Jones DC, Brennan MF: Effect of chronic protein malnutrition on host-tumor composition and growth. Surgical Research 26:pp 79–86, 1979.
28. Lowe CU et al: Committee on Nutrition, American Academy of Pediatrics: Nutritional Management in Hereditary Metabolic Disease Ped 40:pp 289–304, 1967.
29. Knox WE: Phenylketonuria. In: The Metabolic Basis of Inherited Disease, Stanbury JB, Wyngaarde JB and Fredrickson DS (eds). McGraw-Hill, NY, pp 226–295, 1972.
30. Edmund J et al: Reduced intake of phenylalamine and tyrosine as treatment of choroidal malignant melanoma. Mod Prob Ophthal 12:pp 504–509, 1974.
31. Lorincz AB, Kuttner RE: Response of malignancy to phenylalamine restriction; a preliminary report on a new concept of managing malignant disease. Nebr State Med J 50:p 609, 1965.
32. Rogers S: Personal communication.
33. Buzzati-Traverso AA: Paper chromatographic pattern of genetically different tissues: A contribution to the biochemical study of individuality. Proc Nat Acad Sci 36:463, 1950.
34. Rogers S, Berton WM: Application of paper chromatography to some of the general problems of pathology. Lab Invest 6:310, 1957.
35. Rogers S, Woodhall B: Rapid method of metabolically characterizing individual tumors. Proc Soc Exptl Biol Med 98:pp 874–877, 1958.
36. Rogers S, Robertson JT: A method of studying metabolic variation between individual tumors. Nutrition and Cancer, 1981. (In press).
37. Rogers S: Comparative amino acid uptakes of a variety of human tumors and its significance. Nutrition and Cancer, Vol 2 No. 3, p 148, 1981.
38. Rogers S, Banks WL Jr, Young HF, Robertson JT et al: Individually tailored dietary restriction therapy and the treatment of malignant gliomas – A status report. Cancer (submitted 1981).
39. Rogers S, Woodhall B: Rapid method of metabolically characterizing individual tumors. Proc Soc Exptl Biol and Med 98:pp 874–877, 1958.
40. Rose WC: Amino acid requirement of man. Fed Proc 8:pp 546–552, 1949.
41. Leverton RM, Gram MR, Chaloupka M, Brodousky E, Mitchel A: The quantitative amino acid requirements of young women, I. Threonine. J Nutrition 58:pp 59–81, 1955.
42. National Academy of Sciences: Recommended Dietary Allowances, 8th Edition, Washington DC, p 128, 1974.
43. USDA: Handbook No 8 Revised-Composition of Foods, US Govt Printing Office, Washington DC, p 190, 1963.
44. Burgess JB, Mashburn LT, Robertson JT: Safety of essential amino acid restriction in adults with

malignant brain tumors. Nutrition and Cancer, pp 16–21, 1980.
45. Hubner KF, Purvis JT, Mahaley MS Jr, Robertson JT et al: Brain tumor imaging by positron emission computed tomography using 11C-labeled amino acids. Journal of Computer Assisted Tomography 6(3):544–550, June 1982, Raven Press, New York.

12. Pathologic Effects of Chemotherapy

KURT JELLINGER

1. INTRODUCTION

With the growing use of potent antineoplastic agents both systemic and nervous system toxicity have become a more pressing issue, and the advent of increasingly aggressive chemotherapy protocols has created a myriad of complications of cytotoxic therapy that have only recently become important. Rapidly dividing tissues, such as bone marrow and epithelial cells, are most overtly affected by such therapy; the dose in major degree, reflects the tolerance of these organs to the various agents [1, 2, 2a]. Organs with less proliferative activity, such as the central nervous system (CNS), seem less sensitive to the adverse effects of chemotherapy, although cytostatic agents may reach the CNS directly by intrathecal (IT) injections or indirectly via the vasculature in patients treated with oral or parenteral medication. Not only agents that penetrate the CNS, i.e. can cross the blood brain-barrier (BBB), need to be considered for either efficacy or toxicity, since several new factors in cancer therapy combine to enhance the likelihood of CNS toxicity [3, 3a, b]:
a) Drugs with high lipid solubility, such as nitrosoureas, have been introduced into many treatment protocols to maximize CNS penetration;
b) The entire concept of the BBB is under continuing review, and electron-microscopic (EM) studies have shown that the vascular structures in the brain containing malignant tumor are altered and that the BBB in and around tumor tissue varies from relatively intact barrier to areas containing passages open to blood proteins [4];
c) Brain tumor and other cancer protocols which incorporate several cytostatic drugs concurrent with radiotherapy may enhance their separate toxicities on many organs including the CNS and may produce synergistic neurotoxic effects.
d) Patients with all forms of malignancy are surviving longer and relapses are treated with more aggressive cytostatic and multimodality regimens; many of which may directly or indirectly affect the nervous system;
e) Increasing survival allows time for the development and recognition of the deleterious effects of chemotherapeutic modalities on various organ systems which guide the oncologist to modifying dosages of individual agents and

M. D. Walker (ed.), Oncology of the Nervous System. ISBN 0-89838-567-9.
© *1983 Martinus Nijhoff Publishers, Boston/The Hague/Dordrecht/Lancaster. Printed in the Netherlands.*

alterating treatment schedules to avoide acute and long-term complications. Iatrogenic damage due to aggressive anticancer therapy may be either *direct* sequelae of chemotherapy and radiation or *indirect* lesions which are related with, but not strictly resulting from, treatment, but include side effects of various etiology, including iatrogenic immunosuppression, malnutrition, metabolic disorders, etc., while the pathogenesis of some adverse effects of chemotherapy remains obscure.

In this chapter, after initial remarks on the classification of cytostatic agents and general toxicity of chemotherapy, the morphologic effects of this treatment on gliomas, and the CNS complications of systemic and IT chemotherapy and multimodality treatments as well as peripheral neurotoxicity and CNS infections complicating chemotherapy will be reviewed.

2. CLASSIFICATION OF ANTINEOPLASTIC AGENTS

The anticancer agents have been classified with respect to their mode of action, their potential for synergistic toxicities when combined with radiation (Table 1) and their neurotoxicity (Table 2). In addition to diverse mechanisms of action of the various agents, the physiological distribution pattern and cell sensitivities enter into toxic reactions, and must be integrated in order to predict both useful and deleterious combinations.

a) Large molecular weight *antibiotics*, particularly anthracyclines, such as adriamycin and actinomycin B, penetrate only poorly, if at all into the normal brain [1, 5, 51], and toxic interactions with the CNS are unusual except where the BBB is significantly deranged, such as in and around malignant gliomas and metastases, where penetration of these drugs may be clinically useful. In general, however, toxicities of these drugs are limited to structures outside the BBB. However, in structures external to the CNS such agents are highly toxic when combined with ionizing radiation, most probably because their mode of action closely resemble each other. Within this group of drugs, subclassifications can be made in which tissue specific toxicity will predominate. Thus, actinomycin D and adriamycin, while ordinarily excluded from the brain *per se*, can be strikingly toxic to vascular tissues. Both antibiotics are known to be cardiotoxic and also toxic to epidermal structures. Bleomycin is also highly toxic to epidermal structures but does not penetrate the brain in significant levels [1, 31, 5, 5a].

b) *Antimetabolites* inhibiting DNA synthesis, such as methotrexate (MTX), hydroxy-urea (HU), 5-fluorourazil (5-FU), cytosine-arabinoside (CA), penetrate the brain in various degrees [5 – 8]. MTX, an analogue of folic acid, is a potent inhibitor of the enzyme dihydrofolate reductase (DHFR) that is responsible for maintaining the intracellular pool of reduced folate co-factor (tetrahydrofolate) and thus, in addition to inhibition of *de novo* nucleic acid synthesis and thereby

Table 1. Classification of selected antineoplastic agents (modified after [5, 13])

Class	Mechanism	Examples	Basic site of action	'Specific toxicity'
1. DNA-lytic (antibiotics)	a) nuclease activation	adriamycin	inhibition of DNA repair cell arrest in G_2 phase	heart, skin, lung, GI tract
		actinomycin D	toxic to all cells	skin, oesophagus, lung,
		mithramycin	'true radiomimetic'	GI tract, liver, bone marrow
		daunomycin	inhibition of DNA synthesis	epithelium, skin, lung
	b) direct attack	bleomycin	revers. cell arrest in G_2 no impairment of S-phase 'radiomimetic'	oesophagus, GI tract
2. Antimeta-bolites	inhibition of DNA synthesis	methotrexate	irreversible cell arrest inhibition of DHFR	skin, CNS, liver, heart
		hydroxyurea	slow DNA hydrolysis	epithelium, skin, lung
		5-Fluorouracil	cell arrest in S-phase, accumulate in G_2-phase	oesophagus, bone marrow skin, GI, liver, eye
		cytosine arabinosine	inhibition of DNA polymerase inhibition of S-phase	optic nerve, CNS GI tract
		L-asparaginase	deprivation of essential amino acid	anaphylaxis, liver, coagulative system
3. Alkylating agents	a) DNA crosslinks	nitrogen mustard cyclophosphamide nitrosoureas (BCNU, CCNU etc)	inhibits S-phase non radiomimetic inhibit progress S-phase	bone marrow, GI tract lung, bladder, GI tract bone marrow, GI tract
		DTIC	irreversible block in G_2	lung, kidney
		procarbazine	irreversible block in G_2	bone marrow
		cis-platinum	inhibits RNA, DNA synthesis	bone marrow, CNS
		Thio-TEPA	inhibits $S + G_2$ phase	kidney, ototoxicity
	b) DNS monodisordered mitotic spindle	same as above	inhibits DNA synthesis major source of carcinogenesis	GI tract
4. Antimitotics		Vinca alkaloids	arrest of $S + G_2$-phase	peripheral nerve, muscle
		vinblastine	disruption of mitosis	lung, CNS
		vincristine	toxic to structural proteins	bone marrow
		podophyllotoxins VM-26, VP-16	arrest of $S + G_2$-phase	GI tract

Table 2. Classification of selected chemotherapeutic agents with regard to neurotoxicity (Modified after [3a, 3b])

Class	Drug	Synonyms, abbreviation	Mode of administration	Common systemic side effects	Crossing of BBB	Neurotoxicity CNS	Neurotoxicity PNS
1. *Antibiotics*	Adriamycin	Doxorubicin	i.v.	GI, cardiac, bone marrow, skin necrosis	0	0 (+ exper.)	0
	Actinomycin D	Dactinomycin	i.v.	GI, hematolog., dermat.	0	0	0 (+ exper.)
	Bleomycin	Blenoxane	i.v., i.m.	GI, dermatol., pulmonary	0/±	+ (?)	+ (?)
	Daunarubicin	Daunomycine, Cerubidine	i.v.	GI, hematol., cardiac	0	0	0
	Mithramycin	Mithracin	i.v.	GI, dermat., hematol.	±/++	0	0
	Mitomycin	–	i.v.	GI, hematologic	0	0	0
2. *Antimetabolits*							
a) Folic acid antagonists	Methotrexate	MTX (amethopterin)	p.o., i.v., i.a. IT	GI, hematol, renal neurologic	±/++	+ (HD i.v.) ++ (IT)	+ (HD i.v.)
b) Antipyrimidines	Triazinate	TZT	i.v.	GI, hematol, skin	++	0	0
	5-Fluorourazil	5-FU	i.v.	GI, mucosa, cardiac	++	+	0
	Ftorafur	–	i.v.	GI, hematologic	++	++	0
	Cytosine arabinoside	Ara-C, CA	iv., IT	GI, hematol., dermatologic	±	±/+ (IT) + (HD i.v.)	+ (?)
	5-Azacytidine	–	i.v.	GI, hematologic	0/±	+	+
c) Antipurines	6-Mercaptopurine	6-MP	p.o.	GI, hematologic	+	0	0
	Thioguane	TG, 6-TG	p.o., i.v.	hematologic	±	0	0
3. *Alkylating agents* ('classic')	Mechlorethamine	Nitrogen mustard, Mustagon	i.v., i.a.	GI, hematol., dermat.	± (?)	(+) HD i.a.	(+) HD pelvic
	Phenylalanine mustard	Mephalan, Alkeran	p.o.	GI, hematologic	?	0	0

Table 2. (Continued).

	Cyclophosphamide	Cytoxan	p.o., i.v.	GI, hematol., bladder	±		0	0
	Thiotepa	–	i.v., IT	GI, hematologic	± (?)	(+) IT	0	(+) IT
	Chlorambuzil	Leukeran	p.o.	GI, hematologic	?	0	0	0
	Busulfan	Myleran	p.o.	hematolog., pulmon.	?	0	0	0
4. *Plant Alcaloids*	Vincristine	Oncovin, VCR	i.v.	GI, dermatol., neurol.	0	+	+	+
	Vinblastine	Velban, VLB	i.v., i.a.	hematol., dermatol.	?	+	+	+
	Vindesine	DVA	i.v.	Hematol., neurol.	?	+ (?)	+	+
	Epipodophyllotoxine	VM-26, Teniposide	i.v.	GI, mild hematol.	±	0	0	0
		VP-16, Etoposide	i.v., p.o.	GI, mild hematol.	?	0	0	0
5. *Synthetic drugs*								
a) Nitrosureas	Bichloroethyl nitrosurea	BCNU, Carmustine	i.v.	GI, hemat., pulm., hepat.	++	(?)	0	0
	Cyclohexylchloroethyl nitros.	CCNU, Lomustine	p.o.	GI, hematologic	++	0	0	0
	Methyl-CCNU	Semustine	p.o.	GI, hematolog.	+++	0	0	0
	Streptotocin	–	i.v.	GI, renal, hepatic	++	0	0	0
b) others	Cis-Platinum	Cisplatin, DDP	i.v.	renal, otolog., vestib., GI	+ (?)	0	0	+
	Dimethyltriazene-imidazole carboximide	Dacarbazine, DTIC	i.v.	GI, hematologic	±	?	+	0
	Hexamethylmelamine	HXM, HMM	p.o.	GI, hematol., neurol.	?	++	++	+
	Hydroxyurea	HU, Hydrea	p.o.	hematologic	+	0	0	0
	L-asparaginase	Elspar	i.v.	hematologic, GI, hepat. neurolog., pancreatitis	0	+	0	0

Table 2. (Continued)

Class	Drug	Synonyms abbreviation	Mode of administration	Common systemic side effects	Crossing of BBB	Neurotoxicity CNS	Neurotoxicity PNS
5b. (Continued)	Procarbazine	Natulan, PCZ	p.o., i.v.	GI, card., dermatol.	+++	+	+
	Mitotane	Lysodren, DIP, DDD	p.o.	GI, skin, neurol.	?	++	+
	N(Phosphoace-tyl) L-aspartic acid.	PALA	i.v.	GI, dermatologic	?	++	0
	Thymidine	–	i.v.	hematologic	+ (?)	(+) with 5-FU	0
	Methyl-glyco-xal-bis guanylhydrazone	MethylGAG	i.v.	GI, mucositis	?	–	+ (?)
	Acridinylamine-methanesulfon-m-anisidide	Amsacrine, AMSA	i.v.	hematolog., cardiac	?	+ (?)	+ (?)

replicative functions, may results in intracellular buildup of oxidized folates which may to toxic in the brain [6]. MTX, a lipid-soluble and highly ionized drug does not cross the BBB or enters the brain only minimally via passive diffusion. After systemic administration the maximum CSF levels are reached after 24 hours, and there are no direct relations between dosage and CSF level [7]. Neurotoxicity after systemic administration of MTX has been observed following intracarotid infusions and high-dose (HD) intravenous (IV) MTX with citrovorum factor rescue for osteogenic sarcoma and brain tumors [6]. Cytotoxic levels of MTX are achieved in the CSF for one or two days after HD-therapy [8], while citrovorum factor probably penetrates the BBB in lower dosage [3]. Highest MTX concentrations in CSF are achieved after continuing intraventricular infusion [7] and, due to active transport from CSF to blood via choroid plexus systemic toxicity may occur after IT MTX administration used for prophylaxis and treatment of CNS leukemia and tumors [3, 9, 10]. Other antimetabolites, such as CA, penetrate relatively poorly into the CNS and are frequently given intrathecally, whereas 5-FU, a halogenated pyrimidine, enters the CNS freely, reaching levels essentially identical to that found in serum [10]. The mechanism of cytotoxic action of these agents relates to inhibition of DNA synthesis, with inhibition of cell progression through the S- and/or G_2 phase. Some of these drugs, like 5-FU, are potent radiation synergizers [1, 3].

c) Alkylating drugs that induce DNA cross-links together with DNA mono-alkylating sites show very similar biochemical properties despite substantial structural differences that dictate water and lipid solubility leading to quantitative differences in physiologic distribution. While the classical alkylating agents, such as cyclophosphamide, penetrate only slightly, nitrosoureas, (BCNU, CCNU, MeCCNU) and procarbazine (PCZ) are highly lipid-soluble and penetrate the CNS freely, hence their usefulness in the treatment of primary and secondary CNS malignancies. Since nitrosoureas are not only toxic for tumor cells but sensitize them to the action of ionizing radiation, their cytostatic effect is enhanced by association with radiotherapy [1, 3b, 5, 10a, 11]. These agents are of special interest in the context of toxicity because their mode of action via alkylation of DNA leads to both DNA cross-links which are very poorly reparable in mammalian cells and must be active against both (bone marrow) stem cells and CNS (glia and neuronal) cells, and also to mono-alkylating sites which presumably dictate their substantial carcinogenic potential [1]. Neurotoxicity of PCZ is particularly attributed to its monoamine oxidase activity [12, 12a].

d) *Vinca alkaloids*, like vincristine and vinblastine, and podophyllotoxin affect structural proteins and uniquely disrupt mitosis. The Vinca alkaloids have primary toxicity as the limiting factor [1 – 3, 3b, 9, 10] and presumably penetrate the CNS slightly. Most of the cell toxicities encountered by these agents can be explained in terms of either DNA depolymerization or alkylation, with the exception of antimitotics, whose mechanism appear to be specifically related to spindle protein denaturation.

In terms of clinical toxicities, representatives from all the major groups have been shown to have significant general toxicity affecting various organ systems (Table 1 and 2) and most of them are known to have a variable spectrum of nervous system toxicities (see Tables 2 and 7).

3. GENERAL TOXICITY OF CHEMOTHERAPY

There is a large number of side effects both vitally essential and less vital organs (Table 3). The damage to the first group may cause fatal outcome as a complication of chemotherapy, while damage to the second group causes less vital dysfunctions [13]. According to the onset of disorders, the side effects are separated into a) acute or early toxicity and b) delayed or chronic adverse effects [3a, b, 10, 13, 14].

3.1. Acute Side Effects

Toxic side effects manifesting immediately or only short time after the onset of chemotherapy include a) substance non-specific and b) substance-specific changes (Table 4).

a) Among the *non-specific* changes the most frequent side effects are CNS and GI reactions, like nausea, loss of appetite, vomiting, vertigo, etc., usually of short duration, occurring after administration of most cytostatic drugs. Other frequent acute effects are rash and other hypersensitive skin reactions, e.g. seen after administration of bleomycine, PCZ, etc., and damage to rapidly proliferating cell-systems, i.e. the stem cells of bone marrow, epithelial cells of GI tract (enteritis, mucositis), skin and its appendages (alopezia) and gonades (oligo- or azoospermia), and to the vascular system causing phlebitis and periphlebitis. The most frequent and essential complication is bone-marrow depression resulting in leukopenia, thrombocytopenia or pancytopenia [1 – 3, 3a, 5, 9, 10, 10a].

b) *Substance-specific side effects* of acute onset include:

Table 3. Critical normal tissues in chemotherapy and combined radiation and chemotherapy (modified from [13])

Class I (vitally essential organs)	Class II (less essential systems)
bone marrow	skin
heart	mucosa (GI tract)
lung	salivary glands
liver	bladder
kidney	cartilage and bone
CNS	eye

Table 4. Acute type complications of chemotherapy and combined radiation and chemotherapy

Organ	Side effect
bone marrow	depression, pancytopenia
skin, mucosa	rash, dermatitis, mucositis, alopezia
esophagus	stricture
heart	cardiomyopathy, cardial insufficiency
lung	pneumonitis, interstitial fibrosis
liver	hepatitis, hepatopathy
kidney	nephrosis, acute kidney failure, calcification
vascular system	phlebitis, phlebothrombosis
GI tract	enteritis
bladder	cystitis
gonades	oligospermia, azoospermia

1. *Cardiotoxicity* related to treatment with some antitumor drugs, chiefly anthracycline antibiotics (daunorubicin, adriamycine, adriblastin) causing cardiomyopathy which may be enhanced by simultaneous endoxan treatment and mediastinal radiation [14, 15], cyclophosphamide, causing acute hypokaliemia, 5-FU causing stenocardia and EKG alterations, while less overt cardiotoxic effects of other agents — MTX, actinomycin D, mithramycine, podophyllotoxin derivatives — mostly are dose-related [15].

2. *Epithelial damage* to mucosa and skin has been observed particularly after administration of bleomycin, actinomycin D, adriblastin, and 5-FU, and can be accentuated by combination with MTX and/or irradiation [16]. Hemorrhagic cystitis is known to occur after cyclophosphamide treatment [14].

3. Early *pulmonary toxicity* has been observed after administration of bleomycin, actinomycin D plus radiation, MTX [14] and H-D BCNU [10a, 17].

4. *Hepatotoxicity* with transient elevation of serum transaminases and alkaline phosphatase levels or toxic hepatitis are observed after administration of many agents, e.g. L-asparaginase, MTX, actinomycin D, 6-mercaptopurine, 5-FU and alkylating agents [14, 16].

5. *Nephrotoxicity* is a dose-limiting adverse effect of Cis-platinum reduced by administration of thiol-compounds and probenecid [18].

3.2. Delayed and Chronic Effects

Late adverse effects of chemotherapy used in the treatment of CNS tumors and leukemia include a wide variety of lesions only some of which are fully reversible (Table 5):

a) Delayed *bone-marrow suppression* has always been emphasized as the major side effect limiting the dose of many agents, particularly alkylating drugs, vinblastine, PCZ and MTX [1–3, 3b, 9, 10, 12, 13]. A 23 to 40 day delay for maximum stem cell depression is usally noted; thrombocytopenia and leukopenia

Table 5. Delayed and/or chronic side effects of chemotherapy

1. damage to stem cell pools (bone marrow, hematopoiesis)
2. pulmonary toxicity (interstitial fibrosis)
3. cardiotoxicity (congestive heart failure)
4. hepatotoxicity (liver fibrosis)
5. nephrotoxicity (chronic renal failure, nephropathy)
6. immunosuppression
7. fertility disorders
8. mutagenous and teratogenous lesions
9. carcinogenesis

reach its nadir at about 28 days after administration of BCNU, CCNU, MeCCNU and at about 6 weeks after onset of PCZ treatment. Following both single and multiple agent therapy of brain tumors with alkylating drugs (BCNU, CCNU, MeCCNU, PCZ, etc.), mild to moderate thrombocytopenia with less than 50.000 platelets/sqm is observed in 25% [19, 20] to 85% of the patients [21] and severe thrombocytopenia (less than 25.000 platelets) in 2% [20] to 11% [22] with an average of 5 – 7% [19]. Mild leukopenia (WBC under 4000) is seen between less than 25% [18] to 70% [19 – 21], and severe leukopenia (less than 1000 WBC) in 5% [19] to 25% [23], while variable degrees of anemia are observed in 5% [19] to 30% [20]. Cumulative myelosuppression is observed with increasing frequency almost proportional to the number of chemotherapy cycles given [20, 22, 24], and is more frequent and more severe after multiple-agent chemotherapy and multimodality treatments with radiation than after single-agent chemotherapy [19 – 24]. Fatal complications during periods of pancytopenia include uncontrolled infections and hemorrhagic disorders including subdural and intracerebral hematomas [20].

b) *Pulmonary toxicity* occurs in patients treated with bleomycin [14] and BCNU [10a, 26] or with BCNU in combination drug protocols [25] as well as in animals without brain tumors and in long survivors of rats with brain tumors treated with radiation and systemic BCNU [26]. The incidence of symptomatic and/or biopsy proven pulmonary toxicity characterized by pulmonary interstitial fibrosis varies from 1,3% [25] to 20% [26], and there is a relationship between its occurrence on one hand and the total cumulative dose of BCNU, the number of BCNU cycles, the history of lung disease, the patient's age and the platelet count after the first course of BCNU on the other [26]. These findings indicate that history of lung disease, total BCNU dose (sqm, and duration of treatment are the most important factors in predicting lung toxicity which often runs a fatal course [25, 26].

c) *Cardiotoxicity* effects induced by MTX, 5-FU and anthracyclines may occur after a lag period between the suspension of treatment which suggests the possibility of a drug-induced self-perpetuing cardiopathy due to autoimmune antiheart reaction [15]. The most important complication is congestive heart failure, refractory to treatment due to decreased myocardiac contractility, morphology showing degenerated heart muscle cells with inclusion bodies [15]. Other delayed or intermediary cardiopulmonary and vascular side effects are thrombophlebitis and pulmonary embolism [24].

d) *Hepatotoxicity* due to MTX and 6-mercaptopurine may develop after periods of variable duration; the incidence of hepatic fibrosis after MTX therapy may be as high as 84 percent [14].

e) Chronic *renal failure* has been observed after treatment with BCNU and MeCCNU [27]; urate neophropathy and renal calcification after combined leukemia therapy [28]. Hemorrhagic cystitis with ensuing fibrosis of the bladder wall may occur as a delayed effect of cyclophosphamide therapy [14].

f) *Skeletal abnormalities* e.g. osteoporosis and fractures secondary to MTX therapy have been reported [14].

g) *Ototoxicity* and *vestibular toxicity* is known for CIS-platinum [3b, 29].

3.3. Immunosuppression

The majority of antitumor drugs are known to interfere with the immune system at different levels and with different effects. A compound can depress some functions of the immune system and enhance others. The result of usual chemotherapy regimens is a more profound depression of B than T cell functions, although much of T-cell capacity is also depressed [30]. Immunosuppression must be viewed as a major undesirable effect because of an increasing number of fatal infections in patients in malignancy remissions [28, 30, 31] including opportunistic infections of the CNS (see, p. 323). *Graft-versus-host-disease* (GVHD) is another, less common complication of intensified supportive therapy with blood compounds in immunosuppressed patients undergoing chemo-or multimodality treatment for lymphomas and gliomas [32 – 34]. GVHD is a clinical syndrome caused by infusion of allogenic immunocompetent cells reacting against antigens of the recipient. Skin, liver, gut and bone marrow are the main target organs, leading to skin rashes, GI disturbances and marrow hypocellularity. GVHD is most often observed after bone marrow transplantation, but has also been reported in patients with malignancies treated with granulocyte concentrates [33] and recently was seen in a glioblastoma patient treated with multimodality therapy who for bone marrow suppression was given transfusion of buffy coat cells from fresh blood units [34]. Skin eruptions appeared 7 days after transfusion, followed by liver dysfunction, GI disturbances, renal failure and death in metabolic coma. Histology changes found in skin biopsy include focal dyskeratosis and eosinophilic degeneration of basal cells; autopsy disclosed a total denudation of lymphoid tissue, bone marrow aplasia, necrosis of GI mucosa, liver and kidney damage [34]. GVHD depends on three essential conditions, immunodeficiency of the host, donation of a sufficient number of immunocompetent cells and still unknown degree of disparity of transplantation antigens. It is not possible at present, to define the state of immunodeficiency in which GVHD is triggered, nor are precised data available about the number of immunocompetent cells necessary to initiate such a reaction, but irradiation of cellular products destined for immunologically compromised hosts may prevent this fatal disorder.

3.4. Fertility Disorders and Teratogenous Lesions

Fertility disorders including azoospermia, testicular atrophy and ovarial dysfunctions with amenorrhea associated with single agent and combination chemotherapy are frequently observed. It would appear that the total dose and duration of treatment are important determinants in the reversibility of such complications [3, 14, 16]. Although many of the antineoplastic agents are teratogenic in laboratory animals, only a small percentage of the pregnancies associated with chemotherapy are complicated by fetal malformations [14, 16]. The presently available data do not constitute a complete assessment of the genetic risks of chemotherapy, but a small number of dominant lethal mutations resulting in early death of the embryo may be unrecognized or autosomal recessive mutagenic effects may be latent and expressed later in life or in subsequent generations.

3.5. Carcinogenesis

The carcinogenic potential of alkylating and other chemotherapy compounds has been recognized for several years. Associations between treatment with chemotherapy and the development of secondary malignancies are appearing with increasing frequency. The incidence of acute leukemia and other neoplasms arising in patients given aggressive chemotherapy for lymphomas is about 1,9% [35, 36] and the finding that about 26% of the secondary neoplasms in chemotherapeutically treated patients are acute leukemia point to the connection between them and treatment [37, 37a]. Acute leukemia complicating therapy of glioblastoma [38] is opposed by 2 cases of glioblastoma following radiation and IT MTX or bone marrow transplantation for childhood ALL [39a].

4. MORPHOLOGIC EFFECTS OF CHEMOTHERAPY ON GLIOMAS

The effects of chemotherapy on brain tumors have been studied both experimentally *in vitro* and *in vivo* as well as in humans. Studies in animal glioma models [41–43] and in human glioma tissue cultures [42, 44] have shown that nitrosoureas and other agents inhibit the *in vivo* and *in vitro* growth of tumors as well as the full development of tumor proliferation transplacentally induced in rats [45]. The cytostatic effect is enhanced by association with radiotherapy [46], since nitrosoureas are not only toxic for tumor cells but sensitize them to the action of ionizing radiation [11]. The morphologic changes induced by cytostatic drugs disturbing the generative cycle are characterized by dystrophic lesions of the cytoplasm and nucleus, nuclear pyknosis and karyorrhexis, the latter change usually indicating cell death [2, 47, 48]. Inhibition of the S-phase (DNA synthesis) caused by alkylating agents and CA, etc. which, however, kill cells in all phases of the proliferative cycle at equal rates, and kill non-proliferating cells as well at a lower rate [48], induces nuclear enlargement and increased condensation of perinuclear granules composed of mitochondria and fat. The clustering of mitochondria around the nucleus is identical to the cells in the late S- and G_2-phase prior to mitosis. Mononuclear giant cells with nuclear changes induced by both radiation [49] and nitrosoureas [46] may also develop from disturbances of mitotic activity during later stages, with reunition of nucleic particles. Fragmentation of the nucleus which is another result of inhibition of the S-phase may also be induced by non-specific mitotic disorders and degeneration of the nuclear envelope, with margination of chromatin followed by breaking up of the nucleus into discrete fragments, diffuse condensation of chromatin, sometimes being an early phase of necrosis [47]. Radiation which mainly acts on the post-DNA synthesis phase G_2 and early mitosis can induce the formation of multinuclear giant cells with or without capacity of further division [48, 49]. When drugs cause cells to be inhibited in the phase G_2 and subsequently killed in the M-phase, their action is almost ineffective due to the small feaction of cells in M-phase [48]. Inhibition of the late mitosis M-phase will cause an increase in the number of multinucleated giant cells which also may occur spontaneously by polyploidy [50]. Postmitotic unification of karyomeres results in the formation of intranuclear 'inclusions' due to cytoplasmic invagination. Inhibition of DNa synthesis induced by numerous compounds results in increased cytoplasmic density, loss of cytoplasmic details, cytoplasmic vacuolization, displacement of nuclear chromatin, variation in the nuclear size or nuclear pyknosis [46 – 50]. Studies in animal glioma models and human tissue culture indicate that characteristic, although not specific morphologic features are likely to result from radiation and/or chemotherapy. However, the evaluation of the morphologic effects of antitumor treatment on human brain tumors is difficult because of their great variability in cell and tissue morphology [49 – 52]. Following radiotherapy, the development of widespread necrosis and of degenerative vascular changes, occurrence of bizarre giant cells and increase of abnormal mitoses have been reported [49, 54]. In gliomas treated with single or multiple-agent chemotherapy a similar development of necroses and giant cells with decrease of mitoses were considered significant [54, 55]. However, these and other morphologic changes are non specific and may also occur spontaneously in untreated gliomas [49, 51]. Schiffer et al. [53] comparing the biopsy and autopsy findings in malignant gliomas failed to confirm alterations attributable to chemotherapy, although in addition to increased necrosis and cell polymorphism there was a tendency for a decrease in the number of mitoses and an increase of monstrous cells in some cases, apparently related to survival. Monstrous cells showed a positive relation with radiation dose and the survival after radiation and a negative relation with the time after chemotherapy. This could mean that monstrous cells may be provoked by chemotherapy and, hence, may be indicative of the previous action of drugs on cell kinetics and particularly on inhibition of cell division [54, 55]. Comparative evaluation of the cytologic and histologic features of biopsy and autopsy material in 153 cases of anaplastic gliomas treated with postoperative supportive care, radiotherapy, CCNU monotherapy and COMP polychemotherapy gave the following results: The cytology of (untreated) anaplastic glioma shows high cellularity with strong cellular and nuclear pleomorphism, marked variation in the nuclear structure with one or several nuclei and many mitotic figures (Figure 1a). Giant cells with large, irregular, hyperchromatic or multiple nuclei and poor margination of the nuclei or with nuclear fragments are present; cytoplasmic invaginations are rare. In the supportive care group no appreciable changes were seen except for increased necrosis and vascular changes apparently related to the duration of the tumor which is in line with the general observation of progressive development of necrosis in malignant gliomas with and without antitumor treatment [49,51 – 53]. After radiation, in addition to increased necrosis and vascular reaction, there was an increased number of giant cells with abundant bizarre cytoplasm, large, hyperchromatic and irregular nuclei, with frequent pyknosis, irregular fragmentation and uneven distribution of nuclear fragments within the cytoplasm that often presents basophilic or vacuolar changes (Figure 1b). After chemotherapy, there was frequent formation of multinucleated giant and monstrous cells and prominent reticular or vacuolar degeneration of the cytoplasm with slight basophilia, loss of structural details and degeneration of nuclear and cellular membranes (Figure 1 c-g). The large, hyperchromatic nuclei often show budding and herniation, fragmentation with dissemination of fragments or micronuclei within the cytoplasm. Large cytoplasmic inclusions, intranuclear cytoplasmic invaginations and isolated giant nuclei without surrounding cytoplasm were abundant (Figure 1f-g). Although the general architecture of gliomas was not changed after chemotherapy, and no differences were found for cellular density, pleomorphism, meningeal invasion and type of vascular reaction, there was a significant post-treatment increase in multinuclear giant and monstrous cells, nuclear hyperchromasia and intranuclear cytoplasmic inclusions; these changes were more pronounced after chemotherapy than after irradiation (Table 6). A difference between multinucleated giant cells introduced by antitumor therapy and those spontaneously arising in gliomas has been observed from the point of view of cell kinetics [56]. Since no definite morphologic variations were found in relation to survival time, the type and number of chemotherapeutic treatments or the length of time after treatment, these changes may indicate late effects of a previous ac-

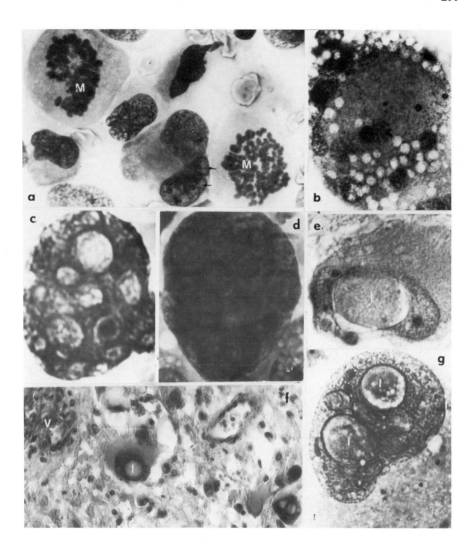

Figure 1. Imprint cytology of glioblastoma. a) Biopsy of untreated tumor showing pleomorphic tumor cells with mitoses (M), lobulated nuclei and micronuclei (arrows) M.G.G. × 665; b) Giant cell with vacuolar degeneration of cytoplasm after CCNU treatment (M.G.G. × 840); c) Degenerating giant nucleus with vacuolation and loss of cytoplasm after polychemotherapy (M.G.G. × 840); d) Degenerating gliant nucleus with loss of chromatin × 945); e) Giant cell with abundant cytoplasm and large intranuclear cytoplasmic invagination (I) after CCNU (M.G.G. × 840); f) Giant cell with large marginated nucleus and intranuclear cytoplasmic invaginations (I) after chemotherapy (× 700); g) Autopsy specimen of anaplastic astrocytoma showing monstrous cells with intranuclear inclusions (I) after HV irradiation H.-E × 210.

Table 6. Histological changes in glioblastomas between time of first and second biopsy or autopsy

Therapy group	Surgery + conventional care Survival 1–14mo n = 62			Surgery + radiotherapy Survival 2–26mo n = 36			Surgery + CCNU Survival 1–16mo n = 15			Surgery + COMP polychemotherapy Survival 3–25mo n = 40		
Histology	↑	↓/=	0	↑	↓/=	0	↑	↓/=	0	↑	↓/=	0
Cellularity	18	44	–	9	27	–	4	11	–	15	25	–
Pleomorphism	14	48	–	5	31	–	4	11	–	8	32	–
Giant cells	16	30	16	14	17	5	10**	5	–	23**	10	7
Monstrous cells	7	13	42	9	12	15	10*	3	2	23*	6	11
Nuclear inclusions	1	8	53	1	5	30	6*	3	6	13*	5	22
Nuclear hyperchromasia	14	42	6	12	23	1	11*	2	2	20*	15	5
Mitoses	20	42	–	14	22	–	1**	14	–	8	32	–
Necroses	34	28	–	30	6	–	10	5	–	31	9	–
Vascular changes	35	27	–	31	5	–	10	5	–	32	8	–
Lymphocyt. infiltrat.	1	18	43	3	13	20	–	3	12	2	9	29
Meningeal invasion	25	24	13	10	16	10	3	2	10	18	16	6
Calcification	–	1	61	3	–	33	1	–	14	–	–	40

* p < 0,01; ** p 0,01 < 0,025. All others not significant ↑ increase ↓ decrease = same 0 not observed.

tion of drugs on cell kinetics. On the other hand, the frequency and intensity of perivascular mononuclear infiltrates considered as an expression of cell-mediated immunity, often decreases in patients with long survival [52, 53, 55]. Their slight, but non-significant reduction after chemotherapy might indicate some immunosuppressive effect of this treatment.

By contrast, there may be a small number of tumors at risk of an adverse response to chemotherapy or radiation with progressive anaplasia that often occurs spontaneously in brain tumors [49, 51]. Among 62 patients anaplastic gliomas treated with postoperative radio- and chemotherapy, Budka et al. [57] observed three cases with increasing preponderance of small anaplastic cells. Comparison of biopsy and autopsy materials from 5 cases with glioblastoma treated with MeCCNU showed increased cellularity and preponderance of small anaplastic cells in the autopsy material in three, while in the two others the tumor had not changed [59]. Necrosis in this group was much less severe than after radiation and radio- plus chemotherapy, and there was no or only minimal fibrinoid necrosis of the tumor vessels which is a prominent feature after irradiation [48, 58, 59]. The occasional increase of anaplasia in treated gliomas, with overgrowth of a primitive, small-cell population suggests enhancement in tumor anaplasia secondary to antitumor treatment. It appears possible that chemotherapeutic agents by altering the cell genome may occasionally 'clone' anaplastic tumor cells with more aggressive behavior, while in the majority of tumors the increased occurrence of both giant cells representing non-dividing terminal stages of proliferation, and of necrosis and vascular response may represent indirect indicators of the previous action of drugs on cell kinetics.

Therapy-induced CNS lesions are observed with increasing frequency in patients with malignant gliomas showing prolongation of life achieved by aggressive multimodality treatment. Neurologic complications possibly related to chemotherapy have been observed in up to 20% of these patients [60]. In malignant gliomas treated by chemo- or radiotherapy, or both, diffuse cerebral edema and extensive tumoral and peritumoral necrosis [20, 49, 52–54, 58, 59], intracranial hemorrhage and other lesions have been observed [20]. Intracranial hypertension which is generally indicative of tumor recurrence, may be caused by extensive necrosis presenting as mass lesion, by fluid accumulation under tension in the operation cavity [60] or by formation of space-occupying necrotic cysts at the site of the original tumor [20, 60, 61]. The incidence of such cystic cavities due to tumor necrosis and regression after chemotherapy or multimodal treatment ranges from 11,8% [60] to 16% [61]. Clinical signs of increased intracranial pressure or neurologic deterioration or CT evidence (Figure 2A, B), or both, of intracranial cysts develop 4 to 12 months after the first craniotomy and 3 to 9 months after radiotherapy after 2 to 4 courses of chemotherapy [61] or after chemotherapy alone [60]. At surgery or autopsy these cysts are limited either by necrotic tissue of superficially by the dura, or they widely communicate with the

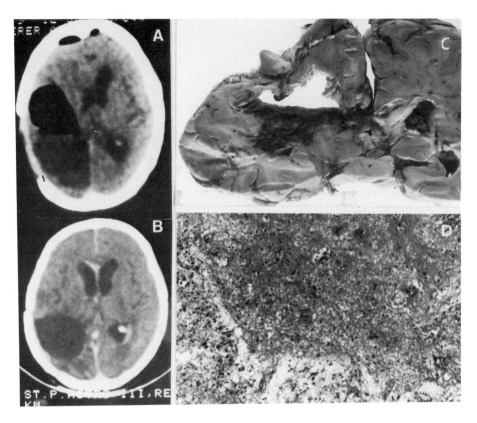

Figure 2. A, B) CT and autopsy specimen showing large fluid-filled cyst in left hemisphere, only small tumor (T) left in callosal splenium; death 14 months after removal of left parietal glioblastoma with radiation (40 Gy) and 5 courses of COMP polychemotherapy; C, D) Large left parietal cyst communicating with ventricle with glio-mesenchymal wall and adjacent glial tissue (biopsy specimen; H.-E × 100) 13 months after surgery, radiotherapy and 4 polychemotherapy courses.

enlarged ventricles (Figure 2A-C). Histology shows coagulative necrosis at or adjacent to the operative site and extending into the white matter; the walls are formed of glio-mesenchymal, hypervascularized tissue with or without fibrinoid vascular necrosis and residual or recurrent tumor (Figure 2D). The pathogenesis of these space-occupying cysts which may also arise spontaneously in anaplastic gliomas and can mimick tumor recurrence is not fully understood. Surgical treatment may produce transient improvement, but usually does not prevent the fatal outcome weeks to months after surgery of cyst formation [61].

Nitrosureas and other chemotherapeutic agents are also synergistic with irradiation delivered to the brain for treatment of malignant gliomas in producing diffuse cortical dysfunction and insidious dementia unrelated to tumor regrowth in more than 10% of the long surviving patients [61a, 61b]. Sequential CT scans in these patients reveal diffuse cerebral atrophy, progressive hydrocephalus, and occasional signs of leukoencephalopathy [42, 61a, b].

Table 7. Neurotoxicity of some chemotherapeutic agents (modified from [3b])

Drug	Route of administration	Brain/cranial nerves	Spinal cord roots/meninges	Periph. nerve/ muscle
Nitrogen mustard	i.v.	0	0	0
	HD i.v., i.a.	edema, necrosis	0	plexo/neuropathy
Cyclophosphamide	i.v., p.o.	0	0	0
Thiotepa	i.v.	0	0	0
	i.t.	?	myelopathy, radiculop.	?
MTX	p.o.	rare leukoencephalopathy	0	0
	i.v. (conv. dose)		0	0
	HD i.v.	dissem. leukoencephalopathy		
	i.t.		asept. meningitis myelo-radiculopathy	0
5-FU	i.v.	encephalopathy cerebell. ataxia	0	0
Ara-C	i.v.	0	0	0
	HD i.v.	encephalopathy cerebell. dysfunct.		
	I.T.		asept. meningitis myelopathy	0
5-Azacytidine	i.v.	encephalopathy	0	"myopathy"
Vincristine	i.v., p.o.	seizures cranial neuropathy	0	per. neuropathy
Bleomycin	i.v., i.m.	encephalopathy	0	?
Nitrosureas	i.v., p.o.	cerebral atrophy encephalopathy	0	0
	intracarotid	focal lesions	0	0
	HD i.v.	multifocal demyelinating necrotic lesions	0	0
Cis-platinum	i.v.	0	0	ototoxicity vestibular les.
DTIC	i.v.	?	0	neuropathy (?)
HXM	p.o.	encephalopathy	0	0
L-asparaginase	i.v.	encephalopathy	0	0
Procarbazine	p.o., i.v.	rare encephalopathy	0	rare neuropathy
Mitotane	p.o.	seizures	0	rare neuropathy
PALA	i.v.	encephalopathy	0	0
Thymidine	i.v. (+ 5-FU)	cerebellar ataxia	0	0
Methy-GAG	i.v.	seizures	0	"myopathy", rare neuropathy

5. NERVOUS SYSTEM COMPLICATIONS OF SYSTEMIC AND IT CHEMOTHERAPY

The neurotoxic effects of cancer chemotherapy which have been extensively reviewed [3, 3a, b, 5, 6, 10, 62 – 64] are divided into four groups:

a) *Acute reactions* ranging from meningeal irritation or chemical arachnoiditis to transient or permanent, occasionally fatal, paraparesis, and transient focal or diffuse cerebral deficits with sudden onset and complete clinical recovery.

b) *Delayed encephalopathy* including transient neurologic impairment and long-term sequelae ranging from abnormal CT in clinically asymptomatic patients to insidious and progressive organic brain syndromes or reduced level of intellectual function to severe disorders associated with a variety of pathologic entities, e.g. disseminated necrotizing leukoencephalopathy (DNLE), periventricular demyelination, diffuse parenchymal damage, dystrophic calcification.

c) *Myelo (radiculo) pathy* with development of transient or permanent paraparesis eventually resulting in death following IT chemotherapy.

d) *Peripheral nervous dysfunction* and polyneuro (myo)pathies chiefly seen after administration of vincristine.

Following *IT administration* of cytostatic drugs 4 types of side effects are seen [64 – 66]: a) acute arachnoiditis reversible within 24 hours; b) acute neurologic dysfunctions reversible after 24 hours; c) paraparesis and other serious dysfunctions; d) encephalopathies. A summary of the neurotoxicity of non-hormonal chemotherapeutic agents is given in Table 7.

5.1. Acute CNS Reactions

Acute neurologic syndromes resulting from chemotherapy can be divided into 3 major groups: a) acute meningeal reactions; b) transient paraparesis particularly following IT chemotherapy; c) acute and transient cerebral dysfunctions (encephalopathies) following both systemic and IT chemotherapy.

5.1.1. Acute Meningeal Reactions (Chemical Arachnoiditis). About 30 to 50% of the patients receiving repeated IT MTX, CA etc. acquire some combinations of fever, headache, nausea, vomiting, dizziness, nuchal rigidity and pain in the back radiating into one or other extremity. This syndrome is generally mild and selflimiting; it occurs soon after instillation and resolves within several days [10, 64, 66 – 70]. It may resemble leukemic meningitis, bacterial or other form of aseptic meningitis [67]. CSF pleocytosis (up to 100 cells/qmm) and elevated protein levels with electrophoretic signs of disturbed BBB present in many cases usually clear without sequelae after therapy [10, 64, 70, 71]. However, both the presence of aseptic meningitis and of blast cells with normal CSF cell counts is occasionally discovered in patients with acute leukemia and lymphomas as a result of routine lumbar puncture [72]. IT chemotherapy of CNS leukemia and lymphomas induces reduction of abnormal CSF cells which show severe damage with pyknotic nuclei and basophilic degeneration of vacuolized cytoplasm with evaginations [72, 73]. Rare forms of intermittent meningitis with severe basophila and/or eosinophilia during IT MTX treatment of acute CSN leukemia or lymphoma have been interpreted as local cell-mediated immune reactions secondary to neoplastic involvement of CSF [74, 75].

5.1.2. Acute Cerebral Dysfunctions (Transient Encephalopathies). A variety of focal and diffuse cerebral deficits with sudden onset and complete restitution has been reported with many drugs [3a, 3b, 5, 10, 62].

a) *L-asparaginase* (LA), used in the treatment of childhood ALL, in about 25 to 50% of the patients causes an acute disorder of consciousness with lethargy, confusion, somnolence, seizures or hallucinations which usually begins within one day after onset of therapy and clears rapidly after its end [64, 76]. Less commonly an organic brain syndrome resembling delirium tremens or Korsakoff syndrome may develop one week or more after treatment. The severity of cerebral dysfunctions which may be accompanied by transient slowing of EEG pattern, is highly variable shows no clear dose-toxicity relations or is related to cumulative dose [3a, b]. It is considered to result from drug-induced metabolic abnormalities and liver dysfunctions [10, 76].

b) *Procarbazine*, a potent monoamine oxidase inhibitor [12] shows its neurotoxicity occurring in 10−31% of the patients, by transient disorders of consciousness ranging from drowsiness, disorientation to stupor associated with diffuse slow wave EEG activity and mental changes including hallucinations, agitation, mania or depression and nervousness accompanied by nausea [3b, 10, 12].

c) *Nitrogen mustard* (mechlorethamine) has produced acute cerebral dysfunctions with seizures, coma and death only after intracarotidal and regional perfusions [10]. Ingestion of overdose of *chlorambucil*, a closely related alkylating agent, caused coma and status epilepticus or lethargia, ataxia and hyperactive jerks clearing rapidly without permanent neurologic damage [10]. Seizures may occur after conventional doses in children with nephrotic syndrome [76a].

d) *Vincristine* after IV administration has been reported to produce seizures in 1−4% of the patients which partly were related to hyponatriemia resulting from inappropriate ADH secretion [10, 77]. Seizures and reversible coma with transient abnormal EEG findings occurring up to 8 days after treatment were observed in up to 15% of the children [10, 64] which was not confirmed by others [78]. Ataxia and athetosis are rare side effects of vincristine [79].

e) *Methotrexate* given in HD therapy for osteogenic sarcoma in 1−2% of the patients produces sudden onset focal cerebral deficits with hemiplegia and focal seizures occurring about 10 days after a course of chemotherapy; symptoms fluctuate or appear on the contralateral side within 72 hours. This syndrome does not recur after similar therapy is reinstituted; neurologic sequelae are minimal, neurologic data are normal and EEG shows diffuse mild abnormalities [80].

Similar type of acute or subacute cerebral dysfunctions including hemiparesis and seizures have been observed in children during combined chemotherapy for ALL [64, 81].

f) *Cis-platinum* may cause acute reversible encephalopathy with transient cortical blindness and occasional seizures, often associated with drug-induced metabolic alterations, e.g. hyponatremia, hypocalcemia, and hypomagnesemia due to renal tubular dysfunction [81a].

g) *5-FU* in less than one percent of patients produces an acute cerebellar syndrome with dysmetria, ataxia, coarse nystagmus and dizzines. This syndrome, the incidence of which increases with high doses or intensive daily-dose regimens is reversible, and can be experimentally reproduced in cats [10, 82]. It is suggested to result from intoxication by fluoroacetate and fluorocitrate, degradation products of alpha-fluoro-beta alanine, the major catabolite of 5-FU which may block the Krebs cycle in the cerebellum [82]. 5-FU causes encephalopathy in up to 40% of the patients, with lethargy or coma, parkinsonism, visual disorders and cerebellar dysfunctions, most of the symptoms showing complete reversibility, and no pathological abnormalities at autopsy [3a, 10]. Reversible encephalopathy has been reported in patients receiving both 5-FU and *thymidine* which are known to produce synergistic cytotoxic effects [81a]. Confusion and lethargy developing within 48 hours of treatment subside within days after cessation of therapy. Cerebellar signs typical of 5-FU are markedly enhanced when combining it with thymidine [3a]. Encephalopathy manifested by somnolence headache, memory impairment and visual illusions, has also been seen in patients treated with HD thymidin infusions alone [3b]. *Bleomycin*, when given with 5-FU or MTX is also known to produce reversible disturbances of consciousness and cognition [3b, 10].

h) *Mitone* (Lysodren), used in the treatment of adenocorticocarcinoma, in 35 – 40% of the patients induces dose-related, usually reversible encephalopathy with somnolence, lethargy, headache and blurred vision [3b].

5.2. Delayed Encephalopathies

Cerebral dysfunctions may not occur until several months after completion of chemotherapy. This delayed type of encephalopathy, usually occurring in patients on long-term treatment receiving large cumulative parenteral and/or IT doses of cytostatic agents [3, 6, 9, 63, 64, 83] with or without cranial irradiation, may include a) transient disorders with apparent clinical recovery [3, 3b, 62 – 64, 84 – 86], b) delayed impairment of mental development [64, 87, 88] and c) permanent and severe disorders with occasional fatal course.

5.2.1. Transient Encephalopathy. Neurologic dysfunctions in long-term survivors of leukemias and lymphomas occurring several months after completion of chemotherapy consist of recent memory loss, incoordination, hyperkinesia, gait ataxia and occasional seizures with EEG abnormalities [10, 63, 64, 87, 88, 90]. Meningoencephalopathy of varying severity with a non-bacterial CSF pleocytosis and complete recovery developed in 18% of leukemic children 2 to 17 months after prophylactic cranial irradiation and IT MTX, while no such symptoms were observed after systemic chemotherapy without CNS treatment [84]. A variety of non-leukemic CNS syndrome ranging from apsetic meningitis [71] to 'somnolence' or 'apathy syndrome' [91, 91a] characterized by lethargy, dizziness, dullness, anorexia, headache, depression and vomiting with slowing of EEG activity and occasional signs of mild intracranial hypertension [64] has been observed in children between one and several months after cranial irradiation and IT MTX [64, 70, 91, 91a], or cranial irradiation without IT chemotherapy [70, 71, 92–94]. All syndromes are reversible within days to weeks. The incidence of this transient late encephalopathy – a counterpart to transient early encephalopathy occurring during chemotherapy – is reported to range from 0 to 79% [64, 91, 91a]. In adults receiving HD-MTX somnolence and confusion accompanying drug administration with complete neurologic recovery is seen in 2 to 20% [3, 80, 90]. The somnolence syndrome may be an early indication for permanent neurologic damage related to combined chemo- and radiation therapy [91a].

After systemic administration of HD CA transient cerebral or cerebellar dysfunction occurring 6–8 days after the first dose and lasting 3–7 days were observed in more than 20% of the patients, CNS toxicity being dose-related [90a]. Intravenous infusion of PALA (N-phosphoacetyl aspartic acid), inhibiting the enzyme aspartate transcarbamylase, an early step in *de novo* pyrmidins biosynthesis, in about 12% of the patients induced delayed onset neurotoxicity with multiple seizures and transient focal neurologic symptoms, occurring between 2 and 6 weeks of cumulative treatment [3b].

5.2.2. Chronic Encephalopathy. Severe forms of long-term sequelae of chemotherapy are clinically featured by insidious onset of irritability, personality changes, agitation, confusion, drowsiness, slurred speech and dementia followed by ataxia, spasticity, dysphagia, altered sensorium, hemianopia, occasional seizures and decerebrate state. The process may be fatal, but most patients survive with signs of permanent neurologic damage [63, 64, 83, 87, 90, 95, 96], and some recover completely [3, 62–64, 80, 85, 86, 90, 97–99, 99a]. Delayed CNS damage may occur after following regimens: a) CNS irradiation and IT chemotherapy or various combinations of oral, IT and low-dose IV MTX without citrovorum factor rescue (CFR) for CNS leukemia or lymphomas [62–64, 83, 85, 90, 96, 100–104]; b) intraventricular MTX with or without cranial irradiation for primary CNS tumors [10, 105–107]; c) cranial radiotherapy and systemic IV MTX with CFR and IT MTX [108] or with systemic HD-MTX for [109]; d) cranial irradiation and systemic chemotherapy plus IT thio-TEPA [110] or CA [110a]; e) IT MTX and/or CA without previous CNS irradiation [63, 89, 111, 112]; f) HD-IV MTX with CFR for osteogenic sarcoma and other neoplasm without cranial irradiation or IT MTX [3, 113, 114]; g) HD BCNV((1500 – 300 mg/m^2/dose) followed by homologous bone marrow transplantation for brain tumors or extracranial metastatic disease without previous CNS irradiation [114a]; h) following DTIC dementia and EEG abnormalities have also been observed [3a, 115].

The *incidence* of this debilitating complication is variable. The syndrome has not been reported with CNS irradiation alone in the 18 to 24 Gy range [83] and rarely with IT MTX and CA alone [64, 86, 90a, 111], and with HD-MTX without either IT MTX or CNS radiation [3, 80, 114]. Rosen *et al.* [114] observed 3 clinical cases among 158 osteosarcoma patients receiving HD-IV MTX (15 g/sqm) with CFR, indicating an incidence of less than 2%. All showed recovery after discontinuition of chemotherapy, but in a later series 7 patients showed severe persistent neurologic sequelae with diffuse white matter hypodensity or atrophy in CT [113, 116]. With combined CNS irradiation and IT and/or IV MTX the incidence is higher; with two modalities clinical encephalopathy may occur in 2 to 15% of the patients [8, 63]. In leukemic patients receiving the standard form of CNS prophylaxis with cranial irradiation and IT MTX, symptomatic encephalopathy may occur in 1 to 10% [83, 99, 102]. In one large series [99], encephalopathy developed in 5% among 248 children prophylaxed with 24 Gy and 5 doses of IT MTX, while Price *et al.* [83] found leukoencephalopathy in 13 of 213 autopsies (5,7%) of children with ALL after combined IT MTX (total dose 42 – 11131 mg) plus IV MTX during and/or after cranial radiation of 20 Gy or more for CNS leukemia. Among 22 children on long-term IV MTX for ALL and lymphoma, four showed severe impairment with seizures, dementia and paraplegia, brain biopsies in three revealing white matter gliosis; another 10 children had mild clinical or EEG changes. None of the 9 normal children had cranial radiation and CNS leukemia, while two of the severely affected had CNS radiation and IT MTX for CNS leukemia [85]. In a personal autopsy series of 31 cases of ALL, mostly children, all treated with systemic and IT MTX during or after CNS radiation, we found two confirmed cases of DNLE (6,4%). Rubinstein *et al.* [96] reported a 45% incidence of LE among patients treated with high doses of CNS radiation (> 35 Gy) and a larger dose of IT MTX plus additional CA and prednisone. Hence, the occurrence of severe encephalopathy appears to be highest when all three modalities, i.e. cranial irradiation, IT and IV MTX or other chemotherapy – are used, particularly if CNS radiation is administered before or during the systemic chemotherapy [63, 63a, 104]. However, similar DNLE was seen after HD systemic BCNU without CNS radiation [114a].

5.2.3. Residual Dysfunctions and CT Findings. While some clinical and neuropsychologic studies did not find any disorders of mental growth and psychic per-

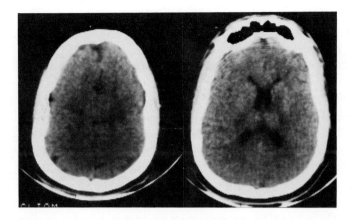

Figure 3. Normal CT scan in male aged 18 years suffering from seizures during and following HD-systemic MTX therapy for metastasizing osteosarcoma (6 cycles 20 g/sqm).

formance after long-term multimodel chemotherapy for childhood leukemia or other neoplasms [3, 99a, 117–119], others observed serious neurologic and neuropsychologic dysfunctions [63, 64, 83, 85, 88, 120]. Since the advent of CT, a number of side effects including decreased density of white matter, calcifications, dilatation of the interior and outer CSF spaces, have been reported after long-term chemotherapy [63, 120–126]. Abnormal CT scans were found in 50 to 66% of asymptomatic long-term survivors with acute leukemia in remission [113, 121, 126] including decreased density of white matter suggesting LE and hydrocephalus or brain atrophy [120–123, 124–126, 126a], while others found no such abnormalities as compared to pretreatment data [127–129], or ventricular dilatation was confined to cases complicated by CNS leukemia treated with IT MTX [3, 129]. From the available data there is no evidence that IT chemotherapy in the absence of CNS leukemia and CNS irradiation may produce CNS changes detectable by CT scan [128, 129]. (See Figure 3). CT scans may show improvement after drug discontinuation [86].

5.3. Neuropathology Syndromes following Chemotherapy

A variety of pathologic changes in the CNS associated with long-term sequelae of chemotherapy has been observed, while in some cases neuropathologic lesions were random postmortem findings unassociated with neurologic signs [96, 110, 130]. The following neuropathology syndromes associated with chemotherapy or combined treatments of malignancies have been observed which often are the anatomical counterparts of the clinical disorders mentioned:

a) Disseminated necrotizing leukoencephalopathy (DNLE)
b) Diffuse parenchymal damage (gliosis, axonal dystrophy)
c) Focal and diffuse subpial necrosis of gray matter

d) Mineralizing microangiopathy and dystrophic calcifications
e) Central pontine myelinolysis
f) Wernicke's encephalopathy
g) Cerebral atrophy and communicating hydrocephalus

5.3.1. Disseminated Necrotizing Leukoencephalopathy (DNLE). Kay *et al.* [90] described 7 children with acute leukemia who after long periods of oral, IV and IT MTX (total doses 944–4458 g) developed neurologic signs, dementia, confusion, tremor, ataxia and somnolence; two had epileptic seizures and one had progression to coma and death. Autopsy showed multiple small focal necroses ('infarcts') in the temporal and parietal lobes associated with fibrinoid vascular necrosis, with no evidence of inflammatory disease or leukemic infiltration. Bresnan *et al.* [105] reported a series of children who received radiation and intraventricular MTX for malignant brain tumors. Six of 9 cases autopsied showed astrogliosis and myelin destruction; only two of the brains showed fibrinoid vascular changes. There appeared to be no correlation between the dosage of MTX and the degree of white matter damage. Shapiro *et al.* [107] observed 3 cases with coagulative necrosis in periventricular white matter with swollen axons and fibrinoid necrosis and vascular thrombosis after cranial irradiation (36–60 Gy) and intraventricular MTX (86–190 mg) to children with recurrent posterior fossa tumors and evidence of ventricular CSF outflow obstruction. They developed signs of acute bilateral cerebral and brainstem dysfunction with rapid progression to decerebrate state; one child developed signs during IT treatment, the others 3 to 5 months after starting MTX; they died 5 to 14 months later. Since 4 other patients with meningeal leukemia treated with IT MTX did not show these periventricular lesions, these were attributed to transependymal absorption of toxic amounts of MTX resulting from CSF outflow obstruction. Since these reports, in addition to a series of clinical and CT proven cases of LE [3a, b, 63, 86, 97, 113, 114, 121–125, 131], a number of morphologically confirmed instances of DNLE have been reported in treated childhood leukemia and lymphomas [3b, 83, 90, 96, 99–104, 110, 132–134] cerebral lymphoid granulomatosis [135], in brain tumors [105–109], soft tissue or osteosarcomas without CNS disease treated with IV HD-MTX [3, 113, 114] and brain tumors treated with HD-BCNU [114a].

The typical clinical course begins 3 to 15 months after CNS irradiation some days after the onset or shortly after completion of MTX or combined IT treatment with the insidious and, rarely rapid, onset of apathy, disorientation, drooling, dysarthria or dysphagia. The dementia is usually progressive and often associated with other neurologic signs including spasticity, ataxia, nystagmus, hemianopia, hemiparesis, seizures, decerebration and coma. CSF shows slightly elevated protein and marked elevation of myelin basic protein [63, 131]. Brain scan reveals increased deposition of nuclides in the periventricular areas, mainly around the frontal horns [99, 102]. In the early phase, CT shows periventricular hypodensity without contrast enhancement, beginning around the frontal and then the occipital horns, and later along the entire length of the ventricles [97, 121–124, 126]. Death occurs one to 7 months after the onset of neurologic disorders, but in some cases DNLE was recognized at autopsy without preceding clinical neurological disorder [94, 96]. The incidence of DNLE in autopsy series ranges from 1% [102] to 25% [96] with an average of 6% [83]. The neuropathology is characterized [83, 96, 101, 111] by:

a) Multiple foci of coagulation necrosis extending by confluence and disseminated in the cerebral white matter in a random manner. In most severely affected cases naked eye observation discloses large necrotic foci in the centrum ovale around the enlarged ventricles (Figure 4), and extensive symmetric demyelination affects both hemispheres. Multifocal necrosis and demyelination may also involve the pons [108], cerebellum and cervical corticospinal tracts [100, 103], while the cortical gray matter and basal ganglia are spared.

b) In the damaged areas there is remarkable absence of inflammatory cellular reaction and a relative paucity of macrophages;

c) The necrotic areas show demyelination, remarkable decrease of astroglial cells and occasional deposits of mineral salts;

d) There is striking axonal damage with axonal swellings within and around the foci which ultrastructurally correspond to reactive and degenerating axons with abundant mitochondria, dense and multivesicular bodies and aggregates of electron-dense crystals [96, 111].

e) Marked status spongiosus and moderate astrocytosis are present in the surrounding white matter;

f) Occasional fibrinoid vascular necrosis was seen in some but not in all cases [83, 96, 99–102]. The initial pathologic changes present as multifocal coalescent areas of spongy necrosis with reactive astroglia; later these changes are more widespread and accompanied by clusters of macrophages reflecting myelin breakdown, while late stages consist of diffuse coagulation necrosis of the white matter with axonal swellings, dystrophic calcifications and reactive gliosis [113]. In the advanced stage the white matter is reduced to a thin gliotic calcified layer with marked enlargement of the ventricles [102]. Multifocal necroses distributed throughout the cerebral cortex and white matter associated with vascular fibrinoid necrosis but without neurofibrillary degeneration was reported in a boy with ALL who, during IV treatment with vincristine (1–2 mg/sqm) developed seizures and, after additional IT MTX treatment without radiotherapy, died in coma 26 days after the onset of chemotherapy [136]. Other patterns of multifocal necrosis with similar histologic features but prominent distribution of the lesions in the brainstem, particularly the pontine basis, was described in 3 patients with neoplasms who received whole brain radiation and systemic chemotherapy including IV MTX-HD and CFR, and occasional IT MTX/CA [108] and in one

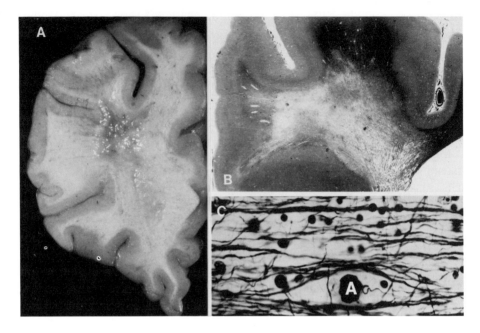

Figure 4. Disseminated necrotizing leukencephalopathy in boy aged 5 years with ALL treated with cranial irradiation (26 Gy), systemic and IT MTX. Necrotic area in left frontal subcortical white matter (A) with flaky demyelination (B) and occasional axonal swellings (C-Bodian × 300).

case of glioblastoma each treated with radiation and PO MeCCNU [108] and procarbazine (58 – case 16).

5.3.2. Parenchymatous CNS Degeneration. Review of 24 autopsy cases of leukemic children who received systemic chemotherapy (L-asparaginase and/or IT MTX) plus cranial radiation (14 – 36 Gy) revealed diffuse subcortical fibrillary gliosis with predilection for the white matter is 95% [95]. Fibrillary gliosis, accentuated in subependymal and subpial regions, olives, cerebellar nuclei, pontine basis, nigra and anterior spinal horns, was independent of myelin pallor and leukemic infiltrations present in 14 brains. There was frequent axonal dystrophy and 22 cases showed prominent Alzheimer type II glia in gray and white matter, most associated with hepatopathy or leukemic liver infiltration. Similar degenerative lesions were seen in 11% among 91 autopsy cases of childhood leukemia, only half of which had received radio- and chemotherapy [137]. Diffuse gray and white matter lesions with abundant Alzheimer type II glia were seen in 35,6% of 115 autopsy cases of non-Hodgkin's lymphomas [138] and in 20% of a personal autopsy series of acute leukemias and adults, many of them treated with systemic chemotherapy, but only one-quarter with CNS radiation.

5.3.3. Diffuse and Focal Subpial Necrosis of Gray Matter. Other patterns of encephalo(myelo) pathy are characterized by diffuse of sharply demarcated areas of incomplete necrosis of gray matter of the brain and spinal cord without meningeal infiltration or vascular pathology [63, 112, 139]. These patterns are exceptional and occur either in non-irradiated patients treated with systemic chemotherapy and IT MTX or IT MTX plus CA [112, 139] in which rim-like superficial tissue destruction in cortex, brainstem, cerebellum and spinal cord (along entrance zone of the roots), i.e. regions adjacent to the CSF suggest effects of a toxic substance in the CSF, or in a patient whose MTX was administered via an Ommaya reservoir, the cannula tip of which was located in brain parenchyma [63].

5.3.4. Mineralizing Microangiopathy and Dystrophic Calcification. Intracranial calcifications involving the basal ganglia and cerebral cortex in children after combined treatment for ALL and lymphomas have been observed by skull radiography [140 – 142], CT [69, 122 – 124, 128, 141, 142] and neuropathology [103, 137, 139, 143, 144]. The lesions which may or may not be associated with clinical signs of delayed encephalopathy or behavior disorders, chiefly affect the gray matter, primarily the putamen with or without additional involvement of the cerebral and cerebellar cortex, the latter pattern was seen in 10 out of 28 autopsy cases [144]. Histologically, deposition of calcified material containing amorphous calcium phosphate and calcium orthophosphate, most probably precursors of hydroxyapatite [145], is found in and around small blood vessels, with lumen of small vessels totally occluded by mineralized debris. Primarily affected are small arteris, precapillary arterioles and venules [143, 144]. The lesions are accompanied by varying necrosis and calcification of adjacent neuronal tissue; inflammatory lesions are absent. The type and pattern of calcifications resembles Sturge-Weber syndrome [139, 141] or Fahr's syndrome showing similar results of chemical analysis and X-ray diffraction [145]. The incidence of cerebral calcification in clinical and autopsy series of childhood leukemia ranges from 0 to 30%. While in some series of children with ALL in remission after CNS prophylaxis no calcification was found in CT [126, 127], in other CT series there were 2,7 to 5% of treated children [125, 128] as compared to 2% among non-oncologic controls [128] and 2% each in two autopsy series of treated childhood leukemia [137, 138] opposed by none among 31 personal autopsy cases. Among 39 children with treated ALL 26% were found to have subcortical calcification in CT [142], while among 163 autopsy cases of treated ALL 17% had calcification; this incidence increased up to 25% after survival for more than 10 months after treatment [143]. The clinical significance of these lesions is variable. While 8 of 10 patients with cerebral calcifications had neurologic problems including ataxia, percepto-motoric disability and seizures [142], others demonstrate gait abnormalities and focal seizuresor somnolence

[143], while most had no objective neurologic dysfunctions unless there may be coexistence with DNLE [132, 143].

5.3.5. Central Pontine Myelinolysis (CPM). Symmetrical patchy demyelination in the central pons with preservation of neurons and axons, disappearance of oligodendroglia and reactive gliosis, described in chronic alcoholism, malnutrition and liver disease, has been observed in some cases with acute leukemia [136, 146 – 148] and lymphoma [149], all treated with chemotherapy. CPM was seen in 4% of 91 autopsy cases of childhood leukemia including the youngest case ever recorded in 1-year old child with ALL [137]. The clinical picture is a confusional state with cranial nerve disorders, weakness of limbs, occasionally resulting in locked- in syndrome. CPM which is frequently associated with liver dysfunction, electrolyte disorders and iatrogenic factors of delayed rehydration and may be combined with Wernicke encephalopathy [150] is suggested to be related to complex metabolic derangements.

5.3.6. Wernicke's Encephalopathy (WE). Combination of WE and ALL was reported in a patient who was successfully treated with thiamine [151], and in 3 autopsy cases, two children with AML and Hodgkin's disease and an adult with malignant lymphoma, clinically presenting increasing stupor 3 to 15 days before death [130]. Autopsy disclosed petechiae in the regions adjacent to the third and fourth ventricle and around the aqueduct, while histologically the lesions indicated a chronic disorder with signs of acute exacerbation. In a prospective study of 24 patients with leukemia and lymphomas, WE was detected in 33,3% [152]. No correlation was found to the type of malignancy and liver steatosis as a marker of malnutrition. In a personal series of 31 cases of ALL treated with systemic chemotherapy, we found 2 instances of WE (6,4%), one associated with DNLE.

5.3.7. Cerebral Atrophy and Communicating Hydrocephalus. Hydrocephalus is considered by recent CT studies to be a main adverse effect of long-term anticancer therapy [121 – 126] and is said to be a direct consequence of radiation or multi-modality treatment [3a]. However, in none of these studies was dilatation of the CSF spaces verified by pretreatment examination. Kretzschmar *et al.* [128] found hydrocephalus in 31% of children with ALL before treatment and the percentage of abnormal CT findings after therapy was constant. These data were confirmed for children after CNS prophylaxis with IT MTX without radiation [127, 129] or CNS leukemia [3] and in osteosarcoma patients on MTX-HD [153] (Figure 3). On the other hand, Crosley *et al.* [137] reviewing 91 autopsy cases of childhood leukemia found cerebral atrophy in 65% which did not correlate with CNS leukemia but there was a relationship between the degree of atrophy and duration of illness and IT MTX given alone or in combination with CNS irradiation, while radiotherapy alone did not seem to have any effect. It must be emphasized, however, that patients who died with CNS leukemia in an era prior to the use of radiation and chemotherapy has also postmortem evidence of hydrocephalus in up to 73% [154]. Communicating hydrocephalus was mainly due to diffuse infiltration of the arachnoid. In a recent series of 124 autopsies of treated leukemia without CNS involvement 7 cases of communicating hydrocephalus were found, in 4 of which was there clear evidence of invasion of the arachnoid villi and sinuses by leukemic cells, areas which are known to be resistent to antitumor therapy, causing obstruction of CSF drainage [155].

5.4. Myelo-Radiculopathy following IT Chemotherapy

Transient or persistent paraparesis of paraplegia have been reported as infrequent complications of IT chemotherapy with vincristine [156–158], MTX [89, 103, 159–174], CA [175], MTX plus CA [161, 164, 165, 167], and Thiotepa [167a]; in some paraparesis recurred with the initial dose of IT CA after MTX [161, 167]. Most of these patients received therapy for preexisting CNS leukemia or tumor at the onset of paralysis, most had received prior courses of IT MTX (range 5 to 53 doses), the cumulative doses of IT MTX ranging from 10 to 305 mg, while one patient died 30 mins after a dose of IT MTX, probably due to acute hypersensitive reactions, after having previously received multiple dose IT MTX without any difficulty [160]. Complications associating previous IT MTX had included fever, chills, headache, vomiting and transient pain or weakness in one or both legs. In most cases, the onset of the deficit was rapid, and occurred immediately after instillation, while in some it was delayed as much as 24 to 40 hours [168], 2 to 6 weeks [174, 175] or even several months after IT administration [173]. Neurologic impairment improved rapidly in some [162, 165, 166, 169], while in 2 cases after changing to IT CA paraplegia developed again, but with only partial resolution [161, 165] and other developed permanent paraplegia and mid-thoracic sensory level or ascending paralysis persisting until death within 30 minutes [160] to several months after onset of symptoms [110a, 159, 167, 173–175]. Neuropathologic information in this group is limited (Table 8): Sullivan *et al.* [159] in a patient with ALL who after 20 doses of IT MTX developed paraplegia and thoracic sensory level persisting for 7 months until death described focal myelomalacia of the thoracic cord and combined tract degeneration in addition to perivascular leukemic infiltration in the involved area. Since the patient had received radation to the cord 6 months prior to death, factors other than IT MTX may have been important. Saiki *et al.* [167] in a girl with ALL who after IT MTX reported transient sensory T-10 level and paraplegia and after later IT CA developed permanent paraplegia and died one month later, autopsy showed demyelination of the spinal nerve roots in the right lumbosacral region, and rim-like demyelination of the subpial spinal cord white matter with no leukemic infiltration. Breuer *et al.* [175] reported a 65-year old male with AML

Table 8. Myelo(radiculo)pathy following IT chemotherapy

Authors	Diagnosis	Age, sex	Drug	Evolution	Neuropathology findings
159	ALL	3,5 M	MTX	7 months death	focal myelomacia thoracic cord, degeneration + demyelination posterior columns, perivasc. leukemic infiltrates
160	ALL	11 F	MTX	death 30 minutes	no neuropathology
161	chorioca metastas.	20 F	MTX	trans. paraplegia; sensory L 1	0
162	ALL	7 M	MTX	trans. sens. paraparesis	
163	ALL	6 F	MTX		no anatomical report
	ALL	8 M	MTX		
164	ALL	4 cases	MTX+CA	transient paraplegia	
165	ALL	14 F	MTX+CA	paraplegia, part. restitution	
166	ALL	5 childr.	MTX	3 fatal – 2 restitution	no anatomical report
89	AML	?	MTX	death	meningeal leukemia only
	ALL	37 M	MTX	death	no anatomical report
167	ALL	13 F	MTX+CA	13 days death	demyelination lumbosacral spinal roots + superficial parts right spinal cord
168	ALL	12 F	MTX	perm. flaccid paraplegia T-6 sensory level; no autopsy	
169	ALL	5 M	MTX[a]+CA[b]	a)transient paraplegia b)death (shock); no autopsy	
	ALL	7 M	MTX	death, hours no autopsy	
	ALL	16 M	MTX	trans. paraplegia, restitution	
	ALL	4 F	MTX	trans. paraparesis, restitution	
170	CML	29 M	MTX	paraparesis, restitution	
171	?	?	MTX	death	
172	ALL	?	MTX	death	no anatomical report
173	ALL	16 F	MTX	months ascending paralysis death	necrosis of anterior horn gray matter throughout spinal cord and mesencephalon; no leukemic lesion

Table 8. (Continued)

Authors	Diagnosis	Age, sex	Drug	Evolution	Neuropathology findings
174	ALL	16 M	MTX	5 months transv. myelopathy death	necrotizing myelopathy T-8-T-10 leukemic infiltration cerebrospinal meninges + thoracic cord roots
175	AML	65 M	CA	6 months death	microvacuolation, axonal swellings + myelin loss superficial white matter of spinal cord
103	myelo-monoc. L	22 F	MTX	months death	patchy demyelination of cervical spinal cord + necrotizing leukoencephalopathy of cerebral hemispheres
156	ALL	2,5 F	VCR (3 mg)	3 days coma, death	enlarged neurons + neurofilaments + cytoplasmic crystals in anterior horn cells and medullary nuclei
157	ALL	5,5 F	VCR (1,2 mg)	death coma 12 days	ascending paralysis + sensory loss – no autopsy
158	lymphobl. lymphoma	29 F	MTX+CA+ VCR (2 mg)	death 14 days paraparesis	swollen motor neurons spinal cord + brain stem, EM: aggregates of intracytoplasmic neurofilaments, myelin and axonal damage subpial cord + roots
167a	meningeal leukemia	2 pat.	thiotepa	a) lower motor neuron disorder, areflexia b) ascending myelopathy respiratory paralysis. – no autopsy	demyelination, gliosis posterior columns
110a	ALL	33 M	CA + syst. CHT	paraplegia, sensory loss T3–8, death 2 months	demyelination posterior columns, spinal white matter lesions
	Ac. monoc. leukemia	66 F	CA + syst. CHT	paraplegia + encephalopathy, death 5 months	sponginess, myelin loss, axonal swelling, gliosis, EM: vesicular disintegration of myelin
	AML	32 M	CA + thiotepa	paraparesis, death 7 months	in spinal cord, roots and peripheral nerves

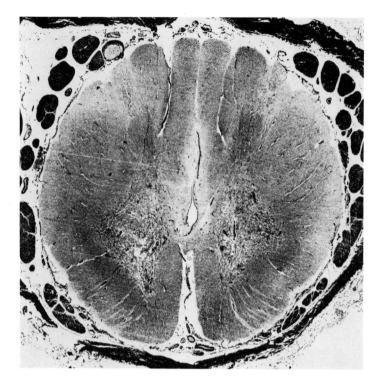

Figure 5. Symmetric anterior horn necrosis in ascending poliomyelo malacia after IT chemotherapy of ALL (courtesy Dr. M. Reznik).

who received systemic chemotherapy and IT CA, and 2 weeks following the last CA administration developed spastic paraparesis that slightly improved within 6 months, when he died from relapsing leukemia. Autopsy showed microvascuolization with scattered axonal swellings, loss of myelin and occasional macrophages throughout the spinal cord, most marked in the superficial white matter of the lower thoracic cord; spinal nerve roots were normal. Reznik [173] reported ascending poliomyelomalacia in a girl with ALL developing several months after IT MTX, with ischemic necrosis of anterior horn gray matter throughout the whole spinal cord (Figure 5) and extending to the mesencephalon, without damage to the white matter, tumor invasion or vascular lesions. Grisold *et al.* [174] in a 16-year old male with ALL who developed a transverse cord lesion during IT MTX treatment 5 months before death observed necrotizing myelopathy of the T-8 to T-10 segments and moderate lymphoblastic infiltration of the cerebrospinal meninges and spinal nerve roots. No parenchymal infiltration or vascular lesions were found. Fatal myeloencephalopathy after accidental IT *vincristine* administration has been observed in 3 cases: A 2,5 year-old girl with ALL who received 3 mg of vincristine sulfate IT, despite saline exchange for CSF died within 3 days incoma. Pathology changes in anterior horn cells and medullary nuclei included enlargement of neurons, broad fields of interwoven neurofilaments, and cytoplasmic crystals [156]. No pathologic examination was performed in a 5-year old child with ALL who received 1,2 mg vincristine IT and despite CSF removal and IT hydrocortisone administration died with ascending paralysis, sensory loss and respiratory failure after 12 days [157]. A woman with lymphoblastic lymphoma treated with combined systemic chemotherapy and IT MTX and CA for CNS involvement was given 2 mg vincristine IT and despite rapid CSF removal developed ascending sensory loss and paraparesis, and afer 14 days died in coma [158]. Histologically, neurons were swollen in rostral spinal cord and brainstem with aggregates of intracytoplasmic neurofilaments similar to those observed in experimental models of vincristine neurotoxicity [156, 176]. In addition, there was shrinkage of neurons in caudal spinal cord and axonal swelling in subpial areas of the cord and brainstem, with myelin damage in the spinal nerve roots. Lower motor neuron disease and ascending myelopathy leading to lethal respiratory paralysis were seen in two among 10 patients with meningeal leukemia or meningeal tumor treated with multiple IT doses of *Thiotepa*. Clinical neurotoxicity occurred after the second and eigth dose, respectively. Autopsy in one patient revealed demyelination and gliosis of the posterior columns extending rostrally to the medulla, while leukoencephalopathy was seen in the brain of another patient who had been given IT MTX plus Thiotepa [167a]. Three adults with leukemia treated with systemic multiple agent chemotherapy, craniospinal irradiation (two) and IT injections of either CA or CA plus Thiotepa (one) developed progressive spinal motor and sensory spinal dysfunctions and one of them encephalopathy in addition to paraplegia. Death occurred 6 weeks to 7 months after onset of neurologic symptoms. Neuropathology showed a central and peripheral myelinopathy with vacuolation, myelin disintegration, axonal swelling, fibrillary gliosis and macrophages in the spinal white matter in the absence of inflammatory and vascular lesions. The earliest changes were vesicular disintegration of the myelin lamellae in both central and peripheral myelin of the cord, roots and peripheral nerves, the relation of which to systemic neoplasia and chemotherapy is unknown [110a].

5.5. Pathogenesis of CNS Complications

For the neurologic side effects of chemotherapy a multifacotrial mechanism is suggested, and there are several etiological factors which are considered to be responsible for the CNS toxicity associated with MTX and other cytostatic agents given systematically or intrathecally with or without combined cranial irradiation.

5.5.1. DNLE and Parenchymal CNS Damage. The etiologic basis of the neurologic sequelae associated with acute and delayed encephalopathies remains to be elucidated, but the following pathogenic factors have been considered:

a) Direct effect of MTX on the replication pathways involving DHFR resulting in cell death;

b) Decreased availability of some neurotransmitters (dopamine-, serotonine) via inhibitory effects in tetrahydrobiopterine metabolism involving biosynthesis of some biogenic amines has been tentatively related to transient mental changes after MTX [6];

c) Folic acid deficiency produced by MTX, a folic acid antagonist, has been causally related to the CNS lesions since treatment with folic and folinic acid arrested the deterioration in some children with MTX encephalopathy [90, 100]; this effect may have been accentuated by blockage of CSF pathways [105 – 107], and the administration of anticonvulsants [90];

d) Action of preservatives or diluents used in MTX, methylhydroxybenzoate, prophylhydroxybenzoate and/or benzoyl alcohol, have been suggested to be responsible for neurotoxicity [164], but CNS lesions are as common in patients receiving preservative-free MTX or other agents (CA) as in those receiving MTX with preservative [65]; hence, preservatives and diluents are not considered as causative factors [83]

e) Some data suggest that neurotoxicity of some agents is related to dosage of the drug administered [102], and an association of increased concentrations and prolonged half-life of MTX in CSF in patients with neurotoxicity has been demonstrated [8, 89]. DNLE with periventricular lesions in patients receiving IT MTX following radiation is suggested to result from obstruction in the outflow of CSF with subsequent accumulation of toxic amounts of MTX which may penetrate via the ventricular ependyma into the brain [106, 107]. While some authors suggest that the duration of treatment as well as the dose appears to be crucial [177], for others there appears to be no correlation between the dosage of MTX or CA and the degree of white matter damage [105, 106], and IT overdose of MTX without neurotoxicity has been reported [178, 179].

f) There appears to be a positive correlation between neurotoxicity of agents and pre-existing CNS leukemia, since most patients in toxic groups had developed CNS leukemia prior to receiving MTX [8, 43, 89, 95, 99, 102], but DNLE developed also without previous CNS leukemia [63, 104, 113, 114].

g) Potentiation of neurotoxic effects by cranial irradiation causing damage to the BBB due to injury of the vascular endothelium is considered an important etiological factor [3, 8, 13, 58, 63, 96, 108]. From the histologic pattern of DNLE it has been suggested that vascular damage produced by the prolonged exposure of the brain to the drugs may play a role, while preexisting CNS leukemia, radiation and MTX may act synergistically in producing the lesions [102]. However, vascular fibrinoid necrosis is observed in both radiated and non-irradiated cases [111], and no correlation has been found between the extent and location of the vascular changes associated with radiation, and PNLE, suggesting that the effects of radiation were different from those due to prolonged concentrations of MTX [105], and that radiotherapy may only increase the vulnerability of the white matter to neurotoxic effects [105].

h) Age at time of radiation, bacterial infections, nutrition, CNS leukemia and dosage of IT MTX are not causally related to development of PNLE [83], which usually occurs only after CNS radiation of more than 20 Gy plus IV MTX. Hence, it has been suggested that the BBB is damaged by radiation and that diffusion of MTX and other drugs IV administered through the damaged BBB results in progressive damage to the CNS white matter. There are basically 3 possible interactions within the CNS between MTX and radiation [63, 63a]: 1. Ionizing radiation and MTX may have synergistic or overlapping mechanisms of neural toxicity [8, 13, 58, 180]; 2. Cytostatic drugs may act as radiosensitizers, increasing the sensitivity of brain tissue to radiation, or recalling radiation-induced injury. Clinical and experimental evidence indicate that MTX may enhance toxicity of radiation on non-CNS tissues [13, 114, 116, 181], and MTX, 5-FU and alkylating agents are known to act as radiosensitizers [11, 46, 63]. MTX may recall radiation injury for lung and skin [3, 8, 181], but whether these phenomena also occur in human CNS remains to be elucidated. Current experience indicates that the incidence of radiation damage to CNS is not altered by pretreatment with MTX [63]. 3. CNS radiation may alter the action and kinetics of MTX and other drugs in the CNS causing accumulation of neurotoxic agents. This may result by a) increasing the permeability of the BBB, allowing more systemically-administered drugs to enter the brain parenchyma; b) by acting on arachnoid and choroid plexus, slowing the turnover of CSF, thus decreasing the clearance of MTX from the CNS; c) radiation may disrupt the ependymal barrier, allowing MTX-containing CSF to enter the brain tissue; d) individual brain cells my accumulate more MTX because of radiation-induced damage on the cellular level. Alterations in cerebrovascular permeability are a well known radiation effect [53, 58, 59, 61], and increased permeability of the BBB for MTX by CNS irradiation [63, 181] and intraarterial mannitol administration was shown experimentally [182]. In summary, for the development of DPLE and other CNS lesions a multifactorial mechanism is suggested, including the malignancy process, gross disturbances in CSF dynamics, and the combined effects of IT and systematic chemotherapy plus CNS irradiation. It is not known which of the many possible factors are operative or predominate in CNS adverse effects of chemotherapy and multimodal treatment. The risk and severity of delayed neurotoxicity are directly proportional to the number of therapeutic modalities, and combinations including CNS radiation appear to be most neurotoxic, particularly when cytostatic drugs are given during or after CNS radiation, while IT or HD systemic chemotherapy followed by CNS radiation is much less likely to produce severe neurologic sequelae [63]. The safest methods are the single modalities, of which HD-MTX may be the least neurotoxic [63a].

5.5.2. Dystrophic Calcification. For calcinosis its relation with cumulative doses of cytostatics has been suggested [141] or a major inductive role was attributed to low-dose radiation-induced damage to microvasculature [143], since similar calcifications are well known sequelae of radiation [49, 51, 58, 63]. Systemic or IT chemotherapy may contribute to the process, while CNS leukemia may not have a major influence. However, identical lesions referred to as 'idiopathic dystrophic calcification' or Fahr's syndrome are occasionally seen in brains of normal or mentally retarded individuals who never had antitumor therapy [137].

5.5.3. Wernicke encephalopathy is not considered to be directly related to cytostatic treatment; nutritional deficiency is the most probably cause, since most patients had gastrointestinal bleeding, hepatic failure and sepsis [132, 152].

5.5.4. Myelo (Radiculopathy) following IT Chemotherapy. Neurotoxicity following IT chemotherapy is related to various factors:
a) Direct toxic effects of CA [175] and MTX due to local folate deficiency [159], decreased tolerance of the CNS to high concentrations of compounds and a prolonged half-life of MTX in the CSF [78, 89, 90, 162], but IT overdose of MTX was also seen without neurotoxicity [178, 179].
b) Preservatives and diluents considered by some as active [167] are unlikely essential causative factors, since paraplegia occurred following both IT MTX and CA [161, 164, 167] and preservative-free MTX administration [65, 67];
c) Physical factors due to exposure to solutions of unphysiological pH and osmolarity [65, 175] which are also implicated for lesions after 'chemical rhizotomy' with distilled water for relief of intractable pain, producing damage to myelin and axons in the nerve roots as high as C-8 and vacuolar changes in the lateral columns [184].
d) Mechanical factors have been implicated [161], since loss of axons and myelin with microvacuolar changes in the peripheral parts of the spinal cord were experimentally produced by exchange of CSF into the cisterna magna [185].

5.5.5. Vincristine Neurotoxicity in the CNS is predominantly directed to the neuron and its axon, since it binds tight with cellular microtubules including mitotic spindle tubules and neurotubules [186]. Experimental IT administration of vincristine or vinblastine produces striking neuronal changes, creating large aggregates of neurofilaments and crystalline masses comprised of neurotubules [156, 176]. Since neurotubules play an important role in axoplasmic flow, their disruption may contribute to the accumulation of neurofilaments in the affected neurons [176].

6. CNS INFECTIONS COMPLICATING CHEMOTHERAPY

An increasing incidence of opportunistic infections with preferential involvement of the CNS is now recognized as a serious and often fatal complication inpatients treated with antineoplastic and immunosuppressive agents [28, 30, 31, 187 – 189]. In the compromised host the clinical symptoms of such infections may have unique features that confound diagnosis [189b]. CNS infections are a potential disadvantage of the management of neoplastic meningeal disease by IT injections or the use of intraventricular reservoirs, infections of the CNS occurring in 6 to 15% of the patients [189c].

In addition to *bacterial* infections frequently affecting the CNS, postmortem studies have established *mycotic infections*, the most common being candidosis [190] and cryptococcosis [191] which, via cerebral and extracerebral involvement in more than half of the patients contribute to their demise. Malignancies were seen in 30 to 50% of cerebral mycoses [191], and antitumor treatment was used in 21 to 31% of patients with cerebral candidosis [190]. *Acquired toxoplasmosis* of the CNS is frequently associated with immunosuppression related to neoplasms (lymphomas) and cancer chemotherapy [192], and relapse of chronic latent toxoplasma encephalitis has been experimentally produced by administration of antineoplastic agents [193]. Since subclinical infections are very common in man, the suppression of cellular immunities by antineoplastic agents seems to be decisive in causing relapsing toxoplasmosis which generally involves the brain exclusively. It may induce diffuse encephalopathy, meningoencephalitis and necrotizing encephalitis or can give rise to a cerebral mass lesion [189]. The neurologic abnormalities are variable and nonspecific and diagnosis is often difficult since antibody titers may be suppressed by chemotherapy or neoplasia. Both mycoses and toxoplasmosis can be successfully treated with appropriate antibiotics even in patients with impaired immunity [189 – 191]. Among *viral infections* known to occur in patients whose immune response is depressed by neoplasm or chemotherapy [28, 30, 182] three types affecting the CNS are of particular interest:

a) *Progressive multifocal leukoencephalopathy* (PMLE) is a clinically progressive CNS disease due to opportunistic infection with papova viruses JC, SV 40 or BK which is frequently superimposed on malignancies with immunodeficiency or after chemotherapy [194 – 196]. Pathologic features are multiple disseminated demyelinating lesions with oligodendroglial inclusions and bizarre malignancy-like astroglia containing large numbers of virions (Figure 6). Clinical signs indicating multifocal cerebral lesions of insidious onset with visual and mental impairment and paralysis are associated with ill-defined low-density CT lesions. The prognosis is poor with progressive deterioration and fatal outcome within 3 to 20 months. While a favorable response to cytarbin and vidarbine or carmustin have been reported in some cases, treatment in others failed [196].

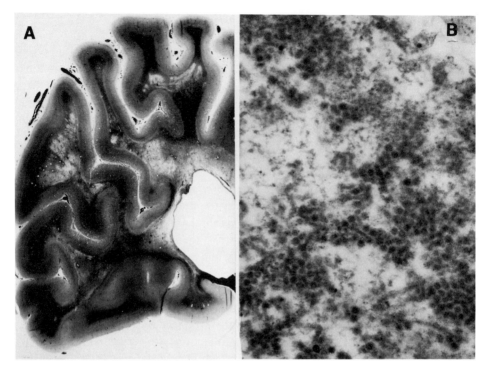

Figure 6. Progressive multifocal leukoencephalopathy in patient with malignant lymphoma treated with systemic polychemotherapy. A) Multiple demyelinating lesions in cerebral hemisphere; B) Virions (33 – 35 nm) in nuleus of oligodendroglia ($\times$ 52700).

Diagnostic confirmation by brain biopsy or immunofluorescent demonstration of JCV or SV 40 antigen in CSF cells are necessary for initiation of therapy. Rare association of PMLE with primary CNS tumors, e.g. multiple glioma [197] and CNS lymphomas [198, 199] may suggest a relationship between the two disorders, since a variety of CNS tumors have been experimentally induced after injection of JC, BK and SV 40 polyoma viruses, and this tumor induction was accomplished more readily in immunosuppressed animals [199].

b) *Subacute immunosuppressive measles encephalitis* (encephalopathy) has been reported after primary measles infection in children and one adult with chemotherapeutically treated malignancies and predominantly cellular immune defects [200 – 212]. Neurological symptoms frequently initiated by seizures, myoclonus or epilepsia partialis continua and hemiparesis occurred within 6 months of exposure to measles virus, and the disease progressed to death within 2 months. Serum antibodies against measles virus were seen in some but not in all cases (Table 9). Neuropathology disclosed necrotic lesions and occasional microglial nodules with no or very little inflammatory reaction, and abundant intranuclear and intracytoplasmic inclusion bodies in neurons and glial cells in cortex and

Table 9. Clinical and pathological aspects of immunosuppressive measles encephalitis

Author	Sex, age	Basic illness	IT chemo-therapy	History of measles	Duration	CSF	Measles antibodies serum titer	CSF inclusions	Intranucl.	EM	IF
200	13 /F	nephrosis	–	+	2 mo	NL	–	–	+	+	–
201	6 /M	ALL	–	+	11 ds	protein	1:8CF	–	+	+	–
	6,5/M	ALL	+	+	3 mo	NL	1:512 (PM)	–	+	+	–
202	2 /M	ALL	+	+	2 wk	NL	–	–	+	+	–
203	5 /M	ALL	–	+	8 ds	NL	no antibody	–	+	+	+
	8 /M	ALL	+	–	19 wk	NL	1:2800CF	1:80CF	no autopsy		
204	9 /M	ALL	+	+	15 wk	NL	1:512CF	1:32CF	+	+	–
	3 /M	ALL	–	+	alive	– –	1:4096CF	1:320CF	–	–	–
205	6 /F	ganglioneuroma	–	+	18 ds	NL	–	–	+	+	–
	5 /M	ALL	+	+	10 ds	protein	1:8CF	–	+	+	–
	6 /M	ALL	+	+	4 wk	NL	1:512CF (PM)	–	+	+	+
206	4 /F	ALL	+	+	2 wk	NL	–	–	– –	– –	+
207	14 /M	rhabdomyosarc.	–	–	4 wk	NL	–	–	+	+	+
208	4,5/F	lymphosarcoma	+	+	2 mo	NL	1:8CF	<1:1CF	–	+	–
	6,6/M	lymphosarcoma	+	+	36 ds	9WBC	– –	–	– –	– –	–
	6,2/M	ALL	+	+	9 ds	NL	1:4 H	–	+	+	–
	1,4/M	ALL	+	+	2 wk	NL	– –	–	–	+	–
209	52 /M	Hodgkin dis.	–	–	6 wk	NL	1:8CF	<1:2CF	+	+	+
210	13 /F	ALL	–	–	alive	– –	1:256CF	not raised	–	–	–
211	2,6/M	neuroblastoma	–	+	alive	NL	1:8CF	1:1024CF	–	–	–
212	6,6/M	ALL	+	+	1 wk	IgG	1:8CF	–	+	+	–

ALL = acute lymphoblastic leukemia; NL = normal; CF = complement fixation; HI = hemagglutination inhibition; EM = paramyxvirus nucleocapsides by EM; IF = specific immunofluorescence for measles; PM = post mortem specimen.

basal ganglia. EM and IF studies [203, 204, 206, 207, 209, 210] revealed paramyxovirusnucleocapsid structures. This disease differs clinically and morphologically from both acute measles encephalitis and subacute sclerosing panencephalitis (SSPE) and seems to be caused by measles virus invading the CNS facilitated by immunosuppressive cytotoxic treatment [201–202, 207, 209]. Some of the cases of fatal subacute measles encephalitis have been reported to develop also herpes zoster during cytostatic treatment [203, 204, 212].

c) Rare instances of inclusion body or necrotizing encephalitis in patients with treated neoplasms have also been noted in association with *herpes simplex* infection [213, 215] or simultaneous infection of herpes simplex and measles virus [216], with possible infection of measles virus and toxoplasma gondii [217] or of unknown etiology [218].

7. PERIPHERAL NEUROTOXICITY (NEUROPATHY, NEUROMYOPATHY)

Of all drugs administered systematically as anticancer agents, *vincristine* has the most severe toxic effects particularly on the peripheral nerves. It is the only anticancer agent whose neurotoxicity is the dose-limiting factor [3, 10]. The clinical features of VC toxicity that can affect all aspects of peripheral nerve function, cranial, motor, sensory and, less commonly, autonomous, are well documented [3, 10, 64, 219–225]. Neuropathy may occur after a single course, although neurotoxicity is cumulative. Infants and elder persons seem to be more susceptible than children; patients in poor nitritional state or with liver dysfunction, those being bedridden and subject to additional pressure injuries of peripheral nerves, and those with preexisting polyneuropathies, such as Charcot-Marie-Tooth syndrome, are particularly susceptible to VC neuropathy [3, 64, 220]. High incidence of neuropathy has also been observed in lymphomas [226]. VC usually produced symmetric polyneuropathy. Loss of the Achilles tendon reflexes or other myotatic reflexes are the earliest and most consistent changes, occurring about two weeks after a single dose, with the reflexes returning to normal in 1 to 3 months [221–223]. Distal paraesthesiae, which are usually the earliest symptoms occurring within 2 to 3 weeks, may involve the hands before the feet and may antedate sensory and motor deficits for long time. Bilateral jaw pain within 24 to 48 hours after a dose and muscle cramps within the tights and calves, particularly after exercise, may be a prominent symptom which accompanies the onset of motor involvement. Weakness impairs the ankle and toe dorsiflectors or may show a predilection for the forearm extensor muscles in some cases [219]. Weakness and muscle atrophy of the lower limbs are initially distal but may develop rapidly such that the patient is unable to stand. Recovery may occur if the drug is stopped or the dose is reduced, but recovery is often slow, and mild sensory impairment and reflex depression often persist. The cranial nerves are occasionally affected. Many patients on long-term VC have 'myopathic facies' with bilateral ptosis without oculomotor dysfunctions and decreased facial expression [3, 64]. Pupillary paralysis with photophobia, diplopia, oculomotor-nerve or bilateral facial nerve palsies and bilateral vocal-cord paresis with mild dysarthria are rare symptoms [3, 10, 221]. Autonomic involvement may occur and abdominal pain, constipation and adynamic ileus or bladder atonia with urinary retention, and orthostatic hypotension may be early symptoms [10]. Electrophysiologic studies demonstrating a mild decrease in motor and sensory nerve conduction velocity which is closely related to dose and duration of therapy, and usually occurs before clinical abnormalities are apparent, with electromyographic signs of denervation of distal muscles, have shown evidence of severe axonal damage involving both motor and sensory fibers; sensory action potential amplitudes remain abnormal despite clinical improvement or recovery [219, 224, 225]. Histologic studies on sural nerve biopsies have confirmed that axonal degeneration of the 'dying-back' type with secondary Wallerian demyelination (Figure 7) is the predominant process with only minor segmental demyelination [224, 227, 228]. EM studies in human and experimental VC neuropathy have shown axonal damage to be due to disturbances of axonal transport [229] related to proliferation of neurofilaments and alteration of the neurotubular system causing a 'dying-back neuropathy' [230, 231]. For *vinblastine* (Velban) and *vindesine* (Desacetyl-Vinblastine amide sulfate) the spectrum of neurological toxicity is similar to that of VC, and includes peripheral neuropathy with paresthesias, distal sensory loss and distal or proximal weakness, jaw pain, urinary retention, paralytic ileus, and hypotension [3a, b, 10, 232].

Peripheral neuropathy has also been observed after treatment of a variety of chemotherapeutic agents: It has been reported in 10 – 20% of the patients treated with *VM – 26* for malignant gliomas [3b, 12a], in 10 – 17% of the patients treated with relative high doses of *procarbazine* [3b, 10]. Neuropathy produced by this drug has been related to depletion of pyridoxal-phosphate [10], but pyridoxine treatment did not prevent neuropathy. On the other hand, MTX, a potent folic acid antagonist, to the best of our knowledge apparently does not induce peripheral nerve dysfunction, although polyneuropathy related to folate deficiency, is well documented [233]. *Cis-platinum* in a small percentage of treated patients causes a pure sensory neuropathy which clinically differs from VC neuropathy [3b, 234]. Rare occurrence of neuropathy has also been reported after treatment with *hexamethylmelamine* [3b], possibly enhanced by prior neuropathy induced by VC, and with *Methyl-GAG*, a drug abandoned because of its GI toxicity [3b]. Peripheral neuropathy has also been related with systemic CA, but the association is not clearly established [236].

Toxic myopathy may be induced by *vincristine*, its clinical features being proximal muscle pain with weakness and wasting [227] often occurring in children [3].

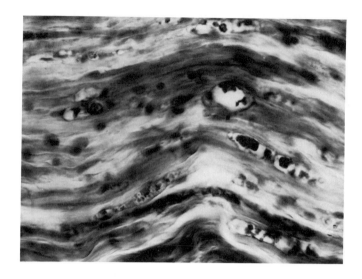

Figure 7. Vincristine polyneuropathy. Sural nerve biopsy showing demyelination with multiple digestion chambers. SSB × 100.

Light and electronmicroscopic studies in man [227] and animals [228, 237] showing segmental necrosis with myofibrillary disruption and accumulation of subsarcolemmal masses have confirmed that the drug has a profound effect on membrane systems and causes severe autophagic degeneration of muscle fibers. Patients with myopathies may be predisposed by vincristine to myotonia [238]. 'Myopathy' induced by 5-*azacytidine*, a non-commonly used chemotherapeutic drug, associated with generalized muscle tenderness and proximal weakness, appearing 4 days of treatment and completely reversed within one week, has been correlated to hypophosphatemia [239]. Rare occurrence of "myopathy" after administration of Methyl-GAG, characterized by reversible proximal weakness, has not been confirmed by muscle enzyme and EMG studies which gave no evidence for myopathy [3b].

8. CONCLUSIONS

The preceding chapter presented an overview of the pathologic effects of chemotherapy with particular reference to the morphologic effects on anaplastic gliomas and to direct and indirect adverse effects on the nervous system. The available data indicate that general and nervous complications are to be anticipated in an increasing number of patients undergoing increasingly aggressive chemotherapy and multimodal treatments where the combination with radiation is synergistically deleterious. A wide spectrum of clinical and morphologic entities may develop to which the oncologists need to be alert, since they represent complication of therapy rather than evidence of disease progression. The differential diagnosis of neurologic disorders occurring in patients treated with chemotherapy has to consider: 1) progression or recurrence of the malignancy process of the CNS or involvement of the CNS by neoplastic process; 2) paraneoplastic syndromes affecting both the peripheral and central nervous system; 3) hemostatic abnormalities producing cerebravascular accidents, hemorrhages, phlebothrombosis, subdural and subarachnoid hemorrhages; 4) opportunistic infections and other concomitant conditions; 5) other iatrogenic damage, particularly radiation lesions. All these changes may be either disease-related or drug-related due to chemotherapy or due to secondary effects resulting from iatrogenic immunosuppression etc. As more and more patients survive longer and longer periods, the full impact of chemotherapy on the nervous system should become apparent, and complications of anticancer therapy will become increasingly important in patients with malignancies, but the etiology and causative factors of many of them remain to be elucidated. Optimum treatment of the individual patient will require cognizance of the various complex factors that enter into the treatment of cancer in order to optimize efficacy and to minimize the adverse effects of antineoplastic chemotherapy.

REFERENCES

1. Guarino AM: Pharmacologic and toxicologic studies of anticancer drugs. In: DeVita VT, Busch H (eds) Methods in Cancer Research, Vol. 17, Cancer Drug Development, New York: Academic Press 1979, pp 91–176.
2. Crooke ST, Prestayko AW: Cancer and chemotherapy. Vol. 3: Antineoplastic agents. New York: Academic Press, 1981.
2a. Dort RT, Fritz WL: Cancer, chemotherapy handbook. New York: Elsevier-North Holland, 1980.
3. Allen JC: The effect of cancer therapy on the nervous system. J Pediatr 93(6):903–909, 1978.
3a. Young DF, Posner JB: Nervous system toxicity of the chemotherapeutic agents. In: Vinken PJ, Bruyn GW (eds) Handbook of Clinical Neurology, Vol. 39, North Holland: Amsterdam 1980, pp 91–129.
3b. Young DF: Neurological complications of chemotherapy. In: Silverstein A (ed.): Neurological complications of therapy. Futura Publ Comp: Mount Kisco, NY, 1982, pp 57–113.
4. Vick NA, Khandekar JD, Bigner DD: Chemotherapy of brain tumors. Arch Neurol (Chic) 34 (9):523–526, 1977.
5. Byfield JE: Central nervous system toxicities from combined therapies. Front Radiat Ther Onc, vol. 13, Basel: Karger, pp 228–240, 1979.
5a. Young RC, Ozols RF, Myers CE: The anthracycline antineoplastic drugs. New Engl J Med 305:139–152, 1981.
6. Abelson HT: Methotrexate and central nervous system toxicity. Cancer Treat Rep 62(8):1999–2001, 1978.
7. Shapiro WR, Young DF, Mehta PM: Methotrexate: Distribution in cerebrospinal fluid after intravenous, ventricular and lumbar injections. New Engl J Med 293:161–166, 1975.
8. Bleyer WA, Poplack DG: Clinical studies on central nervous system pharmacology of methotrexate. In: Clinical Pharmacology of Anti-Neoplastic Drugs. Pinedo HH, Boeslma E,

(ed). Amsterdam/North-Holland 1978, pp 115 – 131.
9. Pochedly C: Neurotoxicity due to CNS therapy for leukemia. Med Ped Oncol 3:101 – 115, 1977.
10. Weiss HD, Walker MD, Wiernik PH: Neurotoxicity of common used antineoplastic agents. New Engl J Med 291:75 – 81, 127 – 133, 1974.
10a. Prestayko, AW, Crooke ST, Baker ST, Carter SK, Schein PS (eds) Nitrosureas. Current status and new developments. New York: Academic Press, 1981.
11. Leenhouts HP, Chadwick KH, Deen DF: An analysis of the interaction between two nitrosourea compounds and X-radiation in rat brain tumour cells. Int J Radiat Biol 37(2):169 – 181, 1980.
12. Weinkam RJ, Shiba DA: Procarbazine. In: Anticancer Drugs. Chabner B (ed). New York: B Saunders, 1978.
12a. Issels BF: The podophyllotoxin derivatives VP16-213 and VM 26. Cancer Chemother Pharmacol 7:73 – 80, 1982.
13. Phillips TL, Fu KK: Acute and late effects of multimodal therapy on normal tissue. Cancer 40(2):489 – 494, 1977.
14. Jaffe N: Late side effects of treatment, skeletal, genetic, central nervous system, and oncogenic. Pediatr Clinics N Amer 23(1):233 – 244, 1976.
15. Ghione M: Cardiotoxic effects of antitumor agents. Cancer Chemother Pharmacol 1(1):25 – 34, 1978.
16. Engelhardt R: Kombinierte Strahlen- und Chemotherapie aus der Sicht des onkologischen Internisten. In: Kombinierte Strahlen- und Chemotherapie. Wannenmacher M (ed), München-Wien-Baltimore: Urban & Schwarzenberg 1979, pp 88 – 95.
17. Litam JR, Dail DH, Spitzer G, et al: Early pulmonary toxicity after administration of high-dose BCNU. Cancer Treat Rep 65(1):39 – 44, 1981.
18. Gale GR, Atkins LM: Cisplatin and diethyldithiocarbamate in treatment of L1210 leukemia. J Clin Hemat Oncol 11(2):41 – 45, 1981.
19. Walker MD, Green SB, Byar DP, Alexander Jr E, Batzdorf U, et al. Randomized comparisons of radiotherapy and nitrosoureas for the treatment of malignant glioma after surgery. New Engl J Med 303(51):1323 – 1329, 1980.
20. Jellinger K, Volc D, Grisold W, Podreka I, Böck P, et al. Multimodality treatment of malignant gliomas. Zbl Neurochir 42(2):99 – 122, 1981.
21. Reagan TH, Bisel HF, Childs Jr DS, Layton DD, Rhoton AL, Taylor W: Controlled study of CCNU and radiation-therapy in malignant astrocytoma. J Neurosurg 44(2):186 – 190, 1976.
22. Levin VA, Wilson CB, Davis R, Wara WM, Pischer TL, Irwin L: A phase III comparison of BCNU, hydroxyurea and radiation therapy to BCNU and radiation therapy for treatment of primary malignant gliomas. J Neurosurg 51(4):526 – 532, 1979.
23. Brisman H, Housepian EM, Chang CM, Duffy P: Adjuvant nitrosourea therapy for glioblastoma. Arch Neurol (Chic) 33(11):745 – 750, 1976.
24. Avellanosa AM, West CR, Tsukada Y, Highby DJ, Barshi S, Reese PA, Jennings E: Chemotherapy of nonirradiated malignant gliomas. Cancaer 44(3):839 – 846, 1979.
25. Durant JR, Norgard NJ, Murad TM, Bartolucci AA, Langford KH: Pulmonary toxicity associated with bischloroethylnitrosoureas (ECNU) Ann Intern Med 90(2):191 – 194, 1979.
26. Aronin PA, Mahaley MS, Rudnick SA, Dudka L, Donohue JD, Selker RG, Moore MN: Prediction of BCNU pulmonary toxicity in patients with malignant gliomas. New Engl J Med 303(39):183 – 188, 1980.
27. Harmon WE: Chronic renal failure in children treated with MethylCCNU. New Eng J Med 300:1200 – 1203, 1979.
28. Thunold S, Moe PJ: Complications of cytostatic therapy in childhood leukemia. Acta path microbiol Scand, A Suppl 236:84 – 96, 1973.

29. Schaefer STD, Wright CG, Post JD, Frenkel EP. Cis-platinum vestibular toxicity. Cancer 47(5):857–859, 1981.
30. Leventhal BG, Cohen P, Triem SC: Effect of chemotherapy on the immune response in acute leukemia. Israel J Med Sci 10(8):866–887, 1974.
31. Simone JV, Holland E, Johnson W: Fatalities during remission of childhood leukemia. Blood 39(5):759–550, 1972.
32. Groff P, Torhorst J, Speck B. Die Graft-versus-Host Krankheit, eine wenig bekannte Komplikation der Bluttransfusion. Schweiz Med Wschr 106:634–639, 1976.
33. Woods WG, Lubin BN: Fatal graft versus host disease following a blood transfusion in a child with neuroblastoma. Pediatrics 67(2):217–221, 1981.
34. Schmidmeier W, Feil W, Gebhart W, Grisold W, Gschnait F, Hinterberger W, Höcker P, Jellinger K, et al: Fatal graft-versus-host reaction following granulocyte transfusions. Blood (in press).
35. Toland DM, Coltan CM, Hall W: Second malignancies complicating Hodgkin's disease. Proc 13th Meet Amer Soc Clin Oncol, Vol. 18, Denver 1977.
36. Kim HD: The development of non-Hodgkin's lymphomas following therapy for Hodgkin's disease. Cancer 46(6):2596–2602, 1980.
37. Louie S, Schwartz RS: Immunodeficiency and pathogenesis of lymphoma and leukemia. Sem Hematol 15(11):117–138, 1978.
37a. Coleman CN, Burke JS, Varghese A, Rosenberg SA, Kaplan HS: Secondary leukemia and Non-Hodgkin's lymphoma in patients treated for Hodgkin's disease. In: Malignant Lymphomas. New York: Academic Press, 1982, pp 259–276.
38. Vogl SE: Acute leukemia complicating treatment of glioblastoma multiforme. Cancer 41(2):333–336, 1978.
39. Chung CK, Stryker JA, Cruse A, Vannucci R, Towfighi J: Glioblastoma multiforme following prophylactic cranial irradiation and intrathecal methotrexate in a child with acute lymphatic leukemia. Cancer 47(11):2563–2566, 1981.
39a. Glioblastoma multiforme in a patient with ALL who received a bone marrow transplant. Transplant Proc 14:770–774, 1982.
40. Merker PC, Wodinsky I, Geran R: Review of selected experimental brain tumor models used in chemotherapy experiments. Cancer Chemother Rep 59(4):729–736, 1975.
41. Rosenblum ML, Knebel KD, Vaquez DA, Wilson CB: Brain tumor therapy. Quantitative analysis using a model system. J Neurosurg 46(2):145–154, 1977.
42. Paoletti P, Walker MD, Butti G, Knerich R (eds): Multidisciplinary Aspects of Brain Tumor Therapy. Amsterdam-New York: Elsevier/North-Holland, 1979.
43. Spence AM, Geraci JP: Combined cyclotron fast-neutron and BCNU therapy in a rat brain tumor model. J Neurosurg 54(2):461–467, 1981.
44. Kornblith PL, Smith BH, Leonard LA: Response of cultured human brain tumors to nitrosoureas. Correlation with clinical data. Cancer 47(2):255–265, 1981.
45. Schiffer D, Giordana MT, Pezzotta S, Paoletti P: Chemotherapeutic effects of some alkylating derivatives of nitrosourea on the development of tumors transplacentally induced in rats by E.N.U.: Acta Neuropath (Berl) 34(1):21–31, 1976.
46. Mealey Jr J, Chen TT, Shupe R: Response of cultured human glioblastomas to radiation and BCNU chemotherapy. J Neurosurg 41(9):339–349, 1974.
47. Searle J, Bawson TA, Abbott PJ, Harmon B, Kerr JFR: An electron microscopical study of the mode of cell death induced by cancer-chemotherapeutic agents in populations of proliferating normal and neoplastic cells. J Path (Edinbg) 116(2):129–138, 1975.
48. Woo KB, Brenkus LB, Wilg KM: Analysis of the effects of antitumor drugs on cell cycle kinetics. Cancer Chemother Rep 59(4):847–860, 1975.
49. Rubinstein LJ: Tumors of the central nervous system. Atlas of Tumor Pathology, 2nd ser., Fasc. 6, Washington, DC.: Armed Forces Institute of Pathology, 1972.
50. Altmann HW, Müller HA: Grundlagen der Karyologie. Verh dtsch Ges Path 57:2–33, 1973.

51. Jellinger K: Glioblastoma multiforme. Acta Neurochir 42(1):5–32, 1978.
52. Schiffer D, Giordana MT, Buoncristiani P, Paoletti P: Human malignant gliomas treated with chemotherapy: a pathological study Neurosurg 3(1):344–347, 1978.
53. Schiffer D, Giordana MT, Paoletti P, Soffietti R, Tarenzi T: Pathology of human malignant gliomas after radiation and chemotherapy. Acta Neurochir 53(3–4):205–216, 1980.
53a. Schiffer D, Giordana MT, Soffiett R, Sciolla R: Histological observations on the regrowth of malignant gliomas after radiotherapy and chemotherapy. Acta neuropath 58:291–299, 1982.
54. Willson N, Duffy PE: Morphologic changes associated with combined BCNU and radiation therapy in glioblastoma multiforme. Neurology (Minneap) 24(1):465–471, 1974.
55. Gerstner L, Jellinger K, Heiss WD, Wöber G: Morphological changes in anaplastic gliomas treated with radiation and chemotherapy. Acta Neurochir 36(2):117–138, 1977.
56. Hoshino T, Wilson CB, Ellis WG: Gemistocytic astrocytes in gliomas. An autoradiographic study. J Neuropath Exp Neurol 34:263–281, 1975.
57. Budka H, Podreka I, Zaunbauer F: Overgrowth of a primitive cell population in operated recurrent gliomas: the possible role of chemo- and radiotherapy. In: Paoletti P, Walker MD et al. (eds). Multidisciplinary aspects of brain Tumor Therapy. Elsevier/North Holland 1979, pp 357–362.
58. Burger PC, Mahaley MS, Dudka L, Vogel FS: The morphologic effects of radiation administered therapeutically for intracranial gliomas. Cancer 44(5):1256–1272, 1979.
59. Jellinger K: Human central nervous system lesions following radiation therapy. Zbl Neurochir 38(3):199–220, 1977.
60. Poisson M, Hauw JJ, Pouillart P, Bataini JP, Mashaly R, Pertuiset BF, Metzger J: Malignant gliomas treated after surgery by combination chemotherapy and delayed irradiation. Acta Neurochir 31(1):27–42, 1979.
61. Volc D, Jellinger K, Flament H, Böck P, Klumair J: Cerebral space-occupying cysts following radiation and chemotherapy of malignant gliomas. Acta Neurochir 57(3–4):177–193, 1981.
61a. Seiler RW: Late results of multimodality therapy of high-grade supratentorial astrozytomas. Surg Neurol 15(2)88–91, 1981.
61b. Lieberman AN, Foo SH, Ransohoff J, Wise A, George A et al: Long term survival among patients with malignant brain tumors. Neurosurgery 19(4)450–453, 1982.
62. Pizzo PA, Poplack DG, Bleyer WA: Neurotoxicities of current leukemia therapy. Amer J Pediatr Hematol 1(2):127–140, 1979.
63. Bleyer WA, Griffin TW: White matter necrosis, mineralizing microangiopathy, and intellectual abilities in survivors of childhood leukemia: Associations of CNS irradiation and methotrexate therapy. In: Gilbert HA, Kagan AR (eds) Radiation Damage to the Nervous System. New York: Raven Press 1980, pp 155–174.
63a. Bleyer WA: Neurologic sequelae of methotrexate and ionizing radiation: A new classification. Cancer Treatm Rep 65, Suppl 1, 89–98, 1981.
64. Hanefeld F, Riehm H: Therapy of acute lymphoblastic leukaemia in childhood: Effects on the nervous system. Neuropaediat 11(1):3–16, 1980.
65. Duttera MJ, Bleyer WA, Pomeroy TC, Leventhal CM, Leventhal BG: Irradiation, methotrexate toxicity, and the treatment of meningeal leukemia. Lancet II, 703–707, 1973.
66. Kölmel HW: Die intrathekale Gabe von Zytostatika. Nervenarzt 49(6):685–696, 1978.
67. Mott MG, Stevenson P, Wood CBD: Methotrexate meningitis. Lancet 2:656, 1972.
68. Haghbin M: Antimetabolites in the prophylaxis and treatment of CNS leukemia. Cancer Treat Rep 61(4):681–687, 1977.
69. Davies-Jones GAB, Preston FE, Timperley WR: Neurological complications in Clinical Haematology. London: Blackwell 1980.
70. Aur RJA, Hustu HO, Verzosa MS et al: Comparison of two methods of preventing CNS leukemia. Blood 42(3):349–357, 1973.
71. Hustu HO, Aur RJA, Verzosa JP et al: Prevention of central nervous system leukemia by irradiation. Cancer 32(7):585–597, 1973.

72. Jellinger K, Budka H: Leukosen und maligne Lymphome. In: Dommasch D, Mertens HG (eds) Cerebrospinalflüssigkeit, CSF. Stuttgart: G. Thieme 1980, pp 34–40.
73. Peter A: Submicroscopic changes in leukemic cells of cerebrospinal fluid following intrathecal methotrexate. Acta Neuropath 29(4):345–352, 1974.
74. Budka HM, Guseo A, Jellinger K, Mittermayer K: Intermittent meningitic reaction with severe basophilia and eosinophilia in CNS leukemia. J Neurol Sci 28(4):459–468, 1976.
75. Glasser L, Corrigan JJ, Payne C: Basophilic meningitis secondary to lymphoma. Neurology (Minneap) 26(9):899–902, 1976.
76. Campbell RHA, Marshall WC, Chessels JM: Neurological complications of childhood leukemia. Arch Dis Childh 52:850–854, 1977.
76a. Williams SA, Makker SP, Grupe WE: Seizures: a significant side effect of chlorambucil therapy in children. J Pediatr 93:516–518, 1978.
77. O'Callaghan M, Ekert H: Vincristine toxicity related to dose. Arch Dis Childh 51(3):289–293, 1976.
78. Dietrich E, Goebel B, Gutjahr P: EEG-Befunde nach Vincristin-Behandlung. Mschr Kinderheilk 126(7):709–712, 1978.
79. Carpenter LI, Lockhart LM: Ataxia and athetosis as side effects of chemotherapy with vincristine in non-Hodgkin lymphoma. Cancer Treatm Rep 62(5):561–562, 1978.
80. Allen JC, Rosen G. Transient cerebral dysfunction following chemotherapy for osteogenic sarcoma. Ann Neurol 3(4):441–444, 1978.
81. Jean R, Navarro M, Marty M et al: Encéphalopathie aigue au cours du traitement des lymphoblastoses de l'enfant. Ann pédiatr 23(7):789–802, 1976.
81a. Berman IJ, Mann MP: Seizures and transient cortical blindness associated with Cis-platinum (II) diamminedichloride (PDD) therapy. Cancer 45:764–766, 1980.
82. Koenig H, Patel A: Biochemical basis of the acute cerebellar syndrome in 5-fluorouracil chemotherapy. Trans Am Neurol Assn 94:290–292, 1972.
82a. Woodcock TM, Martin DS, Damin LAM et al: Combination clinical trials with thymidine and fluorouracil. Cancer 45:1135–1143, 1980.
83. Price RA, Jamieson PA: The central nervous system in childhood leukemia. II. Subacute leukoencephalopathy. Cancer 35(2):306–318, 1975.
84. McIntosh S, Aspnes GT: Encephalopathy following CNS prophylaxis in childhood lymphoblastic leukemia. J Pediatr 91:909–913, 1973.
85. Meadows AT, Evans AE: Effects of chemotherapy on the CNS. A study of parenteral MTX in long-term survivors of leukemia and lymphoma in childhood. Cancer 37(4):1079–1085, 1976.
86. Fusner J, Poplack DG, Pizzo PA, DiChiro G: Leukoencephalopathy following chemotherapy for rhabdomyosarcoma, reversibility of cerebral changes demonstrated by CT. J Pediat 91(1):77–79, 1977.
87. McIntosh S, Klastkin EH, O-Brien RT, Aspnes GT, Klammerer BL et al. Chronic neurologic disturbance in childhood leukemia. Cancer 37(6):853–857, 1976.
88. Eiser C: Intellectual abilities among survivors of childhood leukemia as a functions of CNS irradiation. Arch Dis Childh 53(4):391–395, 1978.
89. Bleyer WA, Drake JC, Chabner BA: Neurotoxicity and elevated cerebrospinal-fluid methotrexate concentration in meningeal leukemia. New Engl J Med 289:770–773, 1973.
90. Kay HEM, Knapton PJ, O'Sullivan JP, Wells DG, Harris RF et al. Encephalopathy in acute leukemia associated with methotrexate therapy. Arch Dis Childh 47(4):344–354, 1972.
90a. Lazarus HM, Herzig RH, Herzig GP, Phillips GL, Roessmann U, Fishman DJ: Central nervous system toxicity of high-dose systemic cytosine arabinoside. Cancer 48:2577–2582, 1981.
91. Terheggen HG, Rado M. Cerebrale Komplikationen der Leukämiebehandlung. I. Das Apathiesyndrom Mschr Kinderheilk 126:693–695, 1978.
91a. CH'ien LT, Aur RJA, Stagner S, Cavallo K. et al.: Long-term neurological implications of

somnolence syndrome in children with acute lymphocytic leukemia. Ann Neurol 8:273 – 277, 1980.
92. Freeman JE, Johnston PGB, Voke JM: Somnolence after prophylactic cranial irradiation in children with acute lymphoblastic leukemia. Brit Med J 4:523 – 525, 1973.
93. Olift A, Bleyer WA,Poplack DG: Acute encephalopathy after initiation of cranial irradiation for meningeal leukemia. Lancet 2:13 – 15, 1978.
94. Garwicz S, Aronson AS, Elmqvist D, Landberg T: Postirradiation syndrome and EEG findings in children with acute lymphoblastic leukaemia. Acta Paediat Scand 64:399 – 403, 1975.
95. Hendlin B, DeVivo DC, Torack R, Lell ME, Ragab AH, Vietti TJ: Parenchymatous degeneration of the central nervous system in childhood leukemia. Cancer 33(2):468 – 482, 1974.
96. Rubinstein LJ, Herman MM, Long TF, Wilburg JR: Disseminated necrotizing leukoencephalopathy: A complication of treated central nervous system leukemia and lymphoma. Cancer 35(2):291 – 305, 1975.
97. Wendling LR, Bleyer WA, DiChiro G, McIlvanie SK: Transient severe periventricular hypodensity after leukemia prophylaxis with cranial irradiation and intrathecal methotrexate. J Comp Ass Tomogr 2:502 – 505, 1978.
98. Shapiro WR, Posner JG: Chronic methotrexate toxicity in the CNS. Clin Bull 10(1):49 – 52, 1980.
99. Aur RJA, Simone JV, Verzosa MS, Hustu HO, Pinkel DP, Barker LF. Leucoencefalopatia en ninos con leucemia linfocitica aguda sometidos a terapeutica preventiva del sistema nervioso central. Sngre 23(1):1 – 12, 1977.
99a. Simone JV: Late complications of treatment of children with leukemia and lymphoma. Bristol-Myers Cancer Symposia, Vol. 3, p 663. New York: Academic Press, 1982.
100. DeVivo DC, Malas D, Nelson JS, Land VJ: Leukoencephalopathy in childhood leukemia. Neurology (Minneap) 27(7):609 – 613, 1977.
101. Rosemberg S: La leucoencéphalopathie, complications du traitement des leucémies infantiles. Arch Franç Pédiat 36(3):291 – 297, 1979.
102. Liu HM, Maurer HS, Vongsvivut S, Conway JJ: Methotrexate encephalopathy. A neuropathologic study. Human Path 9(6):635 – 648, 1978.
103. Ebels EJ: Iatrogenic damage to the CNS in malignant systemic disease. Acta Neuropath (Berl) Suppl VII:352 – 355, 1981.
104. Aur RJA, Simone JV, Verzosa MS, Hustu MD, Barker LF, Pinkel DP et al: Childhood acute leukemia – Study III. Cancer 42(6):2123 – 2129, 1978.
105. Bresnan MD, Gilles FH, Lorenzo AV, Watters GV, Barlow CF: Leukoencephalopathy following combined irradiation and intraventricular MTX therapy of brain tumors in childhood. Trans. Amer Neurol Assn 97:204 – 206, 1972.
106. Norrell H, Wilson CB, Slagel DE, Clark DB: Leukoencephalopathy following the administration of methotrexate into the cerebrospinal fluid in the treatment of primary brain tumors. Cancer 33(4):923 – 932, 1974.
107. Shapiro WR, Chernik NL, Posner JB. Necrotizing encephalopathy following intraventricular instillation of methotrexate. Arch Neurol (Chic) 28(2):96 – 102, 1973.
108. Breuer AC, Blank NK, Schoene WC: Multifocal pontine lesions in cancer patients treated with chemotherapy and CNS radiotherapy Cance 41(6):2112 – 2120, 1978.
109. Pratt RA: Cerebral necrosis following irradiation and chemotherapy for metastatic choriocarcinoma. Surg Neurol 7(3):117 – 120, 1977.
110. Garcia JH, Sandbank U, Gutin P: Multifocal leukoencephalopathy in adult leukemia: Histologic features and etiologic considerations. Acta Neuropath 40(3):273 – 276, 1977.
110a. Mena H, Garcia JH, Valandia F: Central and peripheral myelinopathy associated with systemic neoplasia and chemotherapy. Cancer 48:1724 – 1737, 1981.
111. Nakazato Y, Ishida Y, Morimatsu M: Disseminated necrotizing leukoencephalopathy. Acta pathol Jap 30/4, 659 – 670, 1980.

112. Skullerud K, Halvorsen K: Encephalomyelopathy following intrathecal methotrexate treatment in a child with acute leukemia. Cancer 42(3):1211–1215, 1978.
113. Allen JC, Rosen G, Mehta BM: Leukoencephalopathy following high-dose intravenous methotrexate chemotherapy with citro vorum factor rescue (abst). Ann Neurol 6:179, 1979.
114. Rosen G, Ghavine F, Nirenberg A, Mosende C, Mehta BM: High-dose methotrexate with citrovorum factor rescue for the treatment of central nervous system tumors in children. Cancer Treat Rep 62(00):681–680, 1977.
114a. Schold SC, Fay JW: Central nervous system toxicity from high-dose BCNU treatment of systemic cancer. Neurology 30:429, 1980.
115. Patterson AH: Possible neurologic complications of DTIC. Cancer Treat Rep 61(2):105–106, 1977.
116. Rosen G, Marcove RC, Caparros B, Nirenberg A, Kosloff C, Huvos A: Primary osteogenic sarcoma. Rationale for preoperative chemotherapy and delayed surgery. Cancer 43(7):2163–2177, 1979.
117. Soni SS, Marten GW, Pitner SE, Duenas DA, Pawozek M: Effects of central nervous system irradiation on neuropsychologic functioning of children with acute lymphocytic leukemia. New Engl J Med 293(2):113–118, 1975.
118. Verzosa MS, Aur RJA, Simone JV, Hustu HO, Pinkel DP: Five years after central nervous system irradiation of children with leukemia. Int J Radiat Oncol Biol Psych 1(2):209–215, 1976.
119. Simone JV, Aur RJA, Hustu HD, Verzosa MS, Pinkel DP: Three to ten years after cessation of therapy in children with leukemia. Cancaer 42(3):839–845, 1978.
120. Hübener KH, Treuner J, Voss AC, Metzger HOF: Spätfolgen präventiver Strahlen- und Chemotherapie des Hirnschädels bei Kindern mit akuter Leukämie Strahlenther 156(1):26–29, 1980.
121. Arnold H, Kühne D, Franke H, Grosch I: Findings in CAT after intrathecal methotrexate and radiation. Neuroradiol 16(1):65–68, 1978.
122. Peylan-Ramu N, Poplack DG, Bley CL, Herdt JR, Vermess M, Dichiro G: CAT in methotrexate encephalopathy. J Comp Ass Tomogr 1(4):437–442, 1977.
122a. Peylan-Ramu N, Poplack OG, Pizzo PA, Adornato BT, DiChiro G: Abnormal CT scans in asymptomatic children with ALL after prophylactic treatment of the CNS with radiation and intrathecal chemotherapy. New Engl J Med 298:815–818, 1978.
123. Kingsley DPE, Kendall BE: Cranial computed tomography in leukemia. Neuroradiol 16(6):543–546, 1978.
124. Enzmann DR, Lane B: Enlargement of subarachnoid spaces and lateral ventricles in pediatric patients undergoing chemotherapy. J Pediatr 92:535–539, 1978.
125. Gastaut JA, Gastaut JL, Carcassonne Y: Computerized acial tomography in the study of intracranial complications in hematology. Cancer 21(2):487–501, 1978.
126. Treuner J, Hübner KH, Böhmer H, Küpper U: Auffällige CT-Befunde des Schädels nach präventiver ZNS-Behandlung akuter Leukämien bei Kindern. Onkologie 2(1):83–86, 1979.
126a. Shalen PR, Ostrow PT, Glass PJ: Enhancement of the white matter following prophylactic therapy of the CNS for leukemia: Radiation effects and methotrexate leukoencephalopathy. Radiology 140:409–412, 1981.
127. Day RE, Kingston J, Bullimore JJ, Mott MG, Thomson JLG: CAT brain scans after CNS prophylaxis for ALL. Brit Med J 2:6154, 1978.
128. Kretzschmar K, Gutjahr P, Kutzner J: CT studies before and after CNS treatment for ALL and malignant Non-Hodgkin's lymphoma in childhood. Neuradiol 20(2):173–180, 1980.
129. Ochs JJ, Berger MM, Brecher ML, Sinks LS, Kinkel W, Freeman AI: CT brain scans in children with ALL receiving methotrexate alone as CNS prophylaxis. Cancer 45(9):2274–2278, 1980.

130. DeReuck JL, Sieben GJ, Sieben-Praet MR, Ngendakayo P et al: Wernicke's encephalopathy in patients with tumors of the lymphoid-hemopoietic system. Arch Neurol (Chic) 37(3):338–341, 1980.
131. Gangji D, Reaman GH, Cohen SR, Bleyer WA et al: Elevated basic myelin protein in the CSF of ALL patients with leukoencephalopathy. Proc Amer Ass Cancer Res Amer Assoc Clin Oncol 20:353, 1979.
132. Iinuma K, Hayashi T, Ikuta F: Disseminated necrotizing leukoencephalopathy accompanied with calcium following antineoplastic therapy in a case with ALL. Adv Neurol Sci (Tokyo) 21(2):190–197, 1977.
133. Koga S, Fijimoto T, Hasegawa K, Sueishi K: Disseminated necrotizing leukoencephalopathy following intrathecal methotrexate. Fukuoka Igaku Zasski 67(1):64–71, 1976.
134. Morimatsu M, Hirai S, Ogawa S, Motegi M, Nakazato Y: A case of disseminated necrotizing leukoencephalopathy associated with intrathecal methotrexate therapy. Neurol Med (Tokyo) 9(1):54–62, 1978.
135. Verity MA, Wolfson WL: Cerebral lymphomatoid granulomatosis. A report of two cases, with disseminated necrotizing leukoencephalopathy. Acta Neuropath 36(2):117–124, 1976.
136. Rosemberg S: Encéphalopathie apparve au cours d'un traitement par la vincristine. Arch franç Pédiat 31(4):391–398, 1974.
137. Crosley CJ, Rorke LB, Evans A, Nigro M: Central nervous system lesions in childhood leukemia. Neurology 28(7):678–685, 1978.
138. Jellinger K, Radaszkiewicz TH: Involvement of the central nervous system in malignant lymphomas. Virch Arch Abt A Path Anat 370(4):345–362, 1976.
139. Moir DH, Bale PM: Necropsy findings in childhood leukemia. Pathology 8:247–258, 1978.
140. Borns PF, Rancier LF: Cerebral calcification in childhood leukemia mimicking Sturge-Weber syndrome. Amer J Roentg 122(1):52–55, 1974.
141. Mueller S, Bell W, Seibert J: Cerebral calcifications associated with intrathecal methotrexate therapy in ALL. J Pediatr 88(4):650–653, 1976.
142. McIntosh S, Fisher DB, Rothman S, Rosenfield N, Lobel JS, O'Brien RT: Intracranial calcifications in childhood leukemia. J Pediatr 91:909–913, 1977.
143. Flament-Durand J, Ketelbant-Balasse P, Maurus R, Regnier R, Spehl M: Intracerebral calcifications appearing during the course of ALL treated with methotrexate and X rays. Cancer 35(2):319–325, 1975.
144. Price RA, Birdwell DA: The central nervous system in childhood leukemia. III. Mineralizing microangiopathy and dystrophic calcification. Cancer 42(2):717–728, 1978.
145. Michotte Y, Smeyers-Verbeke J, Ebinger G, Maurus R, Pelsmaekers J et al: Brain calcification in a case of acute lymphoblastic leukemia. J Neurol Sci 25(2):145–152, 1975.
146. Rosman NP, Kakulas BA, Richardson EP: Central pontime myelinolysis in a child with leukemia. Arch Neurol (Chic) 14(3):273–280, 1966.
147 Cadman TE, Rorke LE: Central pontine myelinolysis in childhood and adolescence. Arch Dis Childh 44(2):342–350, 1969.
148. Minauf M, Krepler P: Zentrale pontine Myelinolyse bei einem Kind mit Leukämie. Arch Kinderheilk 180(1):55–65, 1969.
149. Valsamis MP, Peress NS, Wright LD: Central pontine myelinolysis in childhood. Arch Neurol (Chic) 26:307–312, 1971.
150. Goebel HH, Herman-Ben Zur P: Central myelinolysis. In: Handbook of Clinical Neurology, Vinken PJ, Bruyn GW (eds). Vol. 28, Amsterdam-Oxford-New York; Elsevier/North Holland 1976, pp 285–316.
151. Sah N, Wolff JA: Thiamine deficiency: probable Wernicke's encephalopathy successfully treated in a child with ALL. Pediatrics 51:750–751, 1973.
152. Reuck J de, Sieben G, Coster W de, Van der Eecken H: Prospective neuropathologic study on

the occurrence of Wernicke's encephalopathy in patients with tumours of the lymphoid-hemopoietic system. Acta Neuropath Suppl VII, 356–358, 1981.
153. Bowles D, Allen JC, Rosen G: Normal CT of the brain in osteosarcoma patients treated with high-dose methotrexate. Cancer 47:1762–1765, 1981.
154. Moore EW, Thomas LB, Shaw RK, Freireich EJ: The central nervous system in acute leukemia. Arch Intern Med 105(2):141–158, 1960.
155. Reuck J de, Coster W de, Van der Eecken H: Communicating hydrocephalus in treated leukemic patients. Europ Neurol 18(1):8–14, 1979.
156. Schochet SS, Lampert PW, Earle KM: Neuronal changes induced by intrathecal vincristine sulfate. J Neuropath Exp Neurol 27(4):645–658, 1968.
157. Shepherd DA, Steuber CP, Starling KA et al: Accidental intrathecal administration of vincristine. Med Pediatr Oncol 5(1):85–88, 1978.
158. Slyter H, Liwnicz B, Herrick MK, Mason R: Fatal myeloencephalopathy caused by intrathecal vincristine. Arch Neurol (Chic) 30(8):867–871, 1980.
159. Sullivan MP, Vietti TJ, Fernbach DJ, Griffith KM, Haddy TB, Watkins LW: Clinical investigations in the treatment of meningeal leukemia: radiation therapy regimens vs conventional intrathecal methotrexate. Blood 34(3):301–319, 1966.
160. Back EH: Death after intrathecal methotrexate. Lancet 2:1005, 1969.
161. Bagshawe KD, Magrath IT, Golding PR: Intrathecal methotrexate. Lancet 2:1258, 1969.
162. Pasquinacci G, Pardini R, Fedi F: Intrathecal methotrexate. Lancet 1:309–310, 1970.
163. Baum ES, Koch HF, Corby DG, Plunket DC: Intrathecal methotrexate. Lancet 1:649, 1971.
164. Pouillart P, Schwarzenberg L, Schneider M, Amiel JL, Mathé G: Les méningitis lymphoblastiques. Presse Méd 1;387–381, 1971.
165. Thompson SW, Saiki J, Kornfeld M et al: Paraplegia following intrathecal antileukemic therapy. Neurology (Minneap) 21:454, 1971.
166. Bernard J, Boiron M, Jacquillat C, Schaison G, Weil M: Traitement des leucémies aigues. Rapp XXIIIe Congr Ass Pédiatr Lange Franç Vol. III, 1972, p 367.
167. Saiki JH, Thompson S, Smith F, Atkinson R: Paraplegia following intrathecal chemotherapy. Cancer 29:370–374, 1972.
167a. Gutin PH, Levi JA, Wiernik PH et al.: Treatment of malignant meningeal disease with intrathecal Thiotepa. Cancer Treatm Rep 61:885–887, 1977.
168. Luddy RE, Gilman PA: Paraplegia following intrathecal methotrexate. J Pediatr 83:988–992, 1973.
169. Corberand J, Pris J, Robert A, Monnier J, Regnier C: Accidents neurologiques graves secondaires à l'injection intra-rachidienne d'agents cytostatiques. Arch franç Pédiat 30(5):177–188, 1973.
170. Gagliano RG, Costanzi JJ: Paraplegia following intrathecal methotrexate. Cancer 37:1663–1668, 1976.
171. Meyer R, Bergerat JP, Lang JM, Oberling T: Accident neurologique mortel après méthotrexate intrarachidien. Neuv Presse méd 5:149–151, 1976.
172. Tomonaga M, Takeno Y, Ishii N, Okayama M: Paraplegia following intrathecal methotrexate, Adv Neurol Sci (Tokyo) 22:1204–1211, 1978.
173. Reznik M: Acute ascending poliomyelomalacia after treatment of acute lymphocytic leukemia. Acta Neuropathol (Berl) 45(2):153–157, 1979.
174. Grisold W, Lutz D, Wolf D: Necrotizing myelopathy associated with acute lymphatic leukemia. Acta Neuropath 49(3):321–235, 1980.
175. Breuer AC, Pitman SW, Dawson DM, Schoene WC: Paraparesis following intrathecal cytosine arabinoside. Cancer 40(6):2817–2822, 1977.
176. Wisniewski H, Shelanski MI, Terry RD: Effects of mitotic spindle inhibitor on neurotubules and neurofilaments in anterior horn cells. J Cell Biol 38(3):224–229, 1968.

177. Goldie JH, Price LA, Harrap KR: Methotrexate toxicity. Correlation with duration of administration; plasma levels and excretion pattern. Europ J Cancer 8:409–414, 1972.
178. Ettinger LJ, Freeman AI, Creaven PJ: Intrathecal methotrexate overdose without neurotoxicity. Cancer 41(4):1270–1273, 1978.
179. Lampkin BC, Higgins GR, Hammond D: Absence of neurotoxicity following massive intrathecal administration of methotrexate. Case report. Cancer 20(5):1780–1781, 1967.
180. Phillips TL, Fu KK: Quantification of combined radiation therapy and chemotherapy effects of critical normal tissues. Cancer 37:1186–1200, 1976.
181. Rosen G, Tefft M, Martinez A, Chan W, Murphy ML: Combination chemotherapy and radiation therapy in the treatment of metastatic osteogenic sarcoma. Cancer 35(3):622–639, 1975.
182. Griffin TW, Rasey JS, Bleyer WA: The effect of photon irradiation on blood-brain barrier permeability to methotrexate in mice. Cancer 40(5):1109–1111, 1977.
183. Neuwelt EA, Frenkel EP, Barnett HP, Maravilla KR, Rapoport S: Osmotic blood-brain barrier disruption: Computerized tomographic scanning of chemotherapeutic agent delivery. Ann Neurol 6(2):166, 1979.
184. Kelley JM, Asbury AK, King JS: Neuropathologic effects of intrathecal water. J Neuropath Exp Neurol 34(2):388–000, 1975.
185. Bunge PR, Settlage PH: Neurological lesions in cats following cerebrospinal fluid manipulation. J Neuropath Exp Neurol 16(3):471–482, 1957.
186. Shelanski ML, Feit H: Filaments and tubules in the nervous system. In: Bourne GH (ed): The Structure and Function of Nervous Tissue. New York: Academic Press 1972, Vol. 5, pp 47–80.
187. Craft WA, Reid MM, Bruce E, Kernahan J, Gardner PS: Role of infection in the death of children with ALL. Arch Dis Childh 52:752–757, 1977.
188. Bell WE, McCormick WF: Neurologic Infections in Children, Second edition, Philadelphia: Saunders, 1981.
189. Townsend JJ, Wolinsky JS, Baringer JR, Johnson PC: Acquired toxoplasmosis. A neglected cause of treatable nervous system disease. Arch Neurol 23:335–343, 1975.
189a. Henson RA, Urich H: Cancer and the nervous system. London-Edinburgh-New York; Blackwell 1982.
189b. Cairncross JG, Posner JB: Neurological complications of malignant lymphomas. In: Vinken PJ, Bruyn GW (eds) Handbook of Clinical Neurology, Vol. 36. Amsterdam-New York: North Holland/Elsevier 1980, pp 27–62.
189c. Trump DL, Grossman SA, Thompson G, Murray K: CNS infections complicating the management of neoplastic meningitis. Arch. Intern Med. 142:583–586, 1982.
190. Parker JC, McClaskey JJ, Lee RS: Human cerebral candidosis. A postmortem evaluation of 19 patients. Human Path 12(1):23–27, 1981.
191. Weenink HR, Bruyn GW: Cryptococcosis of the nervous system. In: Vinken PJ, Bruyn GW (eds) Handbook of Clinical Neurology, Amsterdam-New York-Oxford: Elsevier/North Holland, Vol. 35, 1978, pp 459–502.
192. Ruskin J, Remington JS: Toxoplasmosis in the compromised host. Ann Int Med 84(2):193–199, 1976.
193. Frenkel JK, Nelson BM, Arias–Stella J: Immunosuppression and toxoplasmic encephalitis. Clinical and experimental aspects. Human Path 6(1):97–111, 1975.
194. Astrom KE, Mancall EL, Richardson EF Jr: Progressive multifocal leukoencephalopathy: a hitherto unrecognized complication of chronic lymphatic leukemia and Hodgkin's disease. Brain 81(2):93–111, 1958.
195. Walker DI: Progressive multifocal leukoencephalopathy: an opportunistic viral in infection of the central nervous system. In: Vinken PJ, Bruyn GW (eds) Handbook of Clinical Neurology.

Amsterdam-New York-Oxford: North Holland/Elsevier, Vol. 34, pp 307 – 329, 1978.
196. Peters ACB, Versteeg J, Bots GTAM, Boogerd W, Vielvoye J: Progressive multifocal leukoencephalopathy. Immunofluorescent demonstration of simian virus 40 antigen in CSF cells and response to cytarabine therapy. Arch Neurol (Chic) 37(4):497 – 980.
197. Castaigne P, Rondot P, Escourolle R, Ribadeau-Dumas JL, Cathala F, Hauw JJ: Leucoencéphalopathie multifocale progressive et 'gliomas' multiples. Rev Neurol 130:379 – 392, 1974.
198. GiaRusso MH, Koeppen AH; Atypical progressive multifocal leukoencephalopathy and primary cerebral malignant lymphoma. J Neurol Sci 35(3):394 – 398, 1978.
199. Ho KC, Garancis JC, Peagle RD, Gerber MA, Borkowski WJ: Progressive multifocal leukoencephalopathy and microglioma in a patient with immunosuppressive therapy. Acta Neuropath 52(1):81 – 83, 1980.
200. Lyon G: Examen ultrastructural du cerveau dans trois encephalopathies aigues et chroniques au cours de déficits immunitaires. Arch franç Pédiatr 29(00):541 – 554, 1972.
201. Breitfeld Y, Yashida Y, Sherman FE, Odagiri K, Yunis EJ: Fatal measles infection in children with leukemia. Lab. Invest 28(3):279 – 291, 19743.
202. Sluga E, Budka H, Jellinger K, Pichler E: SSPE-like inclusion body disorder in treated childhood leukemia. Acta Neuropath Suppl VI:267 – 272, 1975.
203. Pullan CR, Noble TC, Scott TJ, Wisniewski K, Gardner PS: Atypical measles infections in leukemic children on immunosuppressive treatment. Brit Med J 1:1562 – 1565, 1976.
204. Smyth D, Tripp JH, Brett EM, Marshall WC, Almeida J, Dayan AD et al: Atypical measles encephalitis in leukemic children in remission. Lancet 2:574, 1976.
205. Murphy JV, Yunis EJ: Encephalopathy following measles infection in children with chronic illness. J Pedit 88:937 – 942, 1976.
206. Drysdale HC, Jones LF, Oppenheimer DR, Tomlinson AH: Measles inclusion body encephalitis in a child with treated ALL. J Clin Path 29:865 – 872, 1976.
207. Haltia M, Paetau A, Erkkilä H, Donner M, Kaakinen K, Holmström T: Fatal measles encephalopathy with retinopathy during cytotoxic chemotherapy. J Neurol Sci 32(3):323 – 330, 1977.
208. Aicardi J, Goutieres F, Arsenio-Nunes ML, Lebon P: Acute measles encephalitis in children with immunosuppression. Pediatrics 59:232 – 239, 1977.
209. Wolinsky JS, Swoveland P, Johnson KP, Baringer JR: Subacute measles encephalitis complicating Hodgkin's disease in an adult. Ann Neurol 1(5):452 – 457, 1977.
210. Campbell RHA, Marshall WC, Chessells JM: Neurological complication of childhood leukemia. Arch Dis Childh 52:850 – 858, 1977.
211. Petersen FK, Schiotz PO, Valerins NH et al: Immunosuppressive measles encephalopathy. Acta Paediat Scand 67(2):109 – 112, 1978.
212. Spalke G, Eschenbach C: Infantile cortical measles inclusion body encephalitis during combined treatment of ALL. J Neurol 230(3):269 – 277, 1979.
213. Price R, Chernik NL, Horta-Barbosa L, Posner JB: Herpes simplex encephalitis in an anergic patient. Amer J Med 54(3):222 – 228, 1973.
214. Henderson BE, Ziegler JL, Templeton AC: Acute necrotizing encephalitis in a patient with Hodgkin's disease. East Afr Med J 48:592 – 600, 1971.
215. Dayan AD, Bhatti I, Gostling JVT: Encephalitis du to herpes simplex in a patient with treated carcinoma of the uterus. Neurology (Minneap) 17(6):609 – 613, 1967.
216. Dayan AD: Encephalitis due to simultaneous infection by herpes simplex and measles viruses. J Neurol Sci 14(4):315 – 323, 1971.
217. Mellor DH, Purcell M: Unusual encephalitic illness in a child with acute leukemia in remission. Possible role of measles virus and Toxoplasma gondii. Neuropaediatr 7(4):423 – 430, 1976.
218. Vuia O: Micronodular encephalitis and acute leukemia treated with immunosuppressives in a child. Neuropaed 6(3):307 – 312, 1975.

219. Casey EB, Jellife AM, Le Quesne PM, Millet YL: Vincristine neuropathy. Clinical and electrophysiological observations. Brain 96(1):69 – 86, 1973.
220. Rosenthal S, Kaufman S: Vincristine neurotoxicity. Ann Intern Med 80(6):733 – 737, 1974.
221. Sandler SG, Tobin W, Henderson ES: Vincristine-induced neuropathy. A clinical study of fifty patients. Neurology (Minneap) 19:367 – 374, 1969.
222. Daun H, Hartwich G: Die Vincristin-Polyneuritis. Fortschr Neurol Psychiat 39(3):151 – 165, 1971.
223. Caccia MR: Vincristine polyneuropathy in man. J Neurol 216(1):21 – 26, 1977.
224. McLeod JG, Penny R: Vincristine neuropathy: an electrophysiological and histological study. J Neurol Neurosurg Psychiat 32(3):297 – 304, 1969.
225. Heiss WD, Turnheim M, Mamoli B: Combination chemotherapy of malignant glioma. Effect of postoperative treatment with CCNU, vincristine amethopterine and procarbazine. Europ J Cancer 11:1191 – 1202, 1978.
226. Watkins SM: High incidence of vincristine induced neuropathy in lymphomas. Brit Med J 1:610 – 612, 1978.
227. Bradley WG, Lassman LP, Pearce GW, Walton JN: The neuromyopathy of vincristine in man. J Neurol Sci 10(2):107 – 131, 1970.
228. Bradley WG: The neuromyopathy of vincristine in the guinea pig. An electrophysiological and pathological study. J Neurol Sci 10(2):133 – 162, 1970.
229. Green LS, Donoso JA, Hallar-Bettinger IE, Samson FE: Axonal transport disturbances in vincristine-induced polyneuropathy. Ann Neurol 1(3):255 – 262, 1977.
230. Schlaepfer WW: Vincristine-induced axonalalterations in rat peripheral nerve. J Neurpath Exp Neurol 30(3):488 – 506, 1971.
231. Wulfhekel U, Düllmann J: Ein licht- und elektronenoptischer Beitrag zur Vinca-Alkaloid-Polyneuropathie. Virchows Arch Abt A Path Anat 357(2):163 – 178, 1972.
232. Obrist R, Paravicini U, Harmtann D et al.: Vindesine: a clinical trial with special reference to neurological side effects. Cancer Chemother Pharmacol 2:233 – 237, 1979.
233. Botez MI, Peyronnard JM, Bachevealier J, Charron L: Polyneuropathy and folate deficiency. Arch Neurol (Chic) 35(9):581 – 584, 1978.
234. Hemphill M, Pestrank A, Walsh T. et al: Sensory neuropathy in Cis-platinum therapy. Neurology 30:429, 1980.
235. Clark AW, Parhad IM, Griffin JW et al: Neurotoxicity of cis-platinum pathology of the central and peripheral nervous system. Neurology (NY) 30:429, 1980.
236. Russell JA, Powles RL: Neuropathy due to cytosine arabinoside. Brit Med J 2:652 – 653, 1974.
237. Clarke JTR, Karpati G, Carpenter S, Wolfe LS: The effect of vincristine on skeletal muscle in the rat. J Neuropath Exp Neurol 31(3):247 – 266, 1972.
238. Michalek JC, Dibella NH: Exacerbation of myotonic dystrophy by vincristine. New Engl J Med 295:283 – 284, 1976.
239. Ho M, Bear RA, Garvey MB: Symptomatic hypophosphatemia secondary to 5-azacytidine therapy of acute non-lymphatic leukemia. Cancer Treatm Rep 60:1400 – 1402, 1976.

13. The Management of Brain Metastases

J. GREGORY CAIRNCROSS and JEROME B. POSNER

INTRODUCTION

Malignant neoplasms originating outside the central nervous system (CNS), called here 'systemic cancer,' may affect the CNS in a variety of ways (Table 1) [1]. Approximately 20% of patients suffering from systemic cancer develop neurological symptoms at some time during their illness. Although metabolic encephalopathy is probably the most common complication, it usually appears terminally and plays little role in most of the course of the patient's illness.

Table 1. Neurological complications of systemic cancer

Metastatic
  Intracranial
    Brain
    Skull
      calvarium
      base
    Dural
      epidural
      subdural
    Leptomeningeal
    Other structures
      pituitary
      pineal
      choroid plexus
      primary tumor
  Spinal
  Leptomeningeal
  Peripheral nerve & plexus

Non-metastatic
  Metabolic encephalopathy
  CNS infections
  Vascular disorders
  Side effects of therapy
  Paraneoplastic syndromes ('remote effects')

Table 2. Presenting symptoms of brain metastasis

Symptoms	% of 162 Cases (49)	% of 201 Cases (50)
	(1970–73)	(1977–78)
Headache	53	45
Focal weakness	40	21
Behavioral or mental change	31	33
Gait ataxia	20	22
Seizures	15	20
focal motor		7
generalized		13
Speech difficulty	10	14
Visual disturbance		11
Sensory disturbance		10
Limb ataxia		10

Otherwise, the commonest neurological complication affecting patients with systemic cancer is metastatic brain tumor. Brain metastases are found at autopsy in 10–15% of patients dying with systemic cancer [2, 3]. In our experience, 2/3–3/4 of these patients suffer significant neurological symptoms during life [1]. Headache, convulsions, dementia, paralysis and ataxia, all common in patients with brain metastases (Table 2), often render patients unable to care for themselves. With prompt diagnosis and vigorous treatment, neurological symptoms and signs usually improve, allowing patients to resume useful activities until systemic cancer, not CNS disease, causes death.

This chapter reviews the management of brain metastasis, with emphasis on the role of adrenocorticosteroids (steroids), radiation therapy (RT), surgery and chemotherapy. (Metastases to other intracranial structures (Table 1), e.g. the dura, pituitary gland and leptomeninges, occur less frequently and are discussed elsewhere [4, 5, 6].) A brief discussion of biological considerations, clinical features and diagnostic evaluation forms a background for the discussion of management.

BIOLOGY OF METASTASES

Two biological problems that bear directly on the management of brain metastasis are: 1) the pathogenesis of metastasis to the brain and 2) the role of the blood-brain barrier in the diagnosis and therapy of brain metastasis.

Pathogenesis of Metastasis

In the last few years new information has surfaced about basic events leading to the spread of systemic cancer from its primary site to the brain. The biology of the individual tumor cell, the site of origin in the body and interactions between the host brain and the tumor cell all play a role in the development of brain metastasis: the metastatic process begins when tumor cells invade surrounding host tissue. The mechanism of invasion is unknown, although the ability of some malignant cells to release tissue-destructive enzymes, particularly lysosomal hydrolases and collagenolytic enzymes, is a possible factor promoting tumor invasion [7, 8]. The invasive ability of tumor cells may vary not only from tumor to tumor but also from cell to cell within a single tumor [9]. Spread of tumor from its site of origin usually occurs when the tumor penetrates blood vessels and/or lymphatics. (Growth into nerve endings and along nerve sheaths is another, but less common, mode of entry into the CNS [10].) The endothelium of blood vessels within tumors is often defective [11] and may be particularly susceptible to penetration by tumor cells. In addition, there are often areas of ischemia [12] and necrosis [13] in tumors; the blood vessels in these regions show structural alterations that could facilitate invasion by tumor cells. After penetrating blood vessels or lymphatics, tumor cells may be swept downstream or, alternatively, proliferate at the site of invasion and shed tumor emboli into the circulation.

Most tumor cells reach the brain via the arterial circulation, find their way to the penetrating pial blood vessels, and lodge in arterioles or capillaries at the grey matter/white matter junction [14]. (Occasionally tumor emboli are large enough to occlude arteries and produce acute neurological symptoms [15].) If these acute neurological symptoms are transient, a three-phased clinical picture may be observed, consisting of: 1) acute neurological dysfunction produced by the tumor embolus, 2) an asymptomatic period of several months following recovery from the acute symptoms, and 3) subacute onset of progressive symptoms identical to those of the acute episode and indicative of tumor growth from the original embolus.

Since most systemic metastases begin in the venous or lymphatic circulation and enter the brain via the arterial circulation, they must somehow traverse lung capillaries. Entry into the arterial circulation occurs in one of several ways: 1) Usually the metastasis originates in the lung from either a primary lung cancer or from a pulmonary metastasis. Cells are shed from the lung tumor into the pulmonary venous system and find their way through the heart into the systemic arterial circulation. (Since in the resting state the brain receives about 15% of the cardiac output, it is not surprising that the brain is a common site of metastasis from lung tumors disseminating via the arterial circulation.) Clinical observations support this mode of spread: lung cancer is the major source of brain metastases [3]; metastasis to the brain tends to occur early in patients with lung cancer and may be the first indication of the disease [16]; brain metastasis from other cancers (i.e. breast, colon, etc.) occurs later in the illness after the tumor has first spread via the venous circulation to the lungs and metastases have established there (the cascade theory) [17]. 2) Tumor cells occasionally bypass the lung, reaching the systemic arterial circulation directly from the venous side through a patent foramen ovale. Such paradoxical tumor embolism has occasionally been demonstrated pathologically [18]. 3) It is possible that individual tumor cells shed into the systemic venous circulation pass through the lung capillaries into the pulmonary veins uninterrupted, without leading to growth of tumor in the lungs themselves. Fidler's experiments with B16 melanoma (see below) suggest this. This route of spread may explain the occasional patients with non-lung cancer who develop brain metastases in the absence of radiographically or pathologically demonstrable pulmonary metastases [19].

Not all tumor cells enter the brain through the arterial circulation. Another potential hematogenous mode of spread to the brain is via the vertebral venous system (Batson's plexus). When Batson described the plexus [20], he suggested that it provided a pathway of spread to the spinal and intracranial space for cancers of the pelvis and retroperitoneum and might explain why carcinomas such as prostate, uterine and colon at times seed the brain without metastasizing elsewhere in the body. This mode of spread as a cause of brain metastasis remains speculative, but there is substantial clinical evidence to suggest that it represents a major pathway of spread to the bones of the skull, spine and pelvis [21].

Hematogenous spread of cancer to the brain, whether via the arterial circulation or via Batson's plexus, does not explain the tremendous variability in the behavior of different tumor types. For example, melanoma [22] and choriocarcinoma commonly metastasize to the brain (usually after lung metastases are evident clinically), whereas thyroid carcinoma, even when it seeds the lungs, rarely causes brain metastases. Renal and colon carcinomas, although uncommon causes of brain metastasis, are more common than prostate cancer. Prostate cancer is more likely to involve the skull and dura. Hodgkin's disease rarely involves the CNS, whereas 20% of patients with diffuse, poorly differentiated or histiocytic non-Hodgkin's lymphomas develop CNS (usually leptomeningeal) metastases [23]. The interactions between the tumor cell and the target tissue must play a role in determining the distribution of metastases and can be invoked in part to explain the striking differences in the incidence of brain metastasis among different tumor types. Nevertheless, random trapping and arrest of circulating tumor cells, both of which are functions of the organ's blood flow, do influence the distribution of metastases. (Most studies of the distribution of brain metastases suggest that the tumors are found in the hemispheres, cerebellum and brainstem in numbers proportional to the weight and presumably blood flow to that area of the brain. There are exceptions, however. Hunter and Newcastle [24] found a disproportionate number of cerebellar and brainstem metastases when compared to those to the cerebral hemispheres. We have not found this to be the case for all tumor types, but in our experience both renal and colon carcinomas are disproportionately represented in the cerebellum.) It is now clear that mechanical factors alone do not explain the distribution of metastases. Non-random patterns of metastasis have been observed in animal tumor systems [25–28]. Studies in which radio-labeled tumor cells are introduced into the circulation have shown that in many instances circulating cells arrest in a wide variety of organs, but metastases occur only in certain tissues [13, 29].

Present evidence suggests that specific properties of both the circulating tumor cell and the host tissue are important in determining the distribution and growth of metastatic tumors. The importance of the target tissue is illustrated by experiments showing that animal tumor cells that ordinarily arrest in the lung also localize and grow in pieces of lung implanted subcutaneously [25, 29, 30]. By contrast, tumor growth does not occur in subcutaneous implants of other organs. Evidence of the influence of tumor cell properties on the arrest pattern of circulating cells has come from the observation that membrane vesicles shed from a B16 melanoma, which preferentially metastasizes to lung when fused to the surface of a non-metastasizing B16, subline transforms the latter into a lung metastasizing line [31]. Fidler *et al.* have been able by *in vivo* cloning of B16 melanoma to produce cell lines that, when injected intravenously, metastasize to specific organs [32], e.g. the brain but not the lung, suggesting that host interactions with particular clonal lines of tumor cells are important for the development of metastases.

The complexity of the interrelationship between tumor cell and host tissue is magnified by the recent appreciation that malignant primary tumors contain subpopulations of cells with differing metastatic capabilities [33]. The increasing recognition of tumor cell populations with different metastatic capabilities should facilitate experimental efforts to identify the properties which enable malignant cells to metastasize.

Blood-Brain Barrier

Most oncologists believe that brain metastases are refractory to chemotherapy. Although this is not strictly true (see section on chemotherapy), at times a tumor outside the CNS may respond to chemotherapy, but its CNS metastasis may fail to do so. There are 2 possible reasons for the failure of many brain metastases to respond to chemotherapy. First, the brain tumor may have grown from a subpopulation or clone of cells in the primary tumor differing from others in both drug resistance and ability to metastasize to brain. Differences have been observed between the primary cancer and those metastases to organs other than brain, both with respect to biochemical properties [34] and responsiveness to therapy [35, 36]. Second, the blood-brain barrier may exclude water-soluble chemotherapeutic agents that might otherwise be effective in the treatment of brain metastases. For chemotherapy to be effective, the agents must not only be toxic to cancer cells but must be delivered to those cells in effective concentrations. Lipid-soluble chemotherapeutic agents pose no problem since they readily traverse plasma membranes. However, in the normal situation the blood-brain barrier substantially restricts the brood-brain exchange of water-soluble drugs [37, 38].

The blood-brain barrier exists because brain capillaries possess a continuous lining of endothelial cells that are connected by tight junctions [39]. The endothelial cells of brain are highly active metabolically [40] but do not support transendothelial vesicular transport, and the tight junctions between them restrict intercellular diffusion [41]. The blood-brain barrier is structurally and functionally abnormal in animal and human brain tumors, both primary and metastatic [42–45], allowing the entry of substances into the tumor that do not normally enter brain, such as diagnostic markers (e.g. technetium for brain scans and organic iodides for CT scans), plasma components (leading to brain edema) and chemotherapeutic agents. The most characteristic alteration in brain tumor capillaries is the appearance of fenestrated and discontinuous endothelial cells [46]. The fenestrations are similar to those found in the area postrema and other specialized regions of normal brain that are freely permeable to plasma proteins and histologic tracers. Endothelial gaps are also observed but occur less frequently than fenestrations [46]. Autoradiographic methods have been used to demonstrate that the barrier is functionally disrupted in primary and metastatic brain tumors [47]. However, it is not known whether the blood-brain barrier is defective in all areas of brain tumors, or at all times during the course of their development. Preliminary autoradiographic evidence suggests that small metastatic brain tumors may lie behind the blood-brain barrier. If such is the case, then small tumors may escape detection by radionuclide or contrast CT scans and be protected from potentially toxic water-soluble chemotherapeutic agents. Further information on these points may influence the selection of chemotherapeutic agents or lead to the concomitant use of blood-brain barrier opening techniques (i.e. osmotic opening) [48] and regional (i.e. intra-arterial) chemotherapy.

DIAGNOSIS OF BRAIN METASTASES

Symptoms and Signs

Tables 2 and 3 summarize the presenting symptoms and signs of brain metastasis in 2 series of patients evaluated at Memorial Sloan-Kettering Cancer Center (MSKCC) [49, 50]. Several points deserve emphasis: Headache is present in only 50% of patients with a brain metastasis. Headaches may be mild, are often diffuse or located bilaterally in the frontal or occipital regions, and usually have little localizing value. Early morning headache, believed to be highly suggestive of increased intracranial pressure, occurs in only 40% of our patients in whom headache of any kind is a presenting symptom. Headaches are more com-

Table 3. Presenting signs of brain metastasis

Symptoms	% of 162 Cases (49) (1970–73)	% of 201 Cases (50) (1977–78)
Impaired cognitive function	77	42
Hemiparesis	66	54
mild-moderate		48
severe		6
Hemisensory loss	27	16
Papilledema	26	15
Gait ataxia	24	14
Aphasia	19	17
Visual field cut		13
Limb ataxia		10
Depressed level of consciousness		7

mon in patients with multiple metastases or with a single metastasis in the cerebellum. Although headaches are often accompanied by raised intracranial pressure, papilledema is an uncommon finding. Seizures occur in about 15–20% of our patients (others report a higher incidence [51]).

In general, neurological signs are more prominent than neurological symptoms. However, in our most recent series this difference is less striking than it was previously, perhaps reflecting earlier diagnosis in patients with very mild symptoms because of the ease and accuracy of CT scanning. Focal weakness and mental or behavioral changes are by far the most common neurological signs at the time of diagnosis. Focal weakness is usually mild or moderate. Focal weakness does *not* accurately localize the metastatic lesion to the motor area of the contralateral hemisphere. Paillas and Pellet [52] noted that although metastatic tumors are more frequently found in the parietal lobe than in the frontal lobe, motor disturbances are more common than sensory ones. Our own experience is similar: Motor deficits are more common than sensory deficits even when tumors are located posterior to the Rolandic fissure. Mental and behavioral changes are usually caused by multiple brain metastases or by solitary lesions which increase intracranial pressure or cause hydrocephalus. Disturbances of higher function are usually not of localizing value. However, patients with isolated alexia, acalculia, agraphia, etc., represent obvious exceptions.

The course of brain metastasis is usually subacute and easily recognized: Patients with known cancer develop mild, diffuse headache followed over a period of days to weeks by progressive focal neurological symptoms and signs (especially hemipareses). However, there are acute and other unusual clinical presentations of brain metastases that may make clinical diagnosis more difficult: Excluding seizures (i.e. focal motor seizures or grand mal seizures), 5–10% of patients with brain metastasis develop severe neurological disability acutely. In some instances this sudden ictus results from hemorrhage into the tumor, cerebral infarction from embolic or compressive occlusion of a blood vessel [53], or a non-convulsive seizure with profound postictal paralysis [54]; in other instances the cause is unknown. Hemorrhage into a metastasis is particularly common in patients with brain metastasis from choriocarcinoma (both gestational and testicular) and malignant melanoma, although hemorrhage may occur in any metastatic tumor [55]. Occasionally patients present with apoplectic symptoms and signs which resolve completely and mimic transient ischemic attacks. The pathogenesis of these transient episodes is poorly understood. Hypotheses include tumor embolization to the cerebral circulation with reversible ischemia, or a 'silent' focal seizure followed by a rapidly clearing postictal paralysis [54]. Patients with raised intracranial pressure from multiple supratentorial metastases or hydrocephalus may present with a constellation of acute paroxysmal symptoms, including intense headache, visual obscurations, drop attacks and syncope. These attacks are frequently precipitated by postural change and are thought to be related to sudden, severe, reversible increases in intracranial pressure (plateau waves) [56].

Not all atypical presentations of brain metastasis take the form of acute problems. Patients with multiple small brain metastases may develop a confusional state [57]. The absence of headache and the non-focal neurological examination mistakenly suggest a toxic or metabolic encephalopathy. Patients with posterior parietal or occipital lesions and no headache may be totally unaware of an homonymous hemianopia until multiple automobile accidents or other incidents draw attention to the problem [57]. A midline cerebellar metastasis may be overlooked initially because the patient typically complains of gait unsteadiness before objective evidence of cerebellar dysfunction can be demonstrated by examination [57]. Unexplained vomiting, cough headache [58] and cough syncope are unusual symptoms which also raise the suspicion of cerebellar metastasis. The most important point to be made is that patients with brain metastasis develop a variety of neurological problems and that, in a patient with cancer, even unusual neurological findings must be considered to be caused by a brain metastasis until proven otherwise.

DIAGNOSTIC TESTS

The best diagnostic test for brain metastasis is the computerized tomographic (CT) scan. It is by far the most sensitive (6% of neurologically asymptomatic patients with lung cancer have a brain metastasis on screening CT scans [59]) and except for biopsy the most specific test for brain metastasis. It has essentially replaced skull roentgenograms, radionuclide brain scanning, cerebral angiography and pneumoencephalography. The CT scan is fast, accurate, reproducible and safe. Although not all contrast-enhancing mass lesions with surrounding edema are metastatic brain tumors, in the appropriate clinical setting the finding of such a lesion is virtually diagnostic. A detailed discussion of CT scanning is found elsewhere in this volume, but a few points deserve emphasis. The limitations of routine CT scans become apparent in two situations: when the CT scan fails to demonstrate a lesion because it is small (< 5 mm), isodense or enhances poorly, and when the clinical picture is sufficiently complex that the lesion demonstrated may be other than tumor (e.g. cerebral hemorrhage, brain abscess, infarct, etc.) Several special techniques are useful in dealing with these problems. Small metastases, if strategically located (e.g. in the internal capsule, brainstem or cerebellum) may produce devastating neurological symptoms and signs but not be detectable on routine CT scans. Sometimes these lesions can be demonstrated by performing thin and overlapping tomographic sections in the area of clinical concern or by increasing the amount of intravenous contrast material administered ('double-dose' contrast). Double-dose contrast may also be useful in those patients with solitary brain metastases and limited systemic disease who are being considered for brain surgery. The double-dose contrast scan may reveal lesions not detected on the routine scan (Figure 1).

At other times a brain metastasis is suspected on clinical grounds but even the double-dose CT scan with overlapping sections is normal. In this situation we recommend that the CT be repeated in 2 – 4 weeks, especially if symptoms persist. (Only occasionally will a radionuclide brain scan detect a metastasis missed on CT scan.)

Patients with known systemic cancer, a typical history for brain metastasis and one or more contrast-enhancing lesions surrounded by edema are not a diagnostic problem. However, situations arise where the diagnosis is not clear even after a lesion compatible with a brain metastasis has been demonstrated on CT scan. One such situation is when a patient with neurological dysfunction and a contrast-enhancing lesion on CT scan is not known to have systemic cancer. The CT lesion could be either a primary or metastatic tumor (or perhaps a non-neoplastic lesion). Sometimes a double-dose CT scan will reveal other unsuspected lesions, increasing the possibility that one is dealing with a metastatic tumor. Evaluation of the patient for systemic cancer may clarify the diagnosis. A chest xray, sputum cytology (in smokers), rectal examination, stool guaiac, serum carcinoembryonic antigen level, intravenous pyelogram and bone scan, in addition to routine hematology, urinalysis and biochemistry, constitute a reasonable search for systemic disease. If these examinations are normal, it is unlikely that further evaluation with an upper and lower GI series, liver scan, sonogram or CT scan of the abdomen will establish the diagnosis [60]. A biopsy of the brain lesion is then indicated.

Another difficult situation occurs in the patient with a history of cancer and an acute neurological event suggestive of an intracerebral hemorrhage. If a non-contrast CT scan identifies a hemorrhagic lesion, one cannot be certain whether

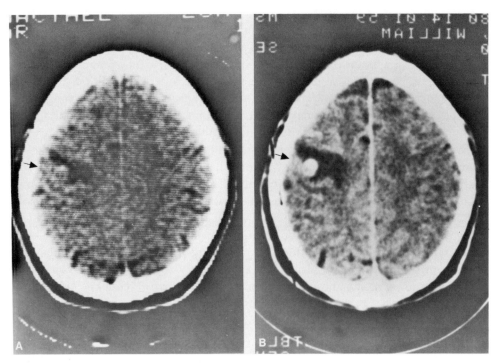

Figure 1. The CT scans of a 56-year old man with adenocarcinoma of the lung who developed clumsiness of the right hand following a three-week history of intermittent headache. The neurological examination revealed mild weakness and clumsiness of the right hand. A CT scan with 100 cc dose of IV contrast (1A) disclosed a non-enhancing area of decreased density in the left posterior frontal region (arrow). The lesion was initially interpreted as an area of infarction. Because the subacute onset of symptoms spoke against a vascular insult, the CT scan was repeated following double-dose (200 cc) IV contrast (1B), revealing a small enhancing lesion with surrounding edema (arrow).

the hemorrhage is primary or into a metastatic tumor (secondary). A contrast scan may provide valuable information, e.g. an area of enhancement adjacent to or remote from the hematoma points to a diagnosis of brain metastasis with secondary hemorrhage. Delaying the contrast scan may lead to confusion because of the tendency for some primary intracerebral hemorrhages to contrast-enhance as healing occurs [61].

Metastatic brain tumors and brain abscesses are often indistinguishable on CT scan and pose another diagnostic problem. Brain abscesses are exceedingly uncommon and not a diagnostic consideration in patients with carcinomas or sarcomas, except in unusual clinical circumstances. However, brain abscesses do occur in patients with lymphoma, particularly Hodgkin's disease. In patients with Hodgkin's disease, toxoplasma abscesses are at least as common, and perhaps more common, than intracerebral metastases. CT criteria such as ring thickness and the average value of CT numbers in the ring center help distinguish abscesses from tumors but are rarely conclusive [62]. If the diagnosis cannot be established on clinical grounds, a biopsy will be necessary.

Occasionally patients with cancer and a history typical of brain metastasis are started on corticosteroids before a CT scan is obtained. A normal CT scan in this setting does not preclude a diagnosis of brain metastasis. The blood-brain barrier reconstitutes with steroids, decreasing the passage of water-soluble radionuclides and iodides into the tumor [63]. At times the lesion may disappear, only to be seen again when steroids are withdrawn.

In addition to its importance as a diagnostic tool, the CT scan provides us with a means to evaluate treatment. The response of a metastasis to radiation can be judged by a repeat contrast CT scan 8 – 10 weeks after treatment. A CT scan with contrast enhancement performed *within* a week of surgical extirpation will often determine whether or not the removal has been complete. Generally speaking, contrast enhancement is not seen along the surgical margin unless tumor remains. However, *after* a week enhancement may be due to postoperative changes or abscess formation, as well as residual tumor, and the differential diagnosis may be difficult [61]. Survival is not a very useful method of evaluating the effectiveness of treatment of brain metastases because most patients, even with brain metastases, die of their systemic cancer. In order to evaluate the various treatments of brain metastases, including radiation dosages, fractionation schedules, radiosensitizers, chemotherapeutic agents, etc., one must rely on the information provided by post-treatment CT scans.

TREATMENT

General Considerations

Table 4 lists the therapeutic options for patients with brain metastasis. Adrenocorticosteroids, radiation therapy and surgery each have a place in the treatment of brain metastasis. The position of chemotherapy, both systemic and intra-arterial, has been considerably less secure, although recent observations suggest that it may be useful in certain situations. At the present time there is no established role for immunotherapy in the treatment of brain metastasis.

Table 4. Treatment options for patients with brain metastases

Adrenocorticosteroids
Whole-brain radiation therapy
Surgery
Chemotherapy
    systemic
    intra-arterial

Several factors must be considered in determining the optimal treatment for each patient with a brain metastasis. These factors include the patient's neurological status, the extent of systemic disease, the number and site of brain metastases, and the sensitivity of the primary tumor to radiation and chemotherapy.

The patient's neurological condition is the most important determinant of emergency treatment. Patients who are acutely deteriorating, with increased intracranial pressure and threatened cerebral herniation, require immediate treatment. A suggested treatment protocol is outlined in Table 5. The protocol includes hyperventilation to lower the pCO_2 and thus the cerebral blood volume, leading to a rapid but transient decrease in intracranial pressure, the use of osmotic agents to decrease the water content of the brain, also leading to a rapid but more sustained decrease in intracranial pressure, and the use of high doses steroids to diminish plateau waves and lower intracranial pressure more slowly but in a more sustained fashion. Radiation therapy is probably ill-advised in the acutely decompensating patient because it may aggravate cerebral edema, further increasing intracranial pressure. In our experience, acute surgical intervention in this situation is seldom helpful unless the patient can first be stabilized by medical means. The exception occurs in patients whose intractable increase in intracranial pressure is the result of obstructive hydrocephalus. In some instances, ventricular drainage can be life-saving when other methods of controlling intracranial pressure fail. However, the majority of acutely decompensating patients can be stabilized by high doses of steroids, osmotic agents or hyperventilation. Further treatment is then dictated by considerations other than their neurological state.

A second factor influencing treatment is the extent of the patient's systemic disease. The expectations of treatment vary with the clinical situation. Brain metastases usually develop in the setting of disseminated and advancing systemic disease. In this situation the goal of treatment is palliation. Occasionally,

Table 5. Management of acutely decompensating patients

I. Hyperventilation
   Maintain the airway
   Intubate stuporous and comatose patients and lower $PaCO_2$ to 25 – 30 mm Hg
II. Hyperosmolar agents
   Mannitol 75 – 100 grams IV stat (20% solution)
III. Diuresis
   Furosemide 80 mg IV stat
IV. Steroids
   Dexamethasone 100 mg IV stat followed by 25 mg qid
IV. Emergency CT scan
V. Ventricular drainage if symptoms are due to hydrocephalus

however, brain metastases appear in a patient whose systemic disease is otherwise controlled or localized. Here palliation is not sufficient. Rather, the aim of treatment must be local (i.e. neurological) 'cure', How this is best achieved is considered in the individual sections below. There is a consensus among neuro-oncologists that surgical extirpation of brain metastases should only be considered if systemic disease is absent (the primary site having been eradicated), minimal, or controllable for an extended period.

The number and site of brain metastases are crucial considerations in selecting treatment. A solitary metastasis in a silent area of brain (e.g. right frontal lobe) in a patient free of systemic disease and potentially 'curable' may be considered for surgical extirpation. This is especially true if the primary tumor is radioresistant. Again, there is a consensus among neuro-oncologists that multiple metastases are not amenable to surgical extirpation. Metastases in surgically less accessible areas of the brain (e.g. thalamus) or in areas of vital function (e.g. speech, language areas) are best treated by non-surgical means first, with surgery to be considered only if medical treatment fails.

The radioresponsiveness of the primary tumor also plays a role in decision-making. A patient with an accessible brain metastasis from malignant melanoma is a more appropriate surgical candidate, because of the recognized radioresistance of this tumor, than a patient with metastatic adenocarcinoma of the lung, who can be expected to respond to RT. We have occasionally treated small asymptomatic brain metastases from tumors sensitive to chemotherapeutic agents (e.g. oat cell carcinoma of lung) with drug alone, withholding RT or surgery while monitoring the response with serial CT scans (see Chemotherapy).

With the aforementioned as background and recognizing that each patient must be considered as a unique problem, Figure 2 outlines our approach to the treatment of patients who are neurologically stable and who have single or multiple brain metastases.

How does one evaluate the efficacy of treatment of metastatic brain tumors? The duration of survival, the usual method of evaluation of cancer therapy, is not very useful. Our own data [50] and those of several other investigators [64, 65] indicate that most patients treated for metastatic brain tumor die of systemic, not brain disease. Since the goal of neurologic treatment is to ameliorate neurological symptoms and signs, a hemiplegic patient who returns to work after therapy only to die within 4 months of pulmonary or liver metastases must be considered a therapeutic success, whereas the same patient who remains hemiplegic but lives a year must be considered a therapeutic failure. Improvement in neurological signs and symptoms thus seems an attractive method of evaluating treatment. There are, however, several problems with this measure. First, neurological improvement, like beauty, is frequently in the eye of the beholder, and enthusiastic therapists may differ substantially from their less enthusiastic counterparts evaluating the clinical outcome in a given patient. Second, brain

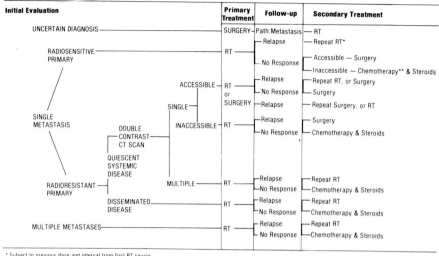

Figure 2. A scheme for the management of neurologically stable patients with brain metastasis. All patients begin steroids prior to the primary treatment. Follow-up includes an assessment of clinical response, steroid requirements and tumor response by CT scan.

edema and its symptoms often respond so well to steroids that many patients begin RT or enter surgery (i.e. definitive treatments) asymptomatic. Consequently, there is no clinical change to evaluate. Lastly, many patients 'successfully' treated for metastatic brain disease develop late neurological symptoms that appear to be due to metastases but are in fact the consequence of vital organ failure (e.g. liver, lung) interfering with brain metabolism and exacerbating the symptoms of a previous disorder.

Serial CT scanning is undoubtedly the most objective way of evaluating the outcome of treatment; if the tumor disappears on CT scan, treatment can be considered a success, whereas if it continues to grow treatment can be considered a failure. At times, however, persistent lesions on CT scan have turned out to be areas of necrosis rather than residual tumor. In general, a combination of the patient's clinical state, post-treatment CT scans and steroid requirements after treatment, represent the best way of evaluating brain metastasis therapy. Survival is useful only if the cause of death can be clearly specified.

The paragraphs below review the current state of each of the common therapeutic modalities and what one might expect from them in the treatment of brain metastases.

No Treatment

Although length of life is a poor measure of treatment efficacy for the majority of patients with brain metastasis, it is clear that life is prolonged by treatment.

The available literature on the natural history of brain metastasis suggests that for untreated patients the median survival is about 4 weeks from the time of diagnosis [66-69]. Symptoms do not remit spontaneously and patients develop increasing intracranial pressure leading to obtundation, stupor and coma. Death results from cerebral herniation with brainstem compression or from intercurrent illness in neurologically devastated patients.

Adrenocorticosteroids

Adrenocorticosteroid hormones (steroids) have been used to treat metastatic brain tumors for 25 years [70]. Prompt clinical improvement is observed in the majority of patients. These medications exert their beneficial effect by decreasing the edema surrounding metastatic tumors [71, 72]. Moreover, experimental evidence in animals and clinical and radiographic observations in patients suggest that steroids may, in a few instances, be oncolytic [73, 74]. Whatever their mechanism of action, more than 70% of patients with brain metastases improve symptomatically when started on steroids [75]. The symptoms most likely to respond to steroid therapy are those which reflect generalized brain dysfunction due to cerebral edema, increased intracranial pressure and brain shifts. Focal symptoms and signs appropriate to the location of the metastasis respond less favorably. Therefore, headache, alterations in consciousness, and signs and symptoms of cerebral herniation are more likely to resolve than seizures, hemiparesis and aphasia. These differences are relative in that neurological symptoms of all types usually improve to some degree with steroid use. The particular steroid used does not seem to make a difference. Most neuro-oncologists favor dexamethasone largely because of its minimal mineralocorticoid activity and because it is less likely to cause steroid psychosis even in large doses [76]. The optimal steroid dose is unknown. The standard starting dose is the equivalent of 16 mg of dexamethasone a day in divided doses. If patients do not respond promptly, higher doses, up to 100 mg a day, can be given with salutary effects [77]. In acutely decompensating patients with incipient herniation, we give 100 mg of dexamethasone as an intravenous bolus.

The median duration of remission for patients treated with steroids alone is approximately one month [78, 79], although much longer remissions have been reported [80]. Although unquestionably of benefit with respect to symptoms, the evidence that steroids alone prolong life is sparse. The few reports addressing this issue suggest that the median survival is about 2 months [68, 79, 81], one month longer than would be expected with no treatment at all. When used in conjunction with RT, steroids hasten neurological improvement [82] but do not appear to alter the likelihood of CNS response, the degree of response, the duration of remission, or the length of survival [82-84]. However, steroids are useful because they promote a rapid amelioration of symptoms and return to independent living, important benefits for patients with a short life expectancy.

In addition to prompt symptomatic improvement, steroids minimize the acute toxicity of RT to the brain (i.e headache, nausea, vomiting, fever, worsening of focal signs, obtundation, herniation, etc.). Some radiotherapists believe that pretreatment with steroids does not reduce or modify the side effects of brain irradiation [85]. However, our experience has been that acute side effects are common in patients who have not been pretreated with steroids and rare in those who have. For patients in whom surgical removal of a brain metastasis is necessary, steroids play an important role in minimizing postoperative edema. Lastly, steroids may be used to extend improvement for a short time in a proportion of patients whose neurological symptoms recur following RT or surgery.

Although of benefit in the management of patients with brain metastasis, prolonged steroid administration is not without risk [86]. The increased catabolism produced by steroids is undesirable, particularly in cachectic patients suffering from cancer. The catabolic effects are not balanced by the increase in appetite and sense of well-being that usually occur with steroid use. In addition, weakness of proximal lower and sometimes upper extremity muscles (steroid myopathy) can appear within days or weeks when high doses of steroids, especially dexamethasone, are used [87]. There is some evidence that the use of steroids promotes dissemination of metastases [88], but this is probably of minor clinical importance in patients who already have metastatic disease.

On balance, the advantages of steroids in the acute treatment of metastatic brain tumors and the amelioration of side effects of RT far outweigh their disadvantages, but steroids must be tapered and stopped wherever possible after definitive treatment (i.e. RT, surgery or chemotherapy). Depending upon the clinical situation, steroid dependence may be viewed as evidence of treatment failure. The steroid taper should be carried out slowly over a period of 3 – 6 weeks, depending in part on how long steroids have been used. If symptoms develop during the tapering of steroids, the dose should be raised until improvement is achieved and then lowered again more gradually. Some patients cannot be tapered off steroids but are asymptomatic on low doses (e.g. 2 – 4 mg a day). Steroid dependency usually implies treatment failure, although psychological dependency on steroids has been described [89]. In some patients, tapering of steroids produces the steroid withdrawal syndrome [90], a constellation of symptoms characterized by headache, malaise, nausea, and at times myalgias, arthralgias, and skin desquamation. Because some of the symptoms are reminiscent of those of cerebral edema and metastatic tumor, the physician may be misled into believing that the withdrawal symptoms represent treatment failure (i.e. failure of the brain metastasis to respond to definitive treatment) and may continue larger doses of steroids than necessary. A CT scan will usually clarify the situation. Because steroids stabilize and reconstitute the blood-brain barrier, they often decrease the amount of contrast material that enters brain metastases. Consequently, when patients are on steroids, metastases may appear smaller than they actually are, or, in some instances, disappear from view entirely.

Radiation Therapy

Radiation therapy (RT) is the preferred treatment for most patients with brain metastasis. The variables usually considered in judging the overall effectiveness of this treatment include initial clinical response, duration of improvement, and length of survival. Table 6 summarizes the findings of the major series published to date. Initial clinical improvement can be anticipated in 39 – 92% of patients [64, 80]. Because most patients who respond to RT are pretreated with steroids and improving already, it may be difficult to judge the response to RT until steroids are discontinued some weeks later. Thus, we define an RT responder as a patient who is asymptomatic or improved off steroids 6 – 8 weeks after RT. Some patients continue to improve both clinically and radiographically (CT scan) after that point. Interseries differences in definition of improvement, inclusion or exclusion of patients starting RT but dying before completion, tumor types treated, and use of steroids account in part for this wide range of response rates. Response rates of 63 – 78% have been reported in 3 series of patients treated with RT but not steroids [68, 91, 92]. The median duration of improvement in most reports is 3 months but ranges from 2 – 6 months. Duration of clinical improvement is not always a reliable indicator of the effectiveness of RT. Patients with successfully treated brain metastases may develop an *apparent* relapse resulting from early-delayed radiation toxicity (see below), metabolic brain disease (a result of advancing systemic cancer) or a non-metastatic neurological complication of systemic cancer that may mimic recurrence of the brain metastasis. In such instances, the clinical findings may be misleading and only the CT scan reveals the true situation. The median length of survival from the onset of treatment varies from 3 – 6 months. Like initial clinical improvement and duration of improvement, survival is an unreliable measurement of response of the brain metastasis to therapy. As the data below indicate, most patients with brain metastasis die not of their brain tumor but of systemic disease. A short median survival usually reflects the advanced nature of the systemic illness at the time the brain metastasis appears, not failure of therapy for the CNS lesion. Although the median survival exceeds the median duration of improvement, 80% of responders alive at 3 months and 60% of those alive at 6 months remain improved with respect to their neurological disease [85, 93]. Between 1/3 [50] and 3/4 [103] of relapsing patients have been reported to improve with a second course of RT. An uncommon but important phenomenon is the development of new brain metastases months or occasionally several years after the successful treatment of the original tumor by RT. The late appearance of brain metastases after prophylactic brain irradiation in patients with oat cell carcinoma is an example. This phenomenon undoubtedly represents re-seeding from the primary site. Many of these patients can be re-radiated, but unfortunately most cannot tolerate the

Table 6. Major studies on the treatment of brain metastasis by radiation therapy (RT)

Study	No. of patients	Tumor types	RT protocol	Steroid use	Initial improvement	Duration of improvement	Length of survival			% Responding to repeat RT
							Median	Mean	> 1 yr	
Chao et al., 1954 [91]	38	All	Variable	None	63%	3.5 mo (mean)		6.5 mo	3%	75%
Chu and Hilaris, 1961 [92]	218*	All	3,000 rad/3 wk	None	78%	5 mo (mean)		5.7 mo	—	70%
Order et al., 1968 [93]	108*	All	Variable	35%	60%	3–6 mo (median)	3–6 mo	6.3 mo	9%	44%
Deeley and Edwards, 1968 [94]	61*	Lung	3,000 rads/4 wk	—	47%	3–6 mo (median)	< 6 mo		14%	—
Hindo et al., 1970 [85]	54	All	1,000 rads/ 1 dose	Some	65%	3 mo (median)		5.6 mo	9%	—
Nisce et al., 1971 [84]	560*	All	3,000 rads/3 wk or 4,000 rads/4 wk	All	80%	5 mo (mean), 3 mo (median)	6 mo		16%	45%
Horton et al., 1971 [79]	19	Lung	4,000 rads/ 3–4 wk	All	74%	13 wk (median)	20 wk	—	—	—
Gottleib et al., 1972 [80]	41*	Melanoma	3,000 rads/2 wk	75%	39%	60 days (median)	86 days	103 days	2%	—
Montana et al., 1972 [95]	62*	Lung	Variable	Most	56%	—	3 mo	—	10%	—
Hazra et al., 1972 [68]	25	Lung	4,000 rads/3 wk	None	76%		23 wk	28 wk	12%	—
Young et al., 1974 [49]	83*	All	1,500 rads/ 2 doses/ 3days	All	57%	2 mo (median)	59 days	103 days	—	—
	79	All	3,000 rads/3 wk	All	62%	3.5 mo (median)	116 days	147 days	—	—

Table 6. (Continued)

Study	No. of patients	Tumor types	RT protocol	Steroid use	Initial improvement	Duration of Improvement	Length of survival Median	Length of survival Mean	>1 yr	% Responding to repeat RT
Posner, 1974 [75]	13*	All	1,500 rads/2 doses 3,000 rads/3 wk	All	—	—	203 days 112 days	— —	0% 10%	— —
Shehata et al., 1974 [83]	81	All	1,000 rads/1 dose or 2,000 rads/1 wk	80%	69%	2 mo (mean)	—	150 days	5–20%	68%
Deutsch et al., 1974 [96]	88*	All	Variable	—	—	—	3–6 mo	—	10%	—
Berry et al., 1974 [97]	124*	All	Variable	51%	63%	3 mo (median)	4 mo	—	10%	—
Hendrickson, 1975 [98]	993	All	3,000 rads/3 wk 3,000 rads/2 wk 4,000 rads/4wk, or 4,000 rads/3 wk	Most	47% (according to functional status assessment)	—	18 wk (for all fractionations)	—	—	—
Hendrickson, 1977 [82]	1,001	All	1,000 r/1 dose 1,200 r/2 doses 2,000 r/1 wk 3,000 r/2 wk, or 4,000 r/3 wk	Most	52% (according to functional status assessment); 80% palliated	—	15 wk	—	15%	—

Table 6. (Continued)

Study	No. of patients	Tumor types	RT protocol	Steroid use	Initial improvement	Duration of improvement	Length of survival Median	Length of survival Mean	Length of survival >1 yr	% Responding to repeat RT
Markesbery et al., 1978 [69]	129	All	3,000 rads/3 wk	All	—	—	15 wk	—	12%	—
Brady and Bajpai, 1979 [99]	343	All	Variable	75%	57%	—	5 mo	—	—	—
Gilbert et al., 1979 [100]	55	Lung	1,300 rads/48 hr	—	50%	—	3–6 mo	—	—	—
	21	Breast	2,000 rads/1 wk		50%	—	3–7 mo	—	—	—
	14	Other	3,000 rads/3 wk or 4,000 rads/4 wk		50%	—	4–5 mo	—	—	—
Nugent et al., 1979 [64]	51**	Lung (small cell)	Variable	Most	92%	—	3 mo	—	—	—
DiStefano et al., 1979 [65]	87	Breast	Variable	—	—	—	4 mo	—	—	—
Cairncross et al., 1980 [50]	183	All	3,900 rads/17 days	All	74%	—	12 wk	—	8%	33%
Carella et al., 1980 [101]	60	Melanoma	Variable	Most	76%	—	10–14 wk	—	—	—
Zimm et al., 1981 [102]	156	All	Variable	All	—	—	3.3 mo	—	12%	35%

*Series includes some patients treated with surgery and RT.
**Number of patients completing a full course of radiation.

large additional radiation doses that would be necessary to eliminate the secondary metastasis.

Further and more helpful information on the effectiveness of RT in the treatment of brain metastasis has been provided by the analysis of steroid requirements after RT, post-treatment CT scans, causes of death in patients with metastatic brain tumors, and neuropathological evaluation. Steroid dependence after treatment points in many instances to radiation failure. However, in a recent report only 1/4 of patients radiated for brain metastasis required post-treatment steroids to maintain neurological improvement [50]. The CT scan offers an ideal way to evaluate treatment. Table 7 summarizes the results of follow-up CT scans in 84 patients with brain metastases treated with RT [50]. Metastatic lesions were judged to be smaller in 78% of patients, with major regression or resolution in a full 60%. The long-standing assumption that brain metastases are a fatal complication of systemic cancer for which treatment is marginally effective is challenged by the observation that 2/3 of patients radiated for brain metastasis die as a result of extraneural disease in the setting of sustained neurological improvement. Table 8 lists the cause of death in 81 patients radiated for brain metastasis [50]. In only 15% of these patients was neurological deterioration the sole cause of death. Zimm *et al* have made very similar observations [102]. Further evidence that RT is effective comes from autopsy studies. Postmortem observations in 187 patients radiated for brain metastasis have been reported [104]. In 5 patients no residual tumor could be found in the brain; in 9 others rare tumor cells were scattered at the periphery of cystic or necrotic lesions, previously the site of large symptomatic metastatic tumors. Steroid requirements after RT and follow-up CT scans are clearly better measures of response and must replace length of survival as the measure of treatment

Table 7. Results of follow-up CT scan in patients radiated for brain metastasis

Primary tumor	# Follow-up scans	Disappearance of lesions	Major improvement	Slight improvement	No change	Worse
Lung	30	9	11	4	2	4
Breast	20	7	8	1	3	1
Melanoma	10	0	0	8	2	0
Colon	3	0	1	0	1	1
Kidney	1	0	0	0	1	0
Testicle	6	1	5	0	0	0
Multiple primaries	3	2	1	0	0	0
Unknown primary	2	0	1	0	1	0
Miscellaneous	9	2	2	2	3	0
Totals	84	21 (25%)	29 (35%)	15 (18%)	13 (15%)	6 (7%)

Table 8. Cause of death in patients radiated for brain metastasis

Primary tumor	Systemic	Neurological	Combined
Lung	22	3	6
Breast	11	3	1
Melanoma	2	3	1
Colon	2	0	1
Kidney	3	0	0
Testicle	6	1	1
Multiple primaries	4	1	1
Unknown primary	1	0	2
Miscellaneous	4	2	0
Totals	55 (68%)	13 (16%)	13 (16%)

efficacy. Future studies comparing radiation protocols and evaluating radiosensitizers must consider these direct measures of treatment response if conclusions are to be meaningful.

The correct radiation portals, total radiation dose, and best fractionation schedule are not established. In most instances the RT port should include the entire brain. At the time of diagnosis 1/2 of patients will have multiple brain metastases visible by CT scan [50]. This observation alone essentially excludes surgery, focal RT or regional chemotherapy as chemotherapeutic alternatives for a great many patients. However, when brain metastases are solitary, alternatives to whole-brain RT may be considered. There is little support for focal RT as a primary treatment modality, principally because of the difficulties in giving a second course of RT should metastases develop in other areas of the brain. Although the radiation portals should include the entire brain, we have recently decided to shield the optic nerves, particularly when delivering high doses of RT, because of potential radiation toxicity to the retina and optic nerves [105]. It is important that the radiation portal is low enough posteriorly to include the entire posterior fossa. On at least one occasion a patient of ours failed to respond to RT because the port did not cover a caudal cerebellar metastasis.

A variety of radiation doses and fractionation procedures have been used in the treatment of patients with metastatic brain tumors. With the possible exception of the rapid-course protocols which may be associated with higher morbidity, more RT-related deaths and earlier relapse [49, 83, 84], it has not been possible to demonstrate a significant difference in therapeutic efficacy among a host of radiation schedules. However, a comparison of RT protocols based largely on initial response rates and length of survival is likely to be misleading, as indicated above.

When considered as a group, patients with brain metastasis respond favorably to RT. When the initial clinical response or the duration of survival are used as criteria, no statistically significant differences emerge between tumors generally

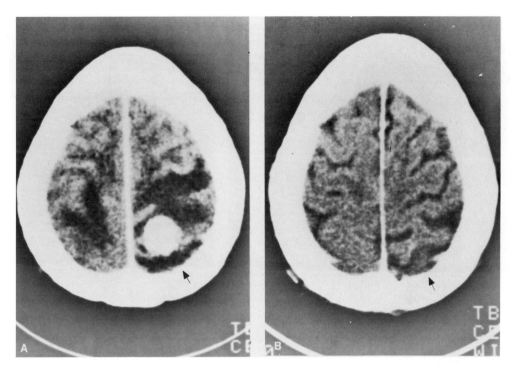

Figure 3. Response to RT of brain metastasis from non-oat cell carcinoma of the lung. A 62 year-old woman with adenocarcinoma of the lung complained of morning headaches for 3 weeks. A CT scan with contrast (A) demonstrated a right posterior parietal brain metastasis with surrounding edema (arrow). A follow-up study 12 weeks after whole-brain RT (B) demonstrates disappearance of the tumor mass. A small low-density area remains (arrow).

considered radiosensitive (e.g. carcinoma of the breast) and those generally considered radioresistant (e.g. malignant melanoma). When CT scans are considered, however, the situation changes. Table 9 summarizes the MSKCC experience accumulated over a recent 18-month period [50]. The primary tumors are listed in decreasing order of radioresponsiveness as judged by post-treatment CT scans. Of the cancers that metastasize to the brain with any regularity, only colon, melanoma and kidney respond poorly to RT. Non-oat cell carcinoma of the lung (Figure 3) and carcinoma of the breast, tumors which account for 60 – 70% of all metastatic brain tumors, respond favorably to radiation alone. To this list must be added highly radiosensitive tumors such as lymphomas and choriocarcinoma, where eradication of the tumor is the rule rather than the exception.

RT emerges as the treatment of choice in patients with multiple brain metastases, solitary metastases and widespread systemic disease, and solitary metastases from lymphoreticular tumors and carcinomas of the lung, breast and

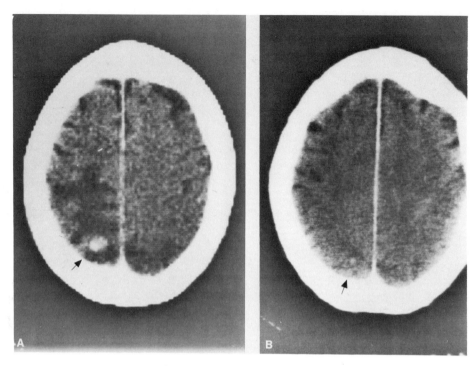

Figure 4. Response to RT of a 'radioresistant' primary tumor metastatic to brain. A 70 year-old woman with metastatic renal carcinoma developed mild headaches. A CT scan with contrast (A) revealed a small left occipital metastasis (arrow). A follow-up study (B) 6 weeks after whole-brain RT (while off steroids) demonstrated a substantial decrease in tumor size (arrow).

Table 9. Radioresponsiveness of brain metastases as judged by follow-up CT scans

Primary tumor	% Showing major improvement or resolution
Lymphoma	100 (1/1)
Testicle	100 (6/6)
Lung (oat cell)	100 (3/3)
Choriocarcinoma (gestational)	100 (1/1)
Endometrium	100 (1/1)
Larynx (epidermoid)	100 (1/1)
Breast	75 (15/20)
Lung (non-oat cell)	63 (17/27)
Colon	33 (1/3)
Melanoma	0 (0/10)
Kidney	0 (0/1)
Sarcoma	0 (0/2)
Other (esophagus, pancreas, prostate)	0 (0/3)

testicle. The best treatment for patients with solitary metastases from melanoma or carcinomas of the colon or kidney remains uncertain. The decision between surgery and RT is based on a number of factors, the 2 most important of which are the extent and progression of disease outside the nervous system and the radioresponsiveness of the primary tumor. Patients with widespread, advancing systemic disease, particularly those who have failed systemic treatment, are best treated with whole-brain RT. The majority of patients fall into this group. However, for the subset of patients with solitary brain metastases and controlled or limited systemic disease, surgery remains an option. At this point the radioresponsiveness of the tumor must be considered. Because we have seen patients with these tumor types respond to whole-brain RT (Figure 4), we recommend RT as the initial therapy and consider surgery if subsequent steroid requirements or CT scans indicate a poor response. It may be argued, however, that patients with limited systemic disease and a single brain metastasis from a radioresistant primary should have the tumor removed surgically with follow-up RT only if the postoperative CT scan demonstrates residual tumor. The location of the metastasis (e.g. motor area or non-dominant frontal pole) is often the deciding factor in the choice between surgery and RT.

In recent years, prophylactic cranial irradiation has become an important treatment modality in patients with oat cell carcinoma of the lung. Irradiation of asymptomatic patients was considered because of the observation that brain metastases developed in up to 50% of patients [64]. Prospective randomized clinical studies have now demonstrated that elective whole-brain RT significantly decreases the incidence of brain metastasis [106–108]. Overall survival has not changed, however [106–108]. Recently, the wisdom of prophylactic RT has been questioned by Baglan and Marks [109] on the grounds that patients who develop symptomatic brain metastases are effectively treated by RT, and those who do not are spared unnecessary brain irradiation. Elective cranial irradiation is also being studied in the management of inoperable non-oat cell carcinoma of the lung. Preliminary results suggest a decrease in the incidence of brain metastases [110].

Radiation Toxicity

Whole-brain RT is not without toxicity (Table 10). There are 3 kinds of radiation encephalopathy: acute, early-delayed and late-delayed. The acute reaction consists of headache, nausea, vomiting, fever, worsening of pre-existing neurological deficits and seizures. These symptoms are transient and characteristically occur after the first dose of radiation. Rarely, patients have developed symptoms and signs of cerebral herniation following high-dose RT [49]. The pathogenesis of the acute reaction is unclear but may be related to changes in the permeability of intracerebral capillaries [111] aggravating cerebral edema or to alterations in cerebral blood flow secondary to disordered

Table 10. Complications of whole-brain radiation therapy

Neurologic
   acute encephalopathy
   early-delayed encephalopathy
   late-delayed encephalopathy (radiation necrosis)
   dementia
   visual loss (optic atrophy)
   stroke syndromes
   cerebral neoplasms

Non-neurologic
   ageusia
   xerophthalmia
   xerostomia
   hearing loss
   acute parotitis
   alopecia

autoregulation [112]. It appears to be dose-related and less frequent if patients are being treated with steroids. Other acute toxicity includes radiation-induced parotitis [113], which is usually asymptomatic and clears within a few days, and radiation-induced loss of taste, which may be very distressing to the patient and may not clear for several weeks.

The early-delayed encephalopathy occurs 4–12 weeks after whole-brain RT and is characterized by transient somnolence and EEG slowing [114]. Pre-existing focal signs may intensify [115]. The pathogenesis of this syndrome is believed to be demyelination resulting from sensitivity of the oligodendroglial cells to RT [116, 117]. Radiation-induced hearing loss is a sometimes serious and always distressing symptom developing 1–3 weeks after treatment. The pathogenesis of hearing loss appears to be a serous otitis media, secondary to swelling and occlusion of the eustachian tubes. Patients should be warned of the possibility of hearing loss and should undergo immediate otologic examination if symptoms develop. Sometimes myringotomy can preserve hearing.

The late-delayed encephalopathy is the most serious complication, being characterized pathologically by necrosis of the brain. It develops months to years after high-dose brain irradiation [118]. The clinical picture suggests recurrent tumor with headaches, papilledema, alterations in consciousness and progressive focal deficits. Although there is no specific treatment, surgical removal of an expanding necrotic mass may be life-saving. Other late complications of brain irradiation include cerebral atrophy [119], learning disabilities [120], dementia [50], visual loss [105], stroke syndromes [121] and, rarely, primary brain tumors [122].

It should be emphasized that the serious complications of whole-brain RT are uncommon and should not restrict its use in the management of patients with brain metastasis.

Surgery

Surgery has a limited role in the management of patients with brain metastasis. Patients with multiple metastases require treatment to the whole-brain and are rarely surgical candidates. Nor is surgery the preferred treatment in patients with solitary metastases and widely disseminated systemic disease. This is especially true of those in whom systemic disease, not CNS disease, is likely to limit survival. The merits of surgical extirpation of accessible solitary metastases in selected patients (i.e. patients with localized systemic disease) have been the subject of several large reviews. These studies are summarized in Table 11. The results of treatment have been similar over the years, with a median survival of 6 months, a one-year survival rate of approximately 25%, and a one-month postoperative mortality rate of less than 10% in recent series. With very careful selection of patients one can achieve a median survival of about 8 months, with one-year survivals as high as 40 – 50% [133], and an operative mortality of less than 5% [128]. These results are superior to those reported after RT alone; however, several important points must be emphasized. First, patients reported in surgical series represent a highly select group. Those with advanced systemic disease or other medical conditions precluding surgery constitute the majority of patients in RT series. Second, median survival and one-year survival rates are indirect and potentially misleading measures of treatment efficacy. The presumption that patients with brain metastasis die neurologic deaths and that shorter survival in RT series must imply less effective treatment of CNS disease is ill-founded. No controlled studies in the literature compare surgery and RT in a randomly selected group of patients.

In our view there are 2 unequivocal indications for surgery. First, patients in whom there is uncertainty about the diagnosis. The CT scan appearance of a metastatic brain tumor is not sufficiently characteristic to permit a diagnosis in the absence of a history of malignancy. Second, patients with a solitary accessible brain metastasis and limited systemic disease who have failed RT. Steroid dependence and CT scan documentation of progression after whole-brain RT are evidence of treatment failure. There is insufficient evidence to support the view that all patients with solitary metastases and localized extraneural disease be treated surgically. However, a case can be made for surgical removal of solitary lesions as the initial treatment in selected patients with radioresistant primary tumors. Tumors unlikely to respond to RT include melanoma, carcinomas of the kidney and colon, and sarcomas. However, even in patients with radioresistant tumors, there is an occasional response, and one could make the case that these patients, like others with solitary metastases, might first be treated with RT followed by surgical extirpation only if radiation fails. In both instances, initial

Table 11. Surgical treatment of cerebral metastases

Series		Tumor types	Number of patients	Postoperative mortality (1 mo.)	Survival		
					Median	One-year	Long-term
Stortebecker	1954 [123]	All	125	25% (20 days)	3.6	21%	3 ≤ 4 years
Richards	1963 [66]	All	108	32%	<5.0	17%	8 > 2 years
Lang	1964 [67]	All	208	22%	4.0	20%	27 > 2 years
Vieth	1965 [124]	All	155	15% (2 weeks)	<6.0	13.5%	12 > 2 years
Raskin	1971 [125]	All	51	12% (2 weeks)	<6.0	30%	4 > 3 years
Haar	1972 [126]	All	167	11%	6.0	22%	7 > 5 years
Ransohoff	1975 [127]	All	100	10%	>6.0	38%	13 > 2 years
Magilligan	1976 [128]	Lung	22	4%	6.0	31%	
Galicich	1980 [129]	All	78	4%, 8%, 32%*	6.0	29%	
Hafstrom	1980 [130]	Melanoma	25	12%	5.0	36%	2 > 3 years
Winston	1980 [131]	All	79	10%	5.0	22%	8 > 2 years
White	1981 [132]	All	122	6%	7.0	30%	18 > 2 years

*For neurological grades I, II and III, respectively.

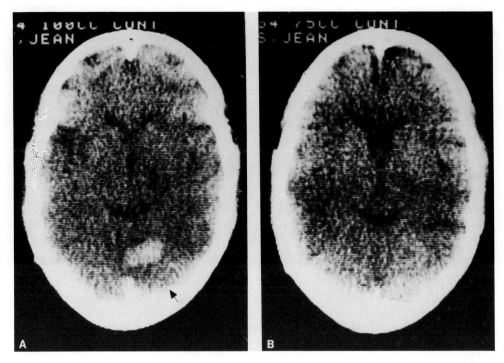

Figure 5. Response of brain metastasis to systemic chemotherapy. A woman with breast cancer developed a malignant pericardial effusion. A CT scan with IV contrast (A), done as a screening procedure, disclosed an unsuspected right superior cerebellar metastasis (arrow). She was treated with adriamycin (water-soluble) and the brain lesion followed with serial CT scans. Coincident with regression of the effusion, the brain metastasis disappeared (B). Subsequently her disease became refractory to adriamycin with reappearance of the effusion and the brain tumor.

treatment with steroids and RT gives the examiner time to assess whether metastatic disease outside the nervous system is going to develop and grow rapidly. Finally, surgical decompression (by a shunting procedure) in acutely deteriorating patients may be life-saving, although in most instances herniating patients respond to high-dose steroids, osmotic agents and hyperventilation.

Chemotherapy

Chemotherapeutic approaches to the treatment of metastatic brain tumors have received surprisingly little attention. It has been the prevailing view that chemotherapy has no role in the management of patients with brain metastases. Certainly chemotherapy cannot be considered a primary treatment modality. However, recent observations in patients with brain metastases from carcinoma of the lung suggest it may be an important adjunct to RT and surgery. Chan *et al.* [134] have reported on the treatment of 24 patients using concomitant adriamycin, CCNU and RT. They have observed an 82% initial response rate, a

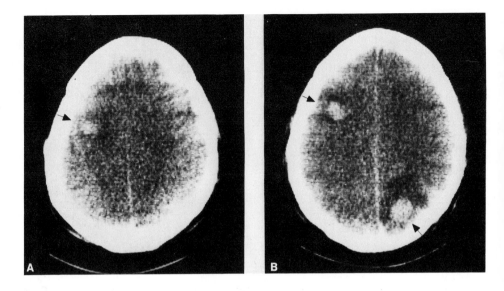

Figure 6. Failure of brain metastasis to respond to chemotherapy. A man with embryonal carcinoma of the testis developed multiple pulmonary metastases followed by mild headache. A CT scan with contrast (A) demonstrated a left frontal metastasis. After two months of combination chemotherapy with predominantly water-soluble agents, the lung metastases had decreased in size, but a second CT scan with contrast (B) demonstrated progression of the CNS disease. The left frontal metastasis is slightly larger (arrow) and a new right occipital tumor has appeared (arrow).

median duration of response of 8 months, complete resolution of lesions by isotope scan in 67%, and deaths due to brain disease in only 15%. Since the blood-brain barrier is defective in at least some areas of many brain metastases, water-soluble chemotherapeutic agents toxic to the tumor cells growing in the brain might be expected to be effective to some degree. Indeed, as Chan's data suggest, this may be the case. Figure 5 illustrates the response of an asymptomatic brain metastasis from carcinoma of the breast to systemic adriamycin. The tumor completely disappeared without additional treatment only to reappear as the tumor relapsed systemically. On the other hand, there are times when tumors are apparently responsive outside the nervous system and yet brain metastases fail to respond. Figure 6 illustrates a patient with testicular carcinoma metastatic to the lungs with a single asymptomatic brain metastasis. Two months of combination chemotherapy, mostly with water-soluble agents, produced a 30% decrease in the size of the pulmonary metastases but did not prevent progression of the brain disease. Whether this represents resistance of a subpopulation of tumor cells growing in the brain or a blood-brain barrier phenomenon is not clear. So little data are available that the percentage of metastatic brain tumors likely to respond to systemic chemotherapy is not established. We reserve systemic chemotherapy for those patients who have an asymptomatic brain metastasis from a systemic tumor which is likely to respond.

Two other chemotherapeutic approaches are currently under investigation. The first is intra-arterial chemotherapy. Provided the drug has a high first pass extraction by the brain tumor, intra-arterial chemotherapy offers 2 advantages over conventional delivery (i.e. intravenous). First, it delivers a high dose of chemotherapeutic agent to the tumor and, second, because the total dose is low, it spares such potentially sensitive organs as the gut and bone marrow. Definite responses in 12 of 25 patients with brain metastasis from lung cancer treated with intra-arterial BCNU have been reported [135]. In this series there were no responses in 10 patients with melanoma. We have observed occasional dramatic responses following intra-arterial BCNU in patients who have failed RT.

A second approach to chemotherapy for brain tumors has been proposed by Neuwalt et al. [136]. These investigators have opened the blood-brain barrier using hypertonic mannitol (in patients with primary brain tumors) in order to increase the delivery of water-soluble chemotherapeutic agents to the tumor. There are no definitive data available on this technique, but our preliminary experiments in animals harboring metastatic brain tumors suggest that the effect of opening the blood-brain barrier is to allow more water-soluble chemotherapeutic agent to enter normal brain, where it may be toxic, without substantially increasing the amount of chemotherapeutic agent that enters the brain tumor [137]. The final position of chemotherapy in the management of patients with brain metastasis awaits further study. At the present time we reserve intra-arterial chemotherapy for patients who have not responded to one of the primary treatment modalities (i.e. RT or surgery).

Anticonvulsants

In a recent review of metastatic brain tumors seizures, either focal or generalized, were the presenting symptom in 20% of patients [50]. An additional 10% developed seizures subsequently [138]. The incidence of seizures is higher in patients with malignant melanoma than in those with brain metastases from other cancers [80], perhaps because melanoma tends to grow in cerebral cortex whereas others tend to be restricted to the white matter. Patients who present with or have had focal or generalized convulsions during their course require anticonvulsant therapy. Phenytoin is the drug of choice. The question of whether anticonvulsants should be used prophylactically is still unresolved. Some investigators choose not to use prophylactic anticonvulsants since it adds yet another drug to the long list of medications taken by these patients, and interactions between anticonvulsants and other agents are common. The exception may be in patients with malignant melanoma, where seizures can be expected. When anticonvulsants, particularly phenytoin, are used in patients with brain metastases, certain precautions are necessary. At times the anticonvulsants are started when the patient is first diagnosed and on high doses of steroids. Steroids may suppress an allergic reaction to the anticonvulsants (i.e. fever, rash, etc.) which only becomes apparent after the patient leaves the hospital, as the steroids are tapered and stopped. Anticonvulsants are microsomal enzyme inducers capable of increasing the metabolism of dexamethasone [139]. Patients on stable doses of steroids may develop new neurological symptoms when anticonvulsants are added. This should not be interpreted as evidence of tumor growth but simply the result of increased metabolism of dexamethasone, requiring higher doses of that drug. We have had the impression that generalized rashes from RT are more common when patients are taking phenytoin than when they are not taking anticonvulsants and that the rash produced by procarbazine is more common in patients taking phenytoin. For these reasons we choose not to use anticonvulsant drugs unless there is a specific indication.

REFERENCES

1. Posner JB: Neurological complications of systemic cancer. Med Clin N Amer 63:783–800, 1979.
2. Posner JB: Management of central nervous system metastases. Sem Oncol 4:81–91, 1977.
3. Posner JB, Chernik NL: Intracranial metastases from systemic cancer. Adv Neurol 19:579–592, 1978.
4. Vonofakos D, Zieger A, Marcu H: Subdural hematoma associated with dural metastatic tumor. Neuroradiology 20:213–218, 1980.
5. Max MB, Chernik NL, Rottenberg DA: Pituitary metastases: incidence in cancer patients and clinical differentiation from pituitary adenomas. Neurology 30:356, 1980.
6. Wasserstrom W, Glass JP, Posner JB: Diagnosis and treatment of leptomeningeal metastasis from solid tumors: experience with 90 patients. Cancer 49:759–772, 1982..
7. Straub P, Weiss L: Cell locomotion and tumor penetration. Eur J Cancer 13:1–12, 1977.
8. Kleinerman J, Liotta L: Release of tumor cells. In Day SB et al (eds): Cancer Invasion and Metastasis: Biologic Mechanisms and Therapy. New York, Raven Press, vol. 5, 1977, pp 135–143.
9. Poste G, Doll J, Hart I et al: In vitro selection of murine B16 melanoma variants with enhanced tissue-invasive properties. Cancer Res 40:1636–1644, 1980.
10. Willis RA: The spread of tumors in the human body. London, Butterworth & Co., 1973.
11. Warren BA: Tumor metastasis and thrombosis. Thromb Diath Haemorr Suppl 59:139–156, 1974.
12. Goldacre RJ, Sylven B: On the access of blood-borne dyes to various tumor regions. Br J Cancer 16:305–314, 1962.
13. Sugarbaker EV: Cancer metastasis: A product of tumor-host interactions. In Hickey RC (ed): Current Problems in Cancer. Chicago, Yearbook Medical Publishers, vol. 3, 1979, pp 3–59.
14. Russell DS, Rubinstein LJ: Pathology of tumors of the nervous system. Baltimore, Williams + Wilkins, 1977.
15. Prioleau PG, Katzenstein AA: Major peripheral arterial occlusion due to malignant tumor embolism: histologic recognition and surgical management. Cancer 42:2009–2014, 1978.
16. Busch E, Christensen E: Treatment of tumor metastatic to the central nervous system. In Pack GT, Ariel IM (eds): Treatment of Cancer and Allied Diseases. New York, P.B. Hoeber Inc., 1959, pp 212–223.

17. Bross IDJ: The role of brain metastases in cascade processes: implications for research and clinical management. In: Weiss L, Gilbert HA, Posner JB (eds): Brain Metastasis. Boston, GK Hall & Co., 1980, pp 66–80.
18. Thompson T, Evans W: Paradoxical embolism. Quart J Med 23:135–150, 1930.
19. Rupp C: Metastatic tumors of the central nervous system: Intracerebral metastases as the only evidence of dissemination of visceral cancer. Arch Neurol Psychiat 59:635–645, 1948.
20. Batson CV: Function of vertebral veins and their role in spread of metastases. Ann Surg 112:138–149, 1940.
21. Del Regato JA: Pathways of metastatic spread of malignant tumors. Sem Oncol 4:33–38, 1977.
22. Walker MD: Brain and peripheral nervous system tumors. In: Holland JF, Frei E III (eds): Cancer Medicine. Philadelphia, Lea & Febiger, 1973, pp 1385–1407.
23. Cairncross JG, Posner JB: Neurological complications of malignant lymphoma. In: Vinken PJ, Bruyn GW (eds): Handbook of Clinical Neurology. Amsterdam, Elsevier North-Holland Publishing Co., vol. 39, 1979, pp 27–62.
24. Hunter KMF, Newcastle NB: Metastatic neoplasms of the brain stem. Canad Med Assoc J 98:1–7, 1968.
25. Sugarbaker EV, Cohen AM, Ketcham AS: Do metastases metastasize? Ann Surg 174:161–166, 1971.
26. Parks RC: Organ-specific metastasis of a transplantable reticulum cell sarcoma. J Nat Cancer Inst 52:971–973, 1974.
27. Pollard M, Luckert PJ: Transplantable metastasizing prostate adenocarcinoma in rats. J Nat Cancer Inst 54:643–649, 1975.
28. Nicolson GL, Brunson KW, Fidler IJ: Specificity of arrest, survival, and growth of selected metastatic variant cell lines. Cancer Res 38:4105–4111, 1978.
29. Fidler IJ, Gersten DM, Hart I: The biology of cancer invasion and metastasis. Adv Cancer Res 28:149–250, 1978.
30. Poste G, Fidler IJ: The pathogenesis of cancer metastasis. Nature 283:139–146, 1980.
31. Poste G, Nicolson GL: Arrest and metastasis of blood-borne tumor cells are modified by fusion and plasma membrane vesicles. Proc Nat Acad Sci 77:399–403, 1980.
32. Brunson KW, Beattie G, Nicolson GL: Selection and altered tumor cell properties of brain colonising metastatic melanoma. Nature 272:543–545, 1978.
33. Fidler IJ, Kripke ML: Metastasis results from preexisting variant cells within a malignant tumor. Science 197:893–895, 1971.
34. Baylin SD, Weisburger WR, Eggleston JC et al: Variable content of histaminase, L-dopa decarboxylase and calcitonin in small-cell carcinoma of the lung: Biologic and clinical implications. New Engl J Med 299:105–110, 1978.
35. Steel GG, Adams K: Stem-cell survival and tumor control in the Lewis lung carcinoma. Cancer Res 35:1530–1535, 1975.
36. Donelli MG, Colombo T, Broggini M et al: Differential distribution of antitumor agents in primary and secondary tumors. Cancer Treat Rep 61:1319–1324, 1977.
37. Rall DP, Zubrod CC: Mechanism of drug absorption and excretion. Passage of drugs in and out of the central nervous system. Ann Rev Pharmacol 2:109–128, 1962.
38. Crone C: The permeability of brain capillaries to non-electrolytes. Acta Physiol Scand 64:407–417, 1965.
39. Brightman MW: Morphology of blood-brain interface. In: Bito LZ, Davson H, Fenstermacher JD (eds): The Ocular and Cerebrospinal Fluids. New York, Academic Press, 1978.
40. Oldendorf WH, Cornford ME, Brown WJ: The large apparent work capability of the blood-brain barrier: A study of the mitochondrial content of capillary endothelial cells in brain and other tissues of the rat. Ann Neurol 1:409–417, 1977.
41. Rapoport SI: Blood-Brain Barrier in Physiology and Medicine. New York, Raven Press, 1976.

42. Hirano A, Matsui T: Vascular structures in brain tumors. Hum Pathol 6:611–621, 1975.
43. Waggener JD, Beggs JL: Vascularization of neural neoplasms. Adv Neurol 15:27–49, 1976.
44. Hirano A, Zimmerman HM: Fenestrated blood vessels in a metastatic renal carcinoma of the brain. Lab Invest 26:465–468, 1972.
45. Long DM: Capillary ultrastructure in human metastatic brain tumors. J Neurosurg 51:53–58, 1979.
46. Vick NA: Brain tumor microvasculature. In: Weiss L, Gilbert HA, Posner JB (eds): Brain Metastasis. Boston, G.K. Hall & Co., 1980, pp 115–133.
47. Blasberg RG, Patlak CS, Jehle JW et al: An autoradiographic technique to measure the permeability of normal and abnormal brain capillaries. Neurology 28:363, 1978.
48. Rapoport SI, Ohno K, Fredericks WR et al: Regional cerebrovascular permeability to [^{14}] sucrose after osmotic opening of the blood-brain barrier. Brain Res 150:653–657, 1978.
49. Young DF, Posner JB, Chu FCH et al: Rapid-course radiation therapy of cerebral metastases: results and complications. Cancer 34:1069–1076, 1974.
50. Cairncross JG, Kim J-H, Posner JB: Radiation therapy for brain metastases. Ann Neurol 7:529–541, 1980.
51. Simionescu MD: Metastatic tumors of the brain: a follow-up study of 195 patients with neurosurgical considerations. J Neurosurg 17:361–373, 1960.
52. Paillas JE, Pellet W: Brain metastasis. In: Vinken PJ, Bruyn GW (eds): Handbook of Clinical Neurology. Amsterdam, Elsevier North-Holland Publishing Co., vol. 18, 1976, pp 201–232.
53. Mori K, Takeuchi J, Ishikawa M et al: Occlusive arteriopathy and brain tumor. J Neurosurg 49:22–35, 1978.
54. Fisher CM: Transient paralytic attacks of obscure nature: the question of non-convulsive seizure paralysis. Canad J Neurol Sci 5:267–273, 1978.
55. Mandybur TI: Intracranial hemorrhage caused by metastatic tumors. Neurology 27:650–655, 1977.
56. Ingvar DH, Lundberg N: Paroxysmal symptoms in intracranial hypertension, studied with ventricular fluid pressure recording and electroencephalography. Brain 84:446–459, 1961.
57. Posner JB: Clinical manifestations of brain metastasis. In: Weiss L, Gilbert HA, Posner JB (eds): Brain Metastasis. Boston, G.K. Hall & Co., 1980, pp 189–207.
58. Symonds C: Cough headache. Brain 79:557–568, 1956.
59. Jacobs L, Kinkel WR, Vincent RG: Silent brain metastasis from lung carcinoma determined by computerized tomography. Arch Neurol 34:690–693, 1977.
60. Voorhies RM, Sundaresan N, Thaler HT: The single supratentorial lesion: an evaluation of preoperative diagnostic tests. J Neurosurg 53:364–368, 1980.
61. Jeffries BF, Kishore PRS, Singh KS et al: Contrast enhancement in the postoperative brain. Radiology 139:409–413, 1981.
62. Coulam CM, Seshul M, Donaldson J: Intracranial ring lesions: can we differentiate by computed tomography? Invest Radiol 15:103–112, 1980.
63. Crocker EF, Zimmerman RA, Phelps ME et al: The effect of steroids on the extravascular distribution of radiographic contrast material and technetium pertechnetate in brain tumors as determined by computed tomography. Radiology 118:471–474, 1976.
64. Nugent JL, Bunn PA, Matthews MJ et al: CNS metastases in small cell bronchogenic carcinoma: increasing frequency and changing pattern with lengthening survival. Cancer 44:1885–1893, 1979.
65. Di Stefano A, Yap HY, Hortobagyi GN et al: The natural history of breast cancer patients with brain metastases. Cancer 44:1913–1918, 1979.
66. Richards P, McKissock W: Intracranial metastases. Br Med J 1:15–18, 1963.
67. Lang EF, Slater J: Metastatic brain tumors: results of surgical and nonsurgical treatment. Surg Clin N Amer 44:865–872.

68. Hazra T, Mullins GM, Lott S: Management of cerebral metastases from bronchogenic carcinoma. Johns Hopkins Med J 130:377–383, 1972.
69. Markesbery WR, Brooks WH, Gupta GD et al: Treatment for patients with cerebral metastases. Arch Neurology 35:754–756, 1978.
70. Kofman S, Garvin JS, Nagamani D et al: Treatment of cerebral metastases from breast carcinoma with prednisolone. JAMA 163:1473–1476, 1957.
71. Long DM, Hartman JF, French LA: The response of human cerebral edema to glucosteroid administration. Neurology 16:521–528, 1966.
72. Weinstein JD, Toy FJ, Jaffe ME et al: The effect of dexamethasone on brain edema in patients with metastatic brain tumors. Neurology 23:121–129, 1973.
73. Shapiro WR, Posner JB: Corticosteroid hormones: effects in an experimental brain tumor. Arch Neurol 30:217–221, 1974.
74. Posner JB, Howieson J, Cvitkovic E: 'Disappearing' spinal cord compression: oncolytic effect of glucocorticoids (and other chemotherapeutic agents) on epidural metastases. Ann Neurol 2:409–413, 1977.
75. Posner JB: Diagnosis and treatment of metastases to the brain. Clin Bull 4:47–57, 1974.
76. Fishman RA: Brain edema. New Engl J Med 293:706–711, 1975.
77. Leiberman A, Le Brun Y, Glass P et al: Use of high-dose corticosteroids in patients with inoperable brain tumors. J Neurol Neurosurg Psychiat 40:678–682, 1977.
78. Ruderman NB, Hall TC: Use of glucocorticoids in the palliative treatment of metastatic brain tumors. Cancer 18:298–306, 1965.
79. Horton J, Baxter DH, Olson KB: The management of metastases to the brain by irradiation and corticosteroids. Am J Roentgenol Radium Ther Nucl Med 3:334–335, 1971.
80. Gottlieb JA, Frei E III, Luce JK: An evaluation of the management of patients with cerebral metastases from malignant melanoma. Cancer 29:701–705, 1972.
81. Posner JB: Brain tumor: current status of treatment and its complications. Arch Neurol 32:781–784, 1975.
82. Hendrickson FR: The optimum schedule for palliative radiotherapy for metastatic brain cancer. Int J Radiat Oncol Biol Phys 2:165–168, 1977.
83. Shehata WM, Hendrickson FR, Hindo WA: Rapid fractionation technique and retreatment of cerebral metastases by irradiation. Cancer 34:257–262, 1974.
84. Nisce LZ, Hilaris BS, Chu FCH: A review of experience with irradiation of brain metastases. Am J Roentgenol Radium Ther Nucl Med 111:329–333, 1971.
85. Hindo WA, De Trana FA III, Lee M-S et al: Large dose increment irradiation in treatment of cerebral metastases. Cancer 26:138–141, 1970.
86. Axelrod L: Glucocorticoid therapy. Medicine 55:39–65, 1976.
87. Lane RJM, Mastaglia FL: Drug-induced myopathies in man. Lancet 2:562–565, 1978.
88. Fidler IJ, Lieber S: Quantitative analysis of the mechanism of glucocorticoid enhancement of experimental metastasis. Res Comm Chem Path Pharm 4:607–613, 1972.
89. McCawley A: Cortisone habituation – a clinical note. New Engl J Med 273:976, 1965.
90. Dixon RB, Christy NP: On the various forms of corticosteroid withdrawal syndrome. Am J Med 68:224–230, 1980.
91. Chao J-H, Phillips R, Nickson JJ: Roentgen-ray therapy of cerebral metastases. Cancer 7:682–688, 1954.
92. Chu FCH, Hilaris BS: Value of radiation therapy in the management of intracranial metastases. Cancer 14:577–581, 1961.
93. Order SE, Hellman S, Von Essen CF et al: Improvement in quality of survival following whole-brain irradiation for brain metastases. Radiology 91:149–153, 1968.
94. Deeley TJ, Edwards JMR: Radiotherapy in the management of cerebral secondaries from bronchial carcinoma. Lancet 1:1209–1213, 1968.

95. Montana GS, Meacham WF, Caldwell WL: Brain irradiation for metastatic disease of lung origin. Cancer 29:1477–1480, 1972.
96. Deutsch M, Parsons JA, Mercado R: Radiotherapy for intracranial metastases. Cancer 34:1607–1611, 1974.
97. Berry HC, Parker RG, Gerdes AJ: Irradiation of brain metastases. Acta Radiol Ther 13:535–544, 1974.
98. Hendrickson FR: Radiation therapy of metastatic tumors. Sem Oncol 2:43–46, 1975.
99. Brady LW, Bajpai D: Intracranial metastatic malignancy: a review of 343 cases. In: Weiss L, Gilbert HA, Posner JB (eds): Brain Metastasis. Boston, G.K. Hall & Co., 1980, pp 269–278.
100. Gilbert H, Kagan AR, Wagner J et al: The functional results of treating brain metastases with radiation therapy. Ibid. pp 269–278.
101. Carella RJ, Gelber R, Hendrickson FR et al: Value of radiation therapy in the management of patients with cerebral metastases from malignant melanoma. Cancer 45:679–683, 1980.
102. Zimm S, Wampler GL, Stablein D et al: Intracerebral metastases in solid-tumor patients: natural history and results of treatment. Cancer 48:384–394, 1981.
103. Kurup P, Reddy S, Hendrickson FR: Results of re-irradiation for cerebral metastases. Cancer 46:2587–2589, 1980.
104. Cairncross JG, Chernik NL, Kim J-H et al: Sterilization of brain metastases by radiation therapy. Neurology 28:1195–1202, 1979.
105. Macdonald DR, Rottenberg DA, Schutz JS et al: Radiation-induced optic neuropathy. Neurology 31 (suppl):43, 1981.
106. Jackson DV, Richards F, Cooper MR et al: Prophylactic cranial irradiation in small cell carcinoma of the lung: A randomized study. JAMA 237:2730–2733, 1977.
107. Moore TN, Livingston R, Heilbrun L et al: The effectiveness of prophylactic brain irradiation in small cell carcinoma of the lung. Cancer 41:2149–2153, 1978.
108. Beiler DD, Kane RC, Bernath AM et al: Low dose elective brain irradiation in small cell carcinoma of the lung. Int J Radiat Oncol Biol Phys 5:941–945, 1979.
109. Baglan RJ, Marks JE: Comparison of symptomatic and prophylactic irradiation of brain metastases from oat cell carcinoma of the lung. Cancer 47:41–45, 1981.
110. Cox JD, Petrovich Z, Paig C et al: Prophylactic cranial irradiation in patients with inoperable carcinoma of the lung: Preliminary report of a cooperative trial. Cancer 42:1135–1140, 1978.
111. Olsson Y, Klatzo I, Carsten A: The effect of acute radiation injury on the permeability and ultrastructure of intracerebral capillaries. Neuropath Appl Neurobiol 1:59–68, 1975.
112. Chapman PH, Young RJ: Effect of cobalt-60 gamma irradiation on blood pressure and cerebral blood flow in the macaca mulatta. Radiat Res 35:78–85, 1968.
113. Cairncross JG, Salmon J, Kim J-H et al: Acute parotitis and hyperamylasemia following whole-brain radiation therapy. Ann Neurol 7:385–387, 1980.
114. Garwicz S, Aronson AS, Elmquist D et al: Postirradiation syndrome and EEG findings in children with acute lymphoblastic leukemia. Acta Paediat Scand 64:399–403, 1975.
115. Boldrey E, Sheline G: Delayed transitory clinical manifestations after radiation treatment of intracranial tumors. Acta Radiol 5:5–10, 1966.
116. Zeman W, Samorojski T: Effects of irradiation on the nervous system. In: Berdjis CC (ed): Pathology of Irradiation. Baltimore, Williams and Wilkins, 1971, pp 213–217.
117. Rider WD: Radiation damage to the brain – a new syndrome. J Canad Assn Radiol 14:67–69, 1963.
118. Rottenberg DA, Chernik NL, Deck MDF et al: Cerebral necrosis following radiotherapy of extracranial neoplasms. Ann Neurol 1:339–351, 1977.
119. Wilson GH, Byfield J, Hanafee WN: Atrophy following radiation therapy for central nervous system neoplasms. Acta Radiol 11:361–368, 1972.
120. Ch'ien LT, Aur RJA, Stagner S et al: Long-term neurological implications of somnolence syndrome in children with acute lymphocytic leukemia. Ann Neurol 8:273–277, 1980.

121. Painter JM, Chutorian AM, Hilal SK: Cerebrovasculopathy following irradiation in childhood. Neurology 25:189–194, 1975.
122. Cohen MS, Kushner MJ, Dell S: Frontal lobe astrocytoma following radiotherapy for medulloblastoma. Neurology 31:616–619, 1981.
123. Stortebecker TP: Metastatic tumors of the brain from a neurosurgical point of view: a follow-up study of 158 cases. J Neurosurg 11:84–111, 1954.
124. Vieth RG, Odom GL: Intracranial metastases and their neurosurgical treatment. J Neurosurg 23:375–383, 1965.
125. Raskin R, Weiss SR, Manning JJ et al: Survival after surgical excision of single metastatic brain tumors. Am J Roentgenol Radium Ther Nucl Med 111:323–328, 1971.
126. Haar F, Patterson RH: Surgery for metastatic intracranial neoplasm. Cancer 30:1241–1245, 1972.
127. Ransohoff J: Surgical management of metastatic tumors. Sem Oncol 2:21–28, 1975.
128. Magilligan DJ, Rogers JS, Knighton PS et al: Pulmonary neoplasm with solitary cerebral metastasis. Results of combined excision. J Thorac Cardiovasc Surg 72:690–698, 1976.
129. Galicich JH, Sundaresan N, Arbit E et al: Surgical treatment of single brain metastasis: factors associated with survival. Cancer 45:381–386, 1980.
130. Hafstrom L, Jonsson P, Stromblad L: Intracranial metastases of malignant melanoma treated by surgery. Cancer 46:2088–2090, 1980.
131. Winston KR, Walsh JW, Fischer EG: Results of operative treatment of intracranial metastatic tumors. Cancer 45:2639–2645, 1980.
132. White KT, Fleming TR, Laws ER: Single metastasis to the brain: surgical treatment in 122 consecutive patients. Mayo Clin Proc 56:424–428, 1981.
133. Galicich JH, Sundaresan N, Thaler HT: Surgical treatment of single brain metastasis: evaluation of results by computerized tomography scanning. J Neurosurg 53:63–67, 1980.
134. Chan PYM, Byfield JE, Campbell T et al: Combined chemotherapy and irradiation in the treatment of brain metastases from lung cancer. In: Weiss L, Gilbert HA, Posner JB (eds): Brain Metastasis. Boston, G.K. Hall & Co., 1980, pp 364–379.
135. Madajewicz S, West CR, Park HC et al: Phase II study: intra-arterial BCNU therapy for metastatic brain tumors. Cancer 47:653–657, 1981.
136. Neuwelt EA, Diehl JT, Vu LH et al: Monitoring of methotrexate delivery in patients with malignant brain tumors after osmotic blood-brain barrier disruption. Ann Int Med 94:449–454, 1981.
137. Hasegawa H, Allen JC, Mehta JC et al: The enhancement of CNS penetration of methotrexate by hyperosmolar intracarotid mannitol and carcinomatous meningitis. Neurology 29:1280–1286, 1979.
138. Cairncross JG, Posner JB: Neurological complications of systemic cancer. In: Yarboro JW, Bornstein RS (eds): Oncologic Emergencies. New York, Grune & Stratton, 1981, pp 73–96.
139. Werk MH, Choi Y, Sholitan L et al: Interference in the effect of dexamethasone by diphenylhydantoin. New Engl J Med 281:32–34, 1969.

14. Prognostic Factors for Malignant Glioma

DAVID P. BYAR, SYLVAN B. GREEN and THOMAS A. STRIKE

1. INTRODUCTION

The word prognosis is derived from two Greek words which taken together mean 'to know before'. When taken in a medical context the word usually means the prediction of the duration, course, and eventual outcome for patients with a given disease based on information about individual patients combined with a knowledge of how the disease behaves generally. Often in studies of cancer, time until death is the major endpoint against which prognostic factors are assessed, but sometimes we may be interested in using other endpoints which may be more difficult to measure. These might include the probability of a response to therapy, the time until recurrence of tumor, or the time until tumor progression. It is important to recognize that factors which may be prognostic for one endpoint may not be for another. Since malignant glioma is a rapidly progressing disease, in this article we shall mainly be interested in prognosis for time until death.

Why should one study prognostic factors? All physicians who have treated patients with cancer, or most other diseases for that matter, realize that patients are quite heterogeneous. The marked variability in the course of the disease for individual patients means that the assessment of treatment differences, determination of the usefulness of diagnostic tests, or the proper interpretation of data from observational studies are often statistical problems. The identification and quantitation of prognostic factors and their influence on the course of the disease is an attempt to explain as much of this variability among patients as possible.

A knowledge of prognostic factors is useful for understanding the natural history of the disease, to predict (albeit somewhat crudely) the expected course for individual patients, and to select appropriate treatment for individual patients. Identification of prognostic factors is important in the analysis of studies comparing treatments because, before evaluating the effect of treatment, we must first assure ourselves that the groups being compared are in fact comparable with respect to important prognostic factors. This requirement is especially important in comparing non-randomized series treated in different ways, but even in randomized studies, tables comparing treatment groups by

M. D. Walker (ed.), Oncology of the Nervous System. ISBN 0-89838-567-9.
© *1983 Martinus Nijhoff Publishers, Boston/The Hague/Dordrecht/Lancaster. Printed in the Netherlands.*

prognostic factors should be prepared and published to demonstrate that comparability has been achieved by randomization. When comparing treatments, a knowledge of prognostic factors allows us to use statistical models which adjust for imbalances in these factors. It is important to adjust treatment comparisons for such imbalances, even when the imbalance itself is not statistically significant.

Statistical models may be used to define groups of patients with comparable risk, and treatments may be compared in these groups which are typically defined by combinations of various prognostic factors. Such analyses may reveal quantitative or qualitative differences in the effects of treatment which may help us understand how the treatment works, what kinds of patients are especially likely to benefit or suffer toxic side effects, and suggest more precise hypotheses to be tested in the future.

Although we recognize that treatment given to glioma patients can affect the importance of prognostic factors, we have not attempted to take that factor into account in this article. During the period covered by most of the work reviewed here, the treatment of malignant gliomas consisted of surgery often followed by radiotherapy with or without various forms of chemotherapy. If any of these treatments were known to cure the disease, then it would be essential to take treatment into account when assessing prognostic factors, but this does not appear to be the case for malignant gliomas.

In most articles about prognostic factors, these variables are examined one at a time for their influence on the outcome variable. This is an important first step, but since it is commonplace for prognostic factors to be correlated substantially with one another, we also need to carry out multivariate analyses to assess their joint effects when acting together. This approach has two advantages. First it allows us to reduce the set of important prognostic variables to the minimum required for reliable prediction, and secondly, in some cases it helps us understand why some variables are prognostic. For example, several authors have suggested that the presence of epileptic convulsions is associated with longer survival for patients with malignant glioma. One possible explanation might be that convulsions call attention to the presence of a tumor even when it is small, and if data on tumor size were available, one would expect to find a strong correlation between the presence of seizures and the size of a tumor if this explanation is correct. Likewise the unfavorable prognostic significance of a history of previous cancer of other sites might be partially or entirely explained by the correlation of this variable with age. In the first part of this article we will review prognostic factors studied one at a time, but in the last section we will present a multivariate analysis and illustrate the concept of forming risk groups based on multiple prognostic factors.

We have not attempted to write an historical account of progress in understanding prognostic factors for malignant gliomas, but instead have placed our emphasis on the recent literature without making any attempts to indicate who first published an important finding. Our goal has simply been to provide a summary of our understanding to date. Hildebrand and Brihaye [1] have presented an excellent summary of earlier work in a review chapter published in 1975. The information in the present work is derived from two sources. First a systematic literature search was carried out using the computerized indexing resources of the National Library of Medicine seeking any articles in English concerned with prognostic factors for brain tumors with special emphasis on malignant gliomas. Secondly we have relied heavily on our own experience in analyzing data for several thousand patients treated under the protocols of the Brain Tumor Study Group (BTSG) of the National Cancer Institute during the past 15 years. The data reported from the BTSG have the advantage of having been obtained prospectively in randomized trials where in most instances the prognostic factors were well balanced across the various treatment groups. In most of the analyses treatment has been ignored, but the prognostic factors we report remain important when adjusted for treatment. Some of these data have been published [2 – 6], but we will also present results of previously unpublished analyses.

We may divide prognostic factors for malignant gliomas into three broad categories: characteristics of the tumor, characteristics of the host, and factors related to the influence of the tumor on the host. Based on our review of the literature and analyses of the results from the BTSG studies we have summarized our impressions (Table 1) as follows: those variables generally agreed to be prog-

Table 1. Summary of impressions of prognostic factors that have been studied

Positive	Negative or equivocal
Tumor characteristics	
Histologic type	Laterality (right versus left)
Presence of giant cells	Location in brain
Extent of surgical removal	Gross findings at operation
	Cellularity
	Lymphocytic infiltration
	Findings on CT scan
Host characteristics	
Age	Sex
	ABO blood type
Effect of tumor on the host	
Performance status	White blood count
Duration of symptoms	Platelet count
History of seizures	
Certain neurologic signs and symptoms (see text)	

nostic (positive), and those either generally found not to be prognostic (negative) or those for which substantial disagreement exists (equivocal).

2. TUMOR CHARACTERISTICS

2.1. Histology of Tumor

Many authors have found that the histologic type of malignant glioma is an important prognostic factor, and this observation has been confirmed in all the studies of the BTSG. Before presenting results concerning this important factor we must first review the taxonomy of malignant gliomas since, unfortunately, different authors use different systems [7 – 10]. Three general types of gliomas are recognized in the various systems of classification. A rough correspondence between the various systems is presented in Table 2. Note that confusion is particularly likely to arise if these tumors are designated as Grade 3 or Grade 4 since these numbers mean different things in the Kernohan [9] and World Health Organization (WHO) [10] systems. It is not surprising that some difficulties in classification exist because the structure of these brain tumors is heterogeneous, mixed or transitional forms are common, and progressive evolution to more anaplastic forms is not uncommon. Pathological characteristics include cytologic pleomorphism, anaplasia, mitotic figures, necrosis, pseudopalisading, invasiveness, presence of multinucleated giant cells, and endothelial hyperplasia. Indeed the histological variety found in many of these tumors is apparent from the name *glioblastoma multiforme*. The tumors to be discussed in this article are of Type 2 and 3 in Table 2. In studies by the BTSG we have generally found that the death rate was 1.5 to 2.0 times higher in the group of patients having glioblastoma multiforme compared to that for other malignant gliomas, but this ratio varies somewhat from study to study.

Burger and Vollmer [4] reviewed pre-treatment biopsy specimens for 184 cases

Table 2. Histologic classifications of gliomas

Authors	Type 1	Type 2	Type 3
Bailey and Cushing [8]	Astrocytoma	Astroblastoma	Spongioblastoma multiforme
Kernohan [9]	Astrocytoma (Grade 1)	Astrocytoma (Grade 2)	Astrocytoma (Grade 3) Astrocytoma (Grade 4)
WHO [10]	Astrocytoma (Grade 2)	Anaplastic astrocytoma (Grade 3)	Glioblastoma multiforme (undifferentiated glioma) (Grade 4)
Others	Astrocytoma	Anaplastic astrocytoma	Glioblastoma multiforme
This article	– –	Other malignant gliomas	Glioblastoma multiforme

of malignant glioma from BTSG study 6901 in order to determine if any of nine histological factors (cellularity, pleomorphism, mitotic figures, necrosis, pseudopalisading, endothelial proliferation, presence of giant cells gliosarcoma, or perivascular infiltration) could be used to subdivide glioblastoma multiforme further into subtypes such as are indicated by Grade 3 and 4 astrocytomas in the Kernohan system. In their study the only histological feature which affected survival was the presence of giant cells. This finding was present in only 6.5% of the cases studied and was associated with prolonged survival even after correcting for the effects of treatment, age, location, and other variables. It is thought that these grotesque giant cells, however alarming in appearance, multiply slowly, if at all [4, 11]. They observed that pleomorphism was an unlikely candidate for further subclassification since it was commonly observed in the presence of giant cells.

Palma *et al*. [12] found that tumors showing definite lymphocytic infiltration were associated with both significantly longer preoperative and postoperative survival than patients with tumors showing only slight or no infiltration, but this finding was not confirmed in the study of Burger and Vollmer. It is possible that the importance of lymphocytic infiltration found by Palma *et al*. was confounded with that for duration of symptoms (see below).

Burger and Vollmer [4] also estimated the cellularity of the tumor as 1 + to 4 + in addition to counting the number of nuclei within fixed areas, both for an area representative of the neoplasm as a whole and for its most cellular region. In a group of patients treated only by surgery and supportive care without radiotherapy or chemotherapy, they found that survival was shortest for patients with the most cellular neoplasms. They suggested that if this observation could be confirmed in a larger sample, the most highly cellular lesions might be a significant subgroup of the glioblastomas. Their failure to find a significant relationship between cellularity and survival in groups treated with chemotherapy or radiotherapy is consistent with the idea that more primitive cells are more susceptible to these treatments. Clearly this interesting observation awaits confirmation.

2.2. Extent of Surgical Removal

Many authors [1, 5, 13 – 15] have found that the extent of surgical removal is an important prognostic variable. Of course this variable is likely to reflect the size, invasiveness, and location of the tumor and particularly its proximity to vital structures. Types of surgical resections may be usefully divided into: biopsy only, subtotal resection, total resection, and total resection plus lobectomy. Using this classification, unpublished data from BTSG study 7501 indicates roughly a doubling of the death rate for patients who had biopsy only compared to those who had a total resection plus lobectomy, with subtotal and total resections falling in between. Gehan *et al*. [5] showed roughly a doubling of the median survival time for patients treated by a partial or total resection compared to those who had biopsy only. Scott and Gibberd [14] present survival curves showing a markedly significant difference for patients having biopsy only compared to those who had partial removal of their tumor.

2.3. Location of Tumor

None of the studies of the BTSG have shown that the laterality (right versus left hemisphere) of the tumor was important for prognosis, even when the handedness of the patient was taken into account. On the other hand, Onoyama *et al.* [13] found a substantially increased survival at two years for patients with tumors in the right cerebral hemisphere (48% survival at two years) compared to the left cerebral hemisphere (18% at two years). Their analysis was not adjusted for other important prognostic factors and this may explain why we failed to confirm it when studying much larger groups of patients.

Although the location of the tumor in the brain is generally assumed to be an important prognostic variable, we have listed it among the negative or equivocal findings because we failed to find convincing data to demonstrate this point. Several authors have suggested that location in the thalamus or basal ganglia is associated with a poorer prognosis, but we are unable either to confirm or contradict this statement with analyses thus far of BTSG data because of the rarity of tumors in these sites. On the other hand, comparisons of survival for tumors in the frontal, temporal, parietal, and occipital lobes have generally not revealed significant differences. An exception to this observation is the finding by Gehan *et al.* [5] that parietal tumors carried a worse prognosis. We conclude that with present information tumor location is of equivocal prognostic significance.

2.4. Gross Findings at Operation

In the earlier BTSG studies [5, 6] a number of gross tumor characteristics were recorded from the operative reports. Included among these were the following adjectives: hypovascular, invasive, solid, necrotic, soft, friable, cystic, hard, vascular, and encapsulated. In their article based on BTSG study 6901 Gehan *et al.* [5] found that hypovascular tumors were associated with improved survival, and Walker *et al.* [6] found higher death rates for necrotic tumors. In the analyses thus far carried out, none of the other features have been demonstrated to be prognostic with convincing significance. This fact combined with the discrepancies in those variables found significant in the two studies just mentioned lead us to classify these gross tumor characteristics as essentially negative, and in fact the BTSG is no longer collecting information on these variables.

2.5. Findings on CT Scan

Although it has been suspected that the size of the tumor is likely to be an important prognostic factor for malignant gliomas, the intracranial location of these tumors has prevented accurate assessment of tumor size. During the past decade CT scans (computerized tomography) have come into general use and with refinements in imaging techniques, studies are beginning to appear which attempt to correlate the findings on CT scan with the patient's course. Marks and Gado [16] have observed that 'the general lack of morphological/pathological correlation makes pathophysiological interpretation of CT scans somewhat conjectural'. Nevertheless, most workers seem to feel that the images obtained by CT scans may be roughly interpreted as consisting of three more or less concentric volumes: the innermost volume of central low density representing tumor necrosis; the next outer volume, often showing enhancement with contrast material, representing actively growing tumor; and the outer low density volume representing edema.

Marks and Gado [16] found that all glioblastomas and 40% of astrocytomas were enhanced by contrast material and suggested that the presence or absence of contrast enhancement might be a more accurate prognostic sign than pathology since it is less subject to sampling error. They claim that CT scan findings correlated with the clinical course in 93% of their cases. In their study of 55 patients, 73% of the tumors persisted, 20% disappeared, and only 7% recurred. However, they did not really find that the CT scan was prognostic. They had hoped to document that CT scan changes would precede clinical deterioration, but in fact for their few recurrences the CT changes coincided with clinical deterioration and therefore only confirmed the presence of tumor regrowth. They did not recommend the routine use of serial CT scans and suggested that such studies were only appropriate on a research basis.

Levin et al. [17] in a study of 61 patients found that the presence of a large peritumoral low density area and a small estimated volume of enhancing tumor were both favorable prognostic signs for the time to documented tumor progression. They point out that their finding with respect to the surrounding peritumoral low density was unexpected but suggest that it might possibly be related to vasogenic edema and breakdown of the blood brain barrier allowing the passage of immune reactive cells or proteins. Clearly, further data are needed to evaluate the importance of this finding and particularly to substantiate or contradict the suggested mechanism.

Reeves and Marks [18] found no relationship between lesion size as estimated from cross-sectional areas on CT scans and survival of patients with glioblastoma multiforme. Unlike Levin et al. [17] they did not subtract out the central low density area thought to represent necrosis when evaluating tumor size. In addition, they assessed the prognostic significance of their measurements against survival rather than time to progression. They concluded that lesion size in glioblastoma multiforme appears to be unrelated to prognosis and therefore should not be used in staging systems for glioblastomas. They suggested that their negative findings might be explained by the very rapid tumor growth immediately prior to presentation and illustrated this concept with a brief case report. Their notion is that the time it takes for a tumor to grow from a small lesion to a large one as seen on CT scan may be short compared to the overall time from diagnosis to death, and thus the effect would not be noted in comparing survival statistics. They also suggested that the lack of correlation might be explained if tumor location were more important than size, but unfortunately they did not have enough data to examine this point.

3. HOST CHARACTERISTICS

3.1. Age of the Patient

There seems to be complete agreement that the age of the patient at the time of diagnosis is inversely correlated with the length of survival. An exception is that patients under age 15 have very short survival [13]. When age at diagnosis is divided into four categories (less than 45, 45 – 54, 55 – 64, greater than or equal to 65) two large studies by the BTSG [6, 19] have been quite consistent in demonstrating that ratios of the death rates for the three higher age categories compared to that for patients under age 45 have been about 1.75, 2.5, and 3.5 respectively. We have observed that age is mildly correlated with three other prognostic variables: duration of symptoms, performance status, and histologic type. However, multivariate analyses adjusted on these three variables and other important prognostic variables with which age is not significantly correlated indicate that the effect of age is so strong that it adds important prognostic information not explained by the other variables.

When examining variables one at a time [19] we found that a history of arteriosclerosis and a history of cancer were significant prognostic factors. However these two variables were highly correlated with age, both being markedly more likely to be positive for older patients, and were not found to add to our ability to predict survival when age was known.

3.2. Sex of the Patient

The ratio of males to females with malignant glioma ranges in various studies from about 1.5:1 to 2:1. This demonstrates that sex is clearly a risk factor for malignant glioma, but we are unaware of any authors who have found that it is a prognostic factor. Of the potential prognostic factors discussed in this article, sex is thus the most clearly negative.

3.3. ABO Blood Type

We are unaware of any reports in the literature indicating that ABO blood type is related to prognosis for malignant glioma. However, unpublished data from BTSG study 7501 indicated that the death rate was higher for patients with

A or AB compared to B or O blood types. The ratio of the two death rates was only 1.37, but this finding was significant at p less than 0.001. This effect persisted with p less than 0.01 even after adjustment for three other important prognostic variables (age, performance status, and histological type). The association of ABO blood type and prognosis did not appear to be mediated through other known prognostic variables since in BTSG study 7501 the correlation coefficient for blood type with other prognostic variables was never greater in absolute value than 0.1.

Since this finding was unexpected, we examined three other unpublished BTSG studies for which information on blood type was available. In one of these we found similar results significant at p less than .01 after adjustment for the three variables mentioned above. However, preliminary analysis of two later studies showed that the ratio of death rates for A or AB versus B or O was only about 1.15 and not statistically significant. We cannot explain these results at present. However, we feel that the finding for ABO blood type is sufficiently provocative to be listed as an equivocal finding, especially since it can easily be checked in other data sets. The mechanism by which such an effect, if true, could operate is a matter of conjecture, but it is conceivable that it might be related to the immunological response of the patient to tumors having A-like tumor antigens.

4. EFFECT OF THE TUMOR ON THE HOST

4.1. Performance Status

Like the histology of the tumor and the age of the patient at diagnosis, the initial performance status has been found to be importantly prognostic by many authors who have studied it [1, 3, 6, 19 – 21] and we are unaware of any reports claiming that it is not. The performance status can be conveniently measured using the Karnofsky scale [22] as given in Table 3. In two large randomized BTSG trials [6, 19] we demonstrated progressively higher death rates as the initial performance status decreased on the Karnofsky scale. Because of the admission criteria to the studies and the number of patients involved, the ten scores on the scale were grouped as follows: 90 – 100, 70 – 80, 50 – 60, and 10 – 40. Taking the highest category as the reference category, the ratios of death rates for the lower categories in order compared to the reference category were roughly 1.4, 2.0, and 2.9. It is interesting to note that the ratio of death rates for the highest and lowest category is roughly 3, about the same ratio obtained for the youngest and oldest age groups discussed above. Gilbert et al. [21] measured performance by a simpler system which they refer to as the 'functional status': full care, partial self-care, and non-self-care. They demonstrated that the probability of being bed-ridden at various points of time after operation was dependent both upon the functional status and on the age of the patient. They also showed that survival time was a function of both variables.

Table 3. Scoring system for assessing performance status adapted from Karnofsky *et al.* [22]

Score	Meaning
100	Normal: no complaints, no evidence of disease
90	Able to carry on normal activity: minor symptoms
80	Normal activity with effort: some symptoms
70	Cares for self: unable to carry on normal activity
60	Requires occasional assistance: cares for most of needs
50	Requires considerable assistance and frequent care
40	Disabled: requires special care and assistance
30	Severely disabled: hospitalized, death not imminent
20	Very sick: active supportive care needed
10	Moribund: fatal processes are progressing rapidly

Because of the unquestionable importance of this factor it is recommended that the performance status be recorded routinely in all studies of malignant gliomas and that data be analyzed taking this variable into account. In reports of single series of patients the distribution of initial performance scores should be reported, and it is particularly important to consider this variable when comparing non-randomized series treated by different methods since differences in survival are likely to be much greater due to maldistribution of initial performance status than they are to differences in treatments received.

4.2. Neurological Symptoms and Signs

In a number of studies [5, 6, 14, 15, 19] a history of seizures (epilepsy) has been associated with a better prognosis. Both the data of Scott and Gibberd [14] and unpublished data from the BTSG indicate that a history of seizure can be associated with tumors in any of the major cerebral lobes. Based on these two studies it seems that this symptom is present in something like 1/3 to 1/2 of patients with malignant glioma. A possible reason that this finding is associated with good prognosis is that symptoms of epilepsy draw attention to the tumor at an earlier point in its course, thus leading to a longer interval between diagnosis and death.

Duration of symptoms is another variable which has consistently been found to be prognostically important [4, 6, 14, 19] with one exception [5]. This exception can be discounted because survival curves based on only 12 and 13 patients respectively were constructed for the two categories with the longest duration of symptoms and such curves are unreliable. In BTSG studies 7201 [6] and 7501, patients whose duration of symptoms was greater than six months had death rates only about 1/2 of those for patients whose duration of symptoms was less than six months. In the analysis of BTSG study 6601 [2] those patients having a duration longer than 25 weeks were divided into four categories and the median survival times following operation were progressively greater as the duration of symptoms increased. A possible explanation for the prognostic significance of this variable is that patients with long duration of symptoms before surgery may have more slowly growing tumors.

In the BTSG studies data were also collected on many additional neurological symptoms, including rapid unconsciousness, headache, personality change, speech disorder, symptoms associated with cranial nerves II, III, IV, or VI, symptoms concerned with other cranial nerves, motor symptoms, sensory symptoms, cerebellar symptoms, and general complaints. These factors were examined by Gehan et al. [5] and none were found significant at p less than 0.05. In BTSG study 7201 [6] headache, personality change, and motor symptoms were found to be significantly prognostic at p less than 0.02 when studied individuality. In a subsequent study [19] the same variables were again examined and three of them were significantly prognostic when studied one at a time – rapid unconsciousness, speech disorder, and motor symptoms. However, in that later study information was also available on the results of the post-operative neurological examination, and most of the variables listed as symptoms have counterparts in the neurological examination. We found that these same three variables were much more significant when based on the results of the neurological examination. In addition, we found evidence that two other variables (sensory deficits, and abnormality of cranial nerves other than III, IV and VI) were significant when examined one at a time. Our impression was that the neurological examination provides more useful prognostic information than does the report of the same abnormalities as symptoms. Possible reasons for this finding are that the neurological exam is not subject to recall bias, it is more objective and more systematic, it is closer to the time of randomization, and the ability to record the information on a five point scale (as was done in this study) may have provided better discrimination of degrees of prognosis than the yes or no questions asked concerning symptoms in this study. Of course a history of seizure and the duration of symptoms are still independently important (see above) and have no counterpart in the neurological examination.

4.3. Hematological Values after Surgery

In the analysis of BTSG study 7501 we were surprised to find that the pre-treatment values of the white blood cell count and platelet count were significantly prognostic. We have not examined these findings in other studies but feel we should report them so that others can either confirm or deny this finding. These measurements were made after the patients had recovered from surgery and before randomization to various chemotherapeutic regimens. The findings were these: for white blood cell count the death rate was increased by a third for patients whose white cell count was outside the range 5,500 to 8,500, whether the abnormal value was high or low; for pre-treatment platelet counts the lowest death rate was found in patients with values greater than 300,000, an intermediate death rate for patients with values between 200,000 and 300,000, and the highest death rate in a group having values below 200,000. For the moment we consider these findings as preliminary and unconfirmed.

5. MULTIVARIATE ANALYSIS

In the introduction of this article we explained that, because of intercorrelations among prognostic variables, it is generally desirable to use multivariate analytic techniques when assessing prognosis. In many articles comparing non-randomized series the authors simply state that the prognostic factors were more or less evenly distributed between the two groups and therefore attribute any differences in survival to the treatments studied. When there are important differences in the distribution of prognostic variables the authors may say that they cannot tell whether the differences in survival are caused by the prognostic factors or by the treatments. It is a common misconception, even among some statisticians, that we need not adjust for prognostic factors unless they differ significantly between the groups being compared. If a variable strongly affects prognosis then even relatively minor imbalances are sufficient to distort analyses appreciably.

Fortunately, modern statistical techniques allow us to adjust for such imbalances when comparing survival experience [23]. There are three methods of adjustments in common use. The Mantel-Haenszel [24] or log rank [25] technique essentially cross-classifies patients in such a way that 'like is compared with like' and then obtains a summary answer. The Cox regression model [26, 27] is a semi-parametric method which has the advantage that the shape of the underlying survival distribution need not be specified, but the effects of the covariates are assumed to act multiplicatively on the underlying, though unspecified, hazard (or death rate). For this reason this technique is often referred to as a 'proportional hazards model'. The Weibull model [28] is a fully parametric method and is a special case of the Cox model in which the shapes of the underlying survival curves are assumed to fall in a certain family of distributions. For any particular analysis a shape parameter for the survival curves is fitted to the data at hand. We have found that with data for malignant glioma, results using the Cox model and Weibull model are essentially the same.

For the purpose of illustration we reproduce here two tables and a figure taken from the published results for BTSG study 7201 [6]. Variables which were significantly prognostic when examined one at a time (using death rates as the measure of prognosis) are presented in Table 4. All these variables were entered into a Weibull regression model and we found that of the original nine variables only five were needed to predict survival. This means that even though all of the other four variables were individually important, they did not significantly in-

Table 4. Effect of individual prognostic variables on death rate

Variable	Category	Number of patients	Number of deaths	Death rate*	P value
Age at randomization (yr)	<45	68	54	0.42	<0.00001
	45–54	108	104	0.76	
	55–64	110	106	1.20	
	≥65	72	72	1.41	
Headache	No	160	150	0.96	0.0186
	Yes	198	186	0.74	
Motor symptoms	No	198	182	0.73	0.0049
	Yes	160	154	0.99	
Seizures	No	228	219	0.92	0.0167
	Yes	130	117	0.70	
Personality change	No	234	215	0.76	0.0188
	Yes	124	121	0.99	
Duration of symptoms (mo)	<4	257	249	0.98	<0.00001
	4–6	25	25	0.96	
	>6	76	62	0.49	
Histopathologic category	Glioblastoma multiforme	302	292	0.96	<0.00001
	Other malignant glioma	56	44	0.44	
Necrotic tumor	No	181	164	0.71	0.0037
	Yes	177	172	0.98	
Performance status (Karnofsky scale)	10–40	66	65	1.52	<0.00001
	50–60	123	120	1.04	
	70–80	78	68	0.70	
	90–100	91	83	0.55	
All patients		358	336	0.83	

*Death rates are expressed as the number of deaths per 10 patient-months.

crease the fit of the model to the data when the other variables were present. Using this model we constructed risk groups based on the predicted survival at two years according to the model. The first risk group consisted of patients whose predicted survival was 25% or greater, for the second group the prediction was between 2.5 and 25%, and for the third group it was less than 2.5%. The predictions of the model as well as actuarial survival curves for the patients in these three risk groups are presented in Figure 1. Although the fit of the model is not perfect, it is quite satisfactory, indicating that we can separate the patients into three groups having markedly different survival experience.

The characteristics of patients falling in these three risk groups are presented in Table 5, along with their observed survival at 24 months. Perhaps the most interesting finding is that fully 50% of the patients in the best risk group had glioblastoma multiforme rather than other malignant gliomas, but these patients were also younger than the other two groups, less frequently had personality

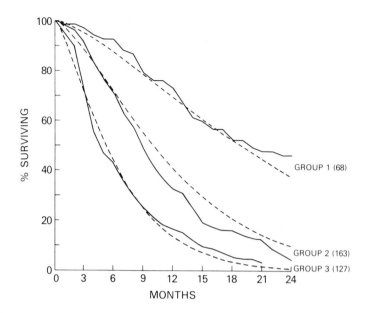

Figure 1. Solid lines are actuarial survival curves and dashed lines are the predicted survival curves for the same groups of patients based on the Weibull model. The numbers of patients are shown in parentheses.

Table 5. Characteristics of risk groups*

Characteristic	Group 1	Group 2	Group 3	All patients
Number of patients	68	163	127	358
Age at randomization				
Mean	36	53	63	53
<45	70.6	11.7	0.8	19.0
45–64	29.4	85.8	45.7	60.9
≥65	0	2.5	53.5	20.1
Personality change	23.5	30.1	46.5	34.6
Duration of symptoms				
>6 mo	58.8	14.1	10.2	21.2
Glioblastoma multiforme	51.5	86.5	99.2	84.4
Performance status (Karnofsky scale)				
Mean	86	68	48	64
10–40	0	6.7	43.3	18.4
50–80	33.8	68.8	52.0	56.2
90–100	66.2	24.5	4.7	25.4
Survival at 24 mo	46.3	4.3	1.6	11.3

*Risk groups are defined in the text.
Figures (except number of patients and means) denote percentages.

changes, had longer duration of symptoms, and had better initial performance status. Analyses such as these can be useful in selecting patients for treatment and they illustrate how the interrelationships between prognostic variables can modify their importance.

6. CONCLUSIONS

In Table 1 we presented a general summary of our impressions based on the analysis of our own data as well as our review of the literature. We combined the two categories 'negative' and 'equivocal' because on present evidence it appears that the only two clearly negative factors studied were laterality of the tumor (right vs left) and the sex of the patient. With most of the other variables in that column it is too early to say whether or not they will emerge as important prognostic factors.

Considering the analyses of the BTSG studies as a whole, age, performance status, and histological category have emerged as the most consistently strong prognostic variables. Not surprisingly, then, these have also been the variables about which there is least dispute in other reports. It is interesting to note that one of them is a tumor factor, one is a host factor, and the third represents the effect of the tumor on the host. The data suggest that these variables are so consistently strong that their significance can be detected even in small series selected in different ways and that they are not likely to have their effects seriously confounded by other variables.

In assessing the importance of any variable (particularly ones obtained from observational studies subject to varying definitions, imprecise measurements, and incompleteness of data recording) replication, that is to say, consistency of findings across different sets of data, is a surer guide than p-values, no matter how significant, in one or a few studies. We may have made mistakes in assigning some variables to the negative or equivocal column in Table 1, but it is much less likely, especially with the three variables just discussed, that we have erred in the other direction. Age, performance status, and histological category are so important and so strong in their effects on survival that they should be regarded as a minimal subset of prognostic factors to be reported when publishing any series, or to be used as adjustment variables in comparing series. We encourage scientists studying malignant gliomas (and most other diseases) to work with experienced biostatisticians familiar with multivariate adjustment techniques when analyzing their data and preparing publications.

REFERENCES

1. Hildebrand J, Brihaye J: Malignant gliomas: prognostic factors and criteria of response. In: Cancer Therapy: Prognostic Factors and Criteria of Response. Staquet M (ed), New York, Raven Press, 1975, pp 307 – 318.
2. Walker MD, Alexander E, Hunt WE, Leventhal CM, Mahaley MS, Mealey J, Norrell HA, Owens G, Ransohoff J, Wilson CB, Gehan EA: Evaluation of mithramycin in the treatment of anaplastic gliomas. J Neurosurg 44:655 – 667, 1976.
3. Walker MD, Alexander E, Hunt WE, MacCarty CS, Mahaley MS, Mealey J, Norrell HA, Owens G, Ransohoff J, Wilson CB, Gehan EA, Strike TA: Evaluation of BCNU and/or radiotherapy in the treatment of anaplastic gliomas. J Neurosurg 49:333 – 343, 1978.
4. Burger PC, Vollmer RT: Histologic factors of prognostic significance in the glioblastoma multiforme. Cancer 46(5):1179 – 1186, 1980.
5. Gehan EA, Walker MD: Prognostic factors for patients with brain tumors. Natl Cancer Inst Monograph 46:189 – 195, 1977.
6. Walker MD, Green SB, Byar DP, Alexander E, Batzdorf U, Brooks WH, Hunt WE, MacCarty CS, Mahaley MS, Mealey J, Owens G, Ransohoff J, Robertson JT, Shapiro WR, Smith KR, Wilson CB, Strike TA: Randomized comparisons of radiotherapy and nitrosoureas for the treatment of malignant glioma after surgery. New Engl J Med 303:1323 – 1329, 1980.
7. Jellinger VK: Pathology of brain tumors with relation to prognosis. Zb Neurochirurgie 39:285 – 300, 1978.
8. Bailey P, Cushing H: A classification of tumors of the glioma group on a histogenetic basis with a correlated study of prognosis. Philadelphia, JB Lippincott, 1926.
9. Kernohan JW, Sayre GP: Tumors of the central nervous system. In: Atlas of Tumor Pathology, Sec 10, Fasc 35. Washington DC, Armed Forces Institute of Pathology, 1952.
10. Zülch KJ: Principles of the new World Health Organization (WHO) classification of brain tumors. Neuroradiology 19:59 – 66, 1980.
11. Jellinger K: Glioblastoma multiforme: morphology and biology. Acta Neurochirurgica 42:5 – 32, 1978.
12. Palma L, DiLorenzo N, Guidetti B: Lymphocytic infiltrates in primary glioblastomas and recidivous gliomas. J Neurosurg 49:854 – 861, 1978.
13. Onoyama Y, Abe M, Yabumoto E, Sakamoto T, Nishidai T, Suyama S: Radiation therapy in the treatment of glioblastoma. Am J Roentgenol 126(3): 481 – 492, 1976.
14. Scott GM, Gibberd FB: Epilepsy and other factors in the prognosis of gliomas. Acta Neurol Scandinav 61:227 – 239, 1980.
15. Horton J, Chang C, Schoenfeld D: Beneficial effect of chemotherapy on survival of patients with malignant glioma treated with surgery and radiation therapy. AACR Abstracts 21:169, 1980.
16. Marks JE, Gado M: Serial computed tomography of primary brain tumors following surgery, irradiation, and chemotherapy. Radiology 125:119 – 125, 1977.
17. Levin VA, Hoffman WF, Heilbron DC, Norman D: Prognostic significance of the pretreatment CT scan on time to progression for patients with malignant gliomas. J Neurosurg 52:642 – 647, 1980.
18. Reeves GI, Marks JE: Prognostic significance of lesion size for glioblastoma multiforme Radiology 132:469 – 471, 1979.
19. BTSG study 7501, unpublished data.
20. Sheline GE: The importance of distinguishing tumor grade in malignant gliomas: treatment and prognosis. Int J Radiation Oncology Biol Phys 1:781 – 786, 1976.
21. Gilbert H, Kagan AR, Cassidy F, Wagner J, Fuchs K, Fox D, Macri I, Gilbert D, Rao A, Nussbaum H, Forsythe A, Eder D, Latino F, Youleles L, Chan P, Hintz BL: Glioblastoma multiforme is not a uniform disease! Cancer Clin Trials 4:87 – 89, 1981.

22. Karnofsky DA, Abelmann WH, Craver LF, Burchenal JH: The use of the nitrogen mustards in the palliative treatment of carcinoma: with particular reference to bronchogenic carcinoma. Cancer 1:634–656, 1948.
23. Byar DP: Analysis of survival data in heterogeneous populations. In: Recent Developments in Statistics. Barra JR et al (eds), Amsterdam, North-Holland Publishing Company, 1977.
24. Mantel N: Evaluation of survival data and two new rank order statistics arising in its consideration. Cancer Chemother Rep 50:163–170, 1966.
25. Peto R, Pike MC, Armitage P, Breslow NE, Cox DR, Howard SV, Mantel N, McPherson K, Peto J, Smith PG: Design and analysis of randomized clinical trials requiring prolonged observation of each patient. II. Analysis and examples. Br J Cancer 35:1–39, 1977.
26. Cox DR: Regression models and life tables. J R Statist Soc (B) 34:187–202, 1972.
27. Kalbfleisch JD, Prentice RL: The Statistical Analysis of Failure Time Data. New York, John Wiley and Sons, 1980.
28. Byar DP, Green SB, Dor P, Williams ED, Colon J, van Gilse HA, Mayer M, Sylvester RJ, Van Glabbeke M: A prognostic index for thyroid carcinoma. A study of the EORTC Thyroid Cancer Cooperative Group. Europ J Cancer 15:1033–1041, 1979.

15. The Challenge to Oncology of the Nervous System

MICHAEL D. WALKER

Brain tumors are among the most lethal, least controlled, and little understood of oncological diseases.

The epidemiology of tumors of the brain indicates a median age-adjusted incidence rate of approximately 4.5/100,000 with a relatively consistent greater occurrence in males. Although there are known genetic factors, such as Von Recklinghausen's disease, tuberous sclerosis and Von Hippel-Lindau disease and there are other associations such as the glioma-polyposis syndrome, little light has been shed on causative factors which might be related to the occurrence of brain tumor. In both man and experimental animals, evidence has been developed demonstrating ionizing irradiation, barbiturates, the nitrosoureas, viruses and possibly head trauma as all being part of a long list of potential etiologic agents for one or another kind of brain tumor. In fact, it is ironic that the nitrosoureas (ethylnitrosoureas) can be utilized to develop uni- and multi-centric tumors in new born rodents which then can be successfully treated by another nitrosourea (CCNU) with demonstrated increased survivorship [1].

Recently an increased incidence of glioma in individuals who have had long-term exposure to petro chemicals has been reported [2]. Much of the material presented in the literature has utilized the term 'brain tumor' without more specific designation as to its histopathologic and biologic characteristics. The association with petro chemicals, however, is primarily with malignant gliomas. Whether brain tumors are the direct effect of some carcinogen on the brain or are involved in a two-step phenomena which requires first initiation and then promotion by either separate etiologic agents or by the same substance at different times, remains to be seen.

If a single etiologic agent causes the development of brain tumors in humans and such an agent can be identified, then clearly methods for either eliminating the agent or altering the effect can be developed. Little research has been carried out in this area and much valuable information in both the prevention and treatment of brain tumors of all kinds could be undertaken.

There are numerous biologic observations which are important to consider in the research designs of the future study and treatment of central nervous system disease. Despite the pathologic appearance of malignant gliomas and their characterization as highly anaplastic pleomorphic cells with numerous mitotic figures, they rarely metastasize [3]. During surgical procedures it is impossible to avoid spilling large quantities of viable tumor cells which come in direct contact with uninvolved surfaces of the brain, venous sinuses with direct continuity to the blood stream and the CSF pathways which circulate around the spinal cord and brain eventually exiting via the Pacchionian granules into the transverse sinus. It has been the observation that tumors elsewhere in the body will often seed a suture line or metastasize in the operative area unless great precautions are taken to avoid such dissemination. Yet, metastasis of primary brain tumor of almost any kind (particularly the highly malignant gliomas) is a curiosity still deserving of a case report in the literature. These tumors which appear to be as highly malignant as any in the body, remain confined to the brain or spinal cord and spread only locally. Even meningiomas, which on occasion dedifferentiate and become more anaplastic in appearance, are rarely found to spread beyond the confines of the brain, skull or immediate local environs. Medulloblastoma, for obscure reasons, tend to have a much higher rate of metastasis, both through the CSF pathways as well as systemically. This was particularly evident in patients who were shunted prior to the time of filters within the shunt system [4]. We have yet to elucidate what the specific characteristics are that make tumor cells or aggregates of tumor cells, which are derived from or arise in the central nervous system, fail to become established as metastasis. As therapies become more effective and patients live longer, we may well learn more of the longterm metastatic potential of these tumors. In the meantime, the cytochemical and immunological characteristics of some CNS tumors which does not allow them to inplant within the brain as the body needs further exploitation and investigation.

The histopathologic classification of brain tumors has been primarily dependent upon the light microscopic description of cytoarchitecture [5]. The nomenclature has been varied in its approach and has been dependent to a degree upon the gestalt of the neuropathologist who is performing the examination. Of the several dozen cytoarchitectual entities which can be described, (such as staining, size, shape, occurrence, and associations), little emphasis has been placed on understanding whether those observations have, in fact, any biologic relevance. The carefully carried out studies of Gilles and his colleagues in posterior fossa astrocytomas of childhood have utilized survival as one major biologic, as well as sociologically important end point [6]. Paired neuropathologists with a dual-headed microscope graded both dependently and independently the degree with which a wide variety of descriptive factors were seen. These factors were then correlated with each other, as well as with survival. Two distinct clusters, Glioma A and Glioma B, were identified. Patients with Glioma A characteristics had a 94% ten-year survival rate and included the following factors: 1) microcysts, 2) Rosenthal fibers, 3) leptomeningial deposits, and 4) focuses of oligodendroglia. Patients characterized as having Glioma B factors had a 29% ten-year survival rate and included such characteristics as: 1) perivascular pseudorosettes,

2) necrosis, 3) high cell density, 4) mitosis, and 5) calcification in the *absence* of any Glioma A characteristics. Thus, a wide variety of other characteristics commonly described do not appear to have a biologic correlation as expressed in what is probably the most important determinant, survival.

The histopathologic description as is commonly seen with both the light microscope and the electronmicroscope must be more carefully brought from a descriptive art to a biologic science wherein the observations could be quantitated and the actual *biological relevance* shown. Such studies need to be carried out for a wide variety of brain tumors and would be most valuable, for example, in determining what characteristics of a meningioma are reproducibly found and predictive of good survivorship versus those which predict toward early recurrence. The physician treating patients can then more knowledgeably design his therapeutic studies to encompass these characteristics.

Tumors of the central nervous system are often afforded a degree of respect which is matched only by the extent of dread. This is particularly true of metastatic disease which occurs several fold more frequently than primary tumors of the brain. More attention thus far has been directed toward primary tumors of the brain, particularly those of the astrocytic series, than toward metastatic disease. This, in part, with the unsubstantiated belief that primary tumors of the brain are comparatively homogeneous, whereas tumors that are of metastatic origin are widely heterogeneic in their biologic characteristics. Only recently has the whole field of cellular heterogeneity become one of importance and may, in fact, be considered as providing some of the significant reasons for therapeutic failure.

Brain tumors in contradistinction to head trauma produce gradual shifts in intracranial contents and gradual build-up of increased intracranial pressure. The brain appears to compensate to a space occupying lesion for a while, but either because of occlusion of venous drainage or of the egress of CSF or just plain the bulk of the lesion, increased ICP is frequently experienced. Much attention has been paid to the ICP of the head injured patient as it is a sensitive index of impending ischemic and destructive events within the brain. The increased ICP of brain tumor, although early recognized, particularly in the prohibition toward performance of lumbar punctures might be a valuable tool for further understanding the mechanics of shifts within the intracranial contents, both rapid and slow. Corticosteroids are extraordinarily effective therapeutic agents, which on CT Scan show a decrease in cerebral edema of brain parenchyma surrounding a tumor and allow in one sense a 'medical decompression'. Fundamental factors of brain shift and movement might be studied profitably in these areas.

Brain tumors, by definition, are abnormal tissue in that they no longer respond to the usual physiological control signals and are aberrantly going through the mitotic cycle. This is seen descriptively by neuropathologic examination but must also have a biochemically relevant component. That is, there must be something aberrant in the biochemistry of glioma. There have been analyses which examined the metabolic oxidative pathways, anaerobic energy metabolism, protein, amino acid and lipid metabolism, and nucleic acid composition of brain tumors [7]. DNA and RNA have been carefully examined and the cell-cycle kinetics have been elucidated [8]. These fundamental and rather basic pieces of information concerning the differences of growing primary neoplasm within the brain, compared to the normal tissue from which they derived, seemed to lie in the literature unaddressed and unused for the design of new appropriate therapy. Are there specific factors, utilizing concepts of controlled nutritional treatment as outlined in the chapter of Robertson *et al.*, which could be more effectively applied for the control and management of these tumors? Despite having defined the cell kinetics of gliomas, how are we going to apply this information for therapeutic gain? The nucleic acid differences are important characterizers of the cell, but can these be utilized to control the growth of these tumors? These and a myriad of other seemingly simple questions should be the basis for very complex research for the next decade.

There is little doubt that immense gain has been made in the accurate location and partial diagnosis of brain tumors through use of the CT Scan and the ability of its computer to reconstruct images in a wide variety of planes. With high resolution and more sophisticated equipment more precise diagnosis can be anticipated. In the past, followup of the patient with brain tumor has been dependent primarily on demonstrating recurrence of signs or symptoms as well as a new or different mass lesion on scanning. The CT Scan has been immediately thrust into the position of authenticating recurrence of tumor. It is being used in an attempt to differentiate between necrotic brain, necrotic tumor, edematous brain and frank viable tumor. Because of its easy pictorial representation, what is actually happening in the brain is thought to appear 'all too obvious'. CT Scanning, however, needs desperately to be correlated with gross neuropathologic observation, as well as histopathologic delineation. What exactly various areas on the CT Scan mean, in terms of the cell content, needs to be more carefully defined, particularly for the patient with recurrent tumor [9]. The whole problem of followup and followup methodology on patients with brain tumor, as well as determining what degree of biologic activity exists within the tumor, remains ill defined.

A number of significant social issues relate to brain tumor. There is a general decline in the autopsy rate throughout the country, thus depriving the clinician of the ultimate knowledge about the results of his treatment of his patient. This seems particularly acute in patients harboring brain tumor as they are often chronically ill, requiring nursing attention and are, therefore, in institutional facilities where autopsies are not frequently performed. Physicians and family find it difficult to take the final step of obtaining autopsies in patients who have been 'through so much'. Nevertheless, the critical information which is needed in order to correlate the early findings and the therapeutic results can only be obtained through these efforts.

There has been a gradual shift in interest in and concern about brain tumor because some researchers are actively treating these tumors, aggressively caring for these patients and showing an interest in their disease process [10]. This is promptly reflected in increased public awareness of brain tumor and acceptance of the fact that it is not just a hopeless disease involving difficult or fatally dangerous surgery. As small advances are made in the treatment, there will be increased concern for these patients. A gratifying occurrence is the development of voluntary agencies who have a specific interest in helping patients with brain tumor and their families understand the many things that are going to happen to them. Such groups as the Association for Brain Tumor Research (Chicago, IL) have recognized the valuable contribution of research in the treatment of this disease by investing their limited and hard-won funds in appropriate research and providing self-help groups for the support of brain tumor patients. They provide a facet of care not normally a part of the medical armamentarium, but which is of great importance to the total well-being of the patient.

With the gradual improvement in the survival of some patients from effective surgery, radiotherapy, and chemotherapy, we are now moving into the position of having patients survive long enough in order for the delayed complications or late effects of treatments to become apparent. The longterm effects may actually be divided into two areas: those that result from the cumulative affects of chemotherapy and/or radiotherapy on both the brain and other tissues, and those that are the effect on the brain of prolonged involvement with compressive and invasive tumor, harboring different biologic properties. Thus, wide-spread invasion of the tumor could be anticipated with involvement of more sensitive areas of the brain leading to progressive dementia. This may be mingled and not easily differentiated from the late effects resulting from the use of highly cytotoxic treatments, such as radiation and chemotherapy. The maximum dose of 6,000 rads of radiotherapy will show, 5 – 10% of patients' who survive long enough, some late effects on the normal adjacent cerebral vasculature which may result in areas of necrosis. More aggressive radiotherapy such as high LET has been shown to apparently sterilize the tumor bed but its devastating effects on the brain led to the ultimate death of the patient [11]. Thus, the therapeutic index of radiation therapy appears to be equally slim. Newer approaches toward the radio-therapy of these tumors using either radio-sensitizers to increase the effect of radiation therapy on tumor but not normal tissue or radio protectors to reduce the effects on normal brain (but hopefuly not the tumor) will be extremely important. Radiation therapy has the luxury of being a locally delivered treatment for a locally occurring tumor. Of all the tumors in which control might be gained through use of appropriate radiation therapy, tumors of the brain are the best candidates because they do not metastasize. Yet, such control has escaped our efforts.

The late effects of chemotherapy are primarily because it is usually systemically delivered treatment applied for local tumor. That is, chemotherapy affects all dividing tissues in the body to a greater or lesser extent and, thus, the bone marrow and gastrointestinal tract become prime targets. More recently, the nitrosoureas have shown a delayed pulmonary fibrosis when used in <1500 mg/m^2 total dose [12]. In a similar fashion, delayed renal toxicity is being uncovered and hepatic dysfunction demonstrated [13]. At least four organ systems are now affected by the toxicity of the nitrosoureas. While it is acceptable to utilize agents which are highly toxic and carry a considerable risk for the treatment of tumors which carry an even greater risk, it would be much more desirable in the long run to have therapeutic agents which have a wide therapeutic index. Thus far, these agents have been hard to find for the field of cancer in general and for the treatment of brain tumor specifically.

The modelling of a disease allows a wide variety of experimental hypotheses to be tested in a model other than the patient. Many tumor models of the central and peripheral nervous systems have been developed and utilized for research in nervous system tumors [14]. The superb monograph by Janisch and Schribers with its English edition by Bigner and Swenberg, details the history of cerebral oncogenic agents in many different hosts. The model chosen for a given series of experiments depends, almost entirely, on the desired outcome. The most desirable model would be one which is autochthonous and of a neural glial cell of origin and histopathologically representative of the tumor of interest. It should demonstrate the same quality of aggressive pathology, be by in large non-metastasizing and have kinetic and biochemical behavior similar to its human equivalent. The definition of human equivalent is often the point of departure, as one must seriously question what part of the human brain-tumor-process one wishes the model to emulate.

There are models which examine carcinogenicity and are relevant to the etiologic agents of brain tumors. Other models are excellent for mechanical or space considerations in that they behave as a well controlled expanding space-occupying lesion. Some models may be designed for biochemical analysis or pharmacologic experimentation. Others may provide insight into the host effect or some of the tumor-host interactions. The most frequently considered purpose for a model is that of identifying effective therapeutic agents. The characteristics of such a model may be divided into those that are of major and of minor concern. The primary requirement is that the model identify effective treatment with a minimum number of false negatives and false positives. Thus, any model must be examined from the point of view of known effective and ineffective treatments for the human condition. Any model that did not respond modestly to radiation therapy and BCNU for example, or did respond to mithromycin, would be highly questionable from the point of view of its ability to identify appropriate therapy as there are treatments which have been carefully tested. Few studies have been carried out where specific models have been systematically evaluated for responsiveness. Other major considerations would include reproducibility and stability in the model such that it could be used repetitively without shift of its intrinsic character.

One unique modelling approach is the *in vitro* chemotherapy assay utilizing the patient's own tumor. Although the cells are removed from their natural environment and are grown in enriched and/or artificial media, greater control over pharmacologic exposure can be achieved than in the experimental animal. Naively one conceives of such a system as being similar to the culture and sensitivity approach utilized for determining the antibiotic of choice for the treatment of bacterial disease. Whereas some bacteria remain remarkably stable over the course of numerous generations in most circumstances, the cells which one is examining in the tissue culture of malignant brain tumors may be very different than their origins after only a few passages. Nevertheless, such approaches have great value as they will allow the therapist to try to individualize therapy for maximum effectiveness against the patient's tumor and to minimum toxicity.

CT Scanning has revolutionized the diagnosis of brain tumor. Whereas CT Scanning defines the gross architecture of the brain and denotes leakage of the blood brain barrier, Positron Emission Tomography (PET) and Nuclear Magnetic Resonance (NMR) have the ability to examine metabolic, biochemical and energy events. Brain tumors would be ideal for intensive study by these methods as they can be anatomically delineated by CT Scan, biopsied at surgery and repetitively studied within a homogeneous environment. The metabolic changes which occur secondary to radiation and chemotherapy could be analyzed in order to maximize the effectiveness of these treatments and devise better treatment schedules. The shift from aerobic to anaerobic metabolic pathways and the so-called hypoxic element within glioma could all be examined. Future studies by both NMR and PET will surely add to the confusion but they may refute or substantiate what are today only hypotheses.

There remains a great deal of work to be done in research leading to understanding the biologic and biochemical events related to brain tumor and its eventual translation into therapeutic action. If a unique etiologic agent, biochemical event, or cellular property can be demonstrated for which agressive therapy can be developed, considerable improvements can be anticipated. Researchers will continue to systematically address the areas of radiation biology, pharmacologic and chemotherapeutic intervention, and basic biology of brain tumor in their efforts to meet the challenge of oncology of the nervous system.

REFERENCES

1. Paoletti P, Giordana MT, Pezzotta S, Schiffer D: Chemotherapy studies on various nitrosoureas: Prevention of development of ethylnitrosourea-induced tumors in the nervous system. Modern Concepts in Brain Tumor Therapy: Laboratory and Clinical Investigations. NCI Monograph 46, DHEW, 1977, pp 161–162.
2. Thomas TL, Decoufle P, Moure-Eraso R: Mortality among workers employed in petroleum refining and petrochemical plants. J Occ Med 22:97, 1980.
3. Bryan P: CSF seeding of intra-cranial tumours: A study of 96 cases. Clin Radio 1 25:355–360, 1974.
4. Kessler LA, Dugan P, Concannon JP: Systemic metastases of medulloblastoma promoted by shunting. Surg Neurol 3:147–152, 1975.
5. Rubinstein LJ (ed): Tumors of the Central Nervous System, Atlas of Tumor Pathology, Second Series Fascicle 6, Armed Forces Institute of Pathology, Washington, D.C., 1972.
6. Winston K, Gilles FH, Leviton A, Fulchiero A: Cerebellar gliomas in children. JNCI 58(4):833–848, 1977.
7. Kirsch WM, Paoletti EG, Paoletti P (eds): The Experimental Biology of Brain Tumors, Charles C. Thomas Publisher, Springfield, IL, 1972.
8. Hoshino T, Wilson CB: Review of basic concepts of cell kinetics as applied to brain tumors. J Neurosurg 42:123–131, 1975.
9. Selker RG, Medelow H, Walker MD, Sheptak PE, Phillips JG: Pathological correlation of CT ring in recurrent, previously treated gliomas. Surg Neurology 17(4):251–254, 1982.
10. Shapiro WR: In: Pinedo HM (ed): Brain Tumors, Chapter 22, Cancer Chemotherapy, EORTC, Excerptia Medica, Amsterdam, 1980, pp 414–431.
11. Cattarell M, Bloom JJ, Ash DV et al: Fast neutron compared with megavoltage x-rays in the treatment of patients with supratentorial glioblastoma: A controlled pilot study. Int J Radiat Oncol Biol Phys 6:261, 1980.
12. Selker RG, Jacobs SA, Moore PB, Wald M, Fisher ER, Cohen M, Bellot P: 1,3-Bis(2-chloroethyl)-1-nitrosourea (BCNU)-induced pulmonary fibrosis. Neurosurgery 7(6):560–565, 1980.
13. Weiss RB, Poster DS, Penta JS: The nitrosoureas and pulmonary toxicity. Cancer Treatment Rev 8:111–125, 1981.
14. Bigner DD, Swenberg JA (eds): Janisch and Schreiber's Experimental Tumors of the Central Nervous System, 1st English Edition, Upjohn Company, Kalamazoo, MI, 1977.

Index

ACNU 37, 54
   treatment of ependymoblastoma group of tumors 48, 49
   treatment of L1210 leukemia 46, 47
Acromegaly 117, 124, 135, 138
   rate of cure 137
   results of transsphenoidal surgery 136
   surgical morbidity 136
Actinomycin D
   classification 287
   penetrates poorly into normal brain 286
   toxicity 293
Adenocarcinoma
   BW10232 275
   C3HBA mammary: phenylalaine rèstriction inhibits tumor growth 274
   lung: metastasis to brain 363
Adenoma
   acidified stem cell 127
   pituitary
      adrenocorticotrophic hormone secretion 139 – 145
      classification 118 – 121
      endocrine-inactive, 146
      growth hormone secreting, 134 – 138
      prolactin-secreting (see Prolactinoma)
      treatment 121 – 124
Adenovirus 12, 41
Adrenocorticosteroid hormone: treatment of brain metastasis 355 – 357
Adriamycin (see Doxorubicin)
Aldosterone 168
Allergic encephalomyelitis 159
Amenorrhea: symptom of prolactinoma 125 – 126, 127, 129 – 130
4-Amino-4-deoxy-10-methyl pteroylglutamic acid (see Methotrexate)

Amphotericin B 49
Antibody-dependent cell mediated cytotoxicity assay 155 – 156
Ara-C (see Cytosine arabinoside)
Armed Forces Institute of Pathology 237
L-Asparaginase
   classification 287
   toxicity 293, 305, 312
Association for Brain Tumor Research 401
Astrocytoma
   assay of amino acids 276
   avian sarcoma virus-induced 51 – 52, 54
   computerized tomography scan 105
   resulting from lead 20
   treatment with radiation therapy 224 – 225
Avian sarcoma virus 40, 75
   induction of astrocytoma 41 – 43
5-Azacytidine: treatment of avian sarcoma virus-induced astrocytoma 51, 52
8-Azaguanine: treatment of ependymoblastoma group of tumors 49
Azo compounds 14
Azoxy compounds 14
Azq
   in vitro assay 251, 261
   treatment of ependymoblastoma group of tumors 48
   treatment of L1210 leukemia 46, 47

Bacillus Calmette-Guérin
   treatment of avian sarcoma virus-induced astrocytoma 51 – 52
   treatment of glioma 156
BCNU 54, 77, 80, 81, 84, 91, 160, 278, 279
   administration via carotid artery 79
   cell cycle nonspecific 73
   classification 287

BCNU (*continued*)
  cross resistance 83
  enhances radiation damage 208
  in vitro assay 72, 250–253, 258, 261
  penetrates central nervous system freely 291
  pulmonary toxicity 83, 274
  structural formula 82
  synergistic effect with irradiation 83
  toxicity 294, 295
  treatment of avian sarcoma virus-induced astrocytoma 51, 52
  treatment of brain metastasis 371
  treatment of ependymoblastoma group of tumors 47, 48, 49
  treatment of glioma 52, 53, 226, 273
  treatment of 9L gliosarcoma 49–50, 51
  treatment of L1210 leukemia 46
  treatment of rat brain tumor 92
Benzoyl alcohol 320
Benzpyrene 33
1,3-Bis(2-chloroethyl)-1-nitrosourea (*see* BCNU)
Bleomycin 90
  classification 287
  penetrates poorly into normal brain 286
  toxicity 292, 293, 294
Blood-brain barrier 66
  cerebral edema 79–80
  malignant gliomas 74–79
  pharmacokinetics of brain tumor chemotherapy 74–80
Bragg peak 216, 217–218
Brain abscess: computerized tomography scan 104
Brain metastasis
  blood-brain barrier 345–346
  diagnostic tests 348–351
  effects of no treatment 354–355
  evaluation of treatment 353–354
  management of acutely decompensating patients 352
  management of neurologically stable patients 352–354
  pathogenesis 342–345
  presenting signs 346–348
  presenting symptoms 342, 346–348
  refractory to chemotherapy 345
  three-phased clinical picture 343
  treatment with adrenocorticosteroid hormones 355–357
  treatment with anticonvulsants 371–372
  treatment with chemotherapy 369–371
  treatment with radiation therapy 357–365
  toxicity 367–369
Brain Tumor Research Center 38
Brain Tumor Study Group 67, 183, 266, 273
Bromocriptine
  treatment of Cushing's disease 145
  treatment of prolactinoma 125, 130, 131, 132, 133–134
Burkitt's lymphoma 80

Calcinosis 322
California Tumor Registry 9
Cancer (*see also* Carcinoma; Tumor)
  breast 21
    metastasis to brain 369
    metastasis to central nervous system 2
    primary site of cerebral metastases 1
  kidney: metastases to central nervous system 2
  lung
    common site of metastasis to brain 343
    metastases to central nervous system 2
    nasopharynx: may spread intracranially 2
  prostate: metastases to central nervous system 2
  skin: iso-effect curves 203–205
  thyroid: metastases to central nervous system 2
Cancer Chemotherapy National Service Center 91
Candidosis 323–324
Carcinoma (*see also* Cancer; Tumor)
  bronchogenic: primary site of cerebral metastases 1
  Ehrlich ascites 31
  gastrointestinal tract: metastases to brain 1
  kidney: metastasis to brain 1
  mammary
    C_3H 174
    TA3 274
  ovarian 69
  renal: metastasis to brain 364
  testicular: metastasis to brain 370
  Walker 256 31, 75–76
Carmustine (*see* BCNU)
Cascade theory 343
CCNU 54, 81, 84, 91, 299
  classification 287

CCNU (*continued*)
  cross resistance 83
  in vitro assay 251
  penetrates central nervous system freely 291
  structural formula 82
  toxicity 293, 294
  treatment of avian sarcoma virus-induced astrocytoma 51, 52
  treatment of brain metastasis 369–370
  treatment of ependymoblastoma group of tumors 47, 48, 49
  treatment of glioma 53, 234
  treatment of 9L gliosarcoma 49
  treatment of L1210 leukemia 46
Central nervous system: metastases 1–2
Central pontine myelinolysis 314
Cerebral edema 66, 79–80
Charity Hospital of Louisiana 9
Chlorambucil 81, 305
1-(2-Chloroethyl)-3-cyclohexyl-1-nitrosourea (*see* CCNU)
1-(2-Chloroethyl)-2-(2,6-dioxo-3-piperidyl)-1-nitrosourea (*see* PCNU)
1-(2-Chloroethyl)-3-(4-methylcyclohexyl)-1-nitrosourea (*see* Methyl-CCNU)
Chlorozotocin 81
Choriocarcinoma 80
Chromophobe adenoma: incidence rates 8
Cisplatinum 52, 80, 85–86
  in vitro study 72
  structural formula 84
  toxicity 293, 294
Citrovorum factor: penetrates the blood-brain barrier 291
Colony forming assay: prediction of drug resistance 92
Complement-dependent assay 155–156
Compton scattering 195
Connecticut
  histologic confirmed primary intracranial neoplasms 5
  incidence rates for primary neoplasms of brain and cranial meninges 6, 7
  primary intracranial tumors 4
Connecticut Tumor Registry 4, 8, 9, 21
Corticosteroid (*see also* specific steroid)
  antiedema effect 172
  anti-inflammatory properties 161
  complications of use 180–184
  diuretic agent 173–174
  effect on cerebral blood flow 173
  effect on cerebrospinal fluid 173
  effect on tumor tissue 174
  effects of diagnostic studies 175
  pharmacology 168–171
  physiology 168–171
  variations in time-dose administration 176–179
Cortisol 168
  biologic half-life 170, 171
  chemical configuration 170
  plasma half-life 171
Cortisone 90
  treatment of Cushing's disease 19
Corynebacterium parvum: treatment of glioma 156
Cox regression model 390
CPA
  treatment of ependymoblastoma group of tumors 47, 48
  treatment of L1210 leukemia 46, 47
Cryptococcosis 323–324
Cyclophosphamide
  classification 287
  penetrates poorly into central nervous system 291
  toxicity 293, 295
  treatment of Burkitt's lymphoma 80
  treatment of Walker 256 carcinoma 75
Cysteamine: radioprotector 211
Cytarbin 323
Cytosine arabinoside 54, 286
  classification 287
  myelopathy following intrathecal chemotherapy 316, 317
  penetrates poorly into central nervous system 291
  toxicity 306, 307, 313, 315, 318, 319, 322
  treatment of ependymoblastoma group of tumors 48
  treatment of L1210 leukemia 46, 47

DAG (*see* Dianhydrogalactitol)
Daunomycin
  classification 287
  toxicity 293
DDMP 90
Dexamethasone 90, 355
  alternate-day therapy 179
  biologic half-life 170, 171

Dexamethasone (*continued*)
    chemical configuration 170
    plasma half-life 171
Dialkyl-aryltriazene 14
2,4,Diamino-5-(3',4'-dichlorophenyl-6-methyl pyrimidine (*see* DDMP)
cis-Diamminedichloroplatinum II (*see* Cisplatinum)
Diamox: diuretic agent 179
Dianhydrogalactitol 54, 80, 85
    structural formula 84
    treatment of ependymoblastoma group of tumors 48, 49
    treatment of 9L gliosarcoma 50
    treatment of L1210 leukemia 46, 47
Diazepam 182
Dibenzanthracene 33
Dibromodulcitol: treatment of ependymoblastoma group of tumors 49
Dibutryl cyclic AMP 249, 257
Dimethyformamide 249, 259
7,12-Dimethylbenz-(a) anthracene 14
5-(3,3-Dimethyl-1-triazeno)imidazole-4-carboxamide (*see* DTIC)
Diphenylhydantoin 182
Disease
    Addison's 167
    Cowden's 12
    Cushing's 117, 122, 124, 139–146, 181
    graft-versus-host 295–296
    Hodgkin's 350–351
        measles encephalitis 325
        treatment with BCNU 83
    von Hippel-Lindau 10, 11, 397
    von Recklinghausen's 10–11, 397
Doxorubicin 90
    classification 287
    enhances radiation damage 208
    penetrates poorly into normal brain 286
    toxicity 293
    treatment of brain metastasis 369–370
DTIC
    classification 287
    treatment of ependymoblastoma group of tumors 48

Eastern Cooperative Oncology Group 227–229
Edema, cerebral
    relieved by adrenocorticosteroid hormones 355

    two forms 172
Electromagnetic spectrum 193–194
EpA: comparative chemotherapeutic results 54
Ependymoblastoma 33, 34, 35, 48
Ependymoblastoma A 33, 34, 35, 37, 48
Ependymoblastoma group 47–49, 54
Ependymoblastoma model 33–34, 36, 91
Ependymoma: treatment with radiation therapy 236–238
N-Ethyl-nitrosourea 37–38
European Organization for Research on the Treatment of Cancer: Brain Tumor Cooperative Group 266

5-Fluorouracil 35, 88–89, 286
    classification 287
    penetrates central nervous system freely 291
    radiosensitizer 321
    structural formula 87
    toxicity 293, 294, 306
    treatment of ependymoblastoma group of tumors 47, 48
    treatment of 9L gliosarcoma 50
    treatment of L1210 leukemia 46, 47
Furosemide: diuretic agent 173, 179

Galactorrhea: symptom of prolactinoma 125–126, 127
Ganglioneuroma: measles encephalitis 325
Gigantism 137, 138
Glioblastoma
    computerized tomography scan 103
    histological changes 300
    imprint cytology 299
    LM cell line 156
    most common intracranial neoplasm in Connecticut series 7
    short survival time 9
    treatment with methyl-CCNU 301
Glioblastoma multiforme 69
    characteristics 65–66
    classification by amino acids uptake 277
    no relationship between lesion size and survival 385–386
    resulting from vinyl chloride 17, 21
    treatment with autologous leukocytes 157
    treatment with high linear energy transfer irradiation, 214–215
    treatment with hydroxyurea 86
    treatment with radiation 210–211

Glioma (*see also* specific type of glioma)
　G26 34, 35, 37, 48
　　comparative chemotherapeutic results 54
　　prediction of human tumor responses 91
　G261 33, 34, 35, 37, 48
　　comparative chemotherapeutic results 54
　avian sarcoma induced 157, 161
　classification 225
　computerized tomography scan 108, 111
　C_6 rat 38
　delayed hypersensitivity to skin tests 152
　dietary therapy 277–281
　effects of steroids 174
　elevation in circulating immune complexes 153
　elevation in IgM levels 153
　evidence of anti-tumor immune response 154–156
　genetic factors 10
　heterogeneity of cells 68–69, 70, 151, 160
　histologic classification 382–383
　humoral immune responsiveness 153
　increased incidence with exposure to petrochemicals 397
　intracranial: signs and symptoms 102–106
　9L 36, 38–40, 49–51
　　comparative chemotherapeutic results 54
　　treatment with BCNU 83, 88
　　treatment with 5-fluorouracil 88
　morphologic effects of chemotherapy 296–303
　neuroectodermal antigen 158
　postoperative management 111–112
　preoperative management 107, 110
　prognostic factors
　　ABO blood type 386–387
　　age of patient 386
　　effect of individual variables on death rate 391
　　extent of surgical removal 383–384
　　findings on computerized tomography scan 384–386
　　gross findings at operation 384
　　hematological values after surgery 389–390
　　histology of tumor 382–383
　　location of tumor 384
　　neurological symptoms and signs 388–389
　　performance status 387–388
　　sex of patient 386
　　three categories 381
　secondary surgery 113
　surgical decision making 107–109
　surgical technique 110–114
　survival rates 225–226, 398–399
　therapy-induced central nervous system lesions 301
　treatment with autologous glioma cells 158–159, 161
　treatment with autologous leukocytes 157
　treatment with bacillus Calmette-Guérin 156
　treatment with BCNU 209, 273
　treatment with CCNU 209
　treatment with *Corynebacterium parvum* 156
　treatment with hydroxyurea 209
　treatment with interstitial radiation therapy 213–214
　treatment with levamisole 156–157
　treatment with procarbazine 209
　treatment with radiation therapy 225–234
　treatment with vincristine 209
　treatment with xenogeneic antiglioma antibody 158
　volume of brain irradiated 230–231

Hammersmith Hospital 215, 233
Heavy charged particle radiation therapy 215–218
Hemangioma: long survival time 9
Hepatoma BW7756: phenylalaine restriction inhibits tumor growth 274
Herpes simplex 326
Herpes zoster 326
HU (*see* Hydroxyurea)
Hydrazo compounds 14
Hydrocephalus 314–315
Hydrocortisone 169
　biologic half-life 170
Hydroxyurea 86, 209, 286
　classification 287
　structural formula 87
Hypopituitarism 133, 137

Immune adherence assay 254
Interphase death: definition 198
Interstitial radiation therapy 213–214
N-Isopropyl-α-(2-methylhydrazino)-p-toluamide hydrochloride (*see* Procarbazine)
Isoproterenol: treatment of C6 glioma 73–74

JC virus 41

Karnofsky scale 388

Leukemia
  acute: treatment with cranial irradiation 208
  acute lymphocytic
    anterior horn necrosis 318
    measles encephalitis 325
    myelopathy following intrathecal chemotherapy 311–312
    treatment with methotrexate 293, 307
    treatment with vincristine 306–307
  acute myelogenous: myelopathy following intrathecal chemotherapy 316–317
  chronic myelogenous: myelopathy following intrathecal chemotherapy 316
  E14 274
  L1210 31, 32–33, 36, 46–47, 54, 274
  meningeal 74
  P388 31, 274
  Rauscher 35
Leukoencephalopathy: progressive multifocal 323–324
Levamisole: treatment of glioma 156–157
Linear energy transfer irradiation: treatment of glioblastoma multiforme 214–215, 232
Lomustine (see CCNU)
Lymphoma, non-Hodgkin's: treatment with BCNU 83
Lymphosarcoma: measles encephalitis 325

Mannitol 371
Mantel-Haenszel model 390
Massachusetts General Hospital 252
Mayo Clinic 83, 226
Measles encephalitis 323–324, 325
MeCCNU (see Methyl-CCNU)
Mechlorethamine (see Nitrogen mustard)
Medical College of Virginia 277, 279
Medulloblastoma
  genetic factors 10
  high rate of metastasis 398
Melanoma
  B16, 31, 344, 345
  hamster 68
  metastasis to brain 1, 371
  neuroectodermal antigen 158
  S91: phenylalaine restriction inhibits tumor growth 274

Memorial Sloan-Kettering Cancer Center 346
Meningioma
  characteristics 65
  computerized tomography scan 106
  elevation in IgM levels 153
  genetic factors 10
  long survival time 9
  most common intracranial neoplasm in Rochester series 7
  occurs more frequently in blacks 8
  relationship to breast cancer 21–22
  resulting from injury 14, 15
  resulting from sodium nitrite 16, 20
  resulting from X-rays 15, 20
6-Mercaptopurine
  toxicity 293, 295
  treatment of L1210 leukemia 47
Methotrexate 35, 54, 77, 78, 89–90, 242, 286
  classification 287
  does not pass blood-brain barrier 291
  enhances radiation damage 208
  etiology of neurologic sequelae 319–322
  implicated in secondary glioblastoma 296
  myelopathy following intrathecal chemotherapy 316, 317
  radiosensitizer 321
  resistance to drug 82
  structural formula 87
  toxicity 293, 294, 295, 305, 306, 307, 309, 312, 313, 315, 318, 319, 322
  treatment of choriocarcinoma 80
  treatment of ependymoblastoma group of tumors 47, 48
  treatment of L1210 leukemia 32, 46
Methyl-CCNU 81, 84
  cross resistance 83
  penetrates central nervous system freely 291
  structural formula 82
  toxicity 83, 294, 295
  treatment of avian sarcoma virus-induced astrocytoma 51, 52
  treatment of ependymoblastoma group of tumors 48, 49
  treatment of glioblastoma 301
  treatment of glioma 52
  treatment of 9L gliosarcoma 49
  treatment of L1210 leukemia 46
Methylcholanthrene 32, 33
  production of tumors 275

Methylhydroxybenzoate 320
1-Methylnitrosourea: treatment of
 L1210 leukemia 47
N-Methylnitrosourea 37–38
Methylprednisolone 90, 91
 alternate-day therapy 178, 179
 bioavailability 177, 180
 biologic half-life 170, 171
 chemical configuration 170
 plasma half-life 171
Metronidazole
 chemical configuration 210
 radiosensitizer 210, 233–234
Microadenoma: prolactin-secreting 117
Minnesota
 Olmsted County: incidence rates for pituitary tumors 8
 Rochester 6–8, 14
Misonidazole
 chemical configuration 210
 radiosensitizer 210–211, 233, 234
Mithramycin 90
 classification 287
 toxicity 293
 treatment of ependymoblastoma group of tumors 47, 48
 treatment of glioma 53
Mitotic death: definition 198
Mouse
 BALB/c 36, 275
 C57, 34
 C57BL 36, 275
 C_3H 33
 DBA 32, 36
 NIH Swiss 36
 nude 52, 53, 55, 69, 70
  introduction of human gliomas 44–45
  model for human brain tumors 91
Mouse mammary tumor virus 68–69
6MP (see 6-Mercaptopurine)
MTX (see Methotrexate)
Murine sarcoma virus 40

National Cancer Institute 8, 91
 Brain Tumor Study Group 381
 Diet, Nutrition and Cancer Program 279
National Institutes of Health 252
Nephrosis: measles encephalitis 325
Neuroblastoma
 measles encephalitis 325

neuroectodermal antigen 158
Neurocutaneous melanosis 11–12
New York University Medical Center 101
Nitrogen mustard 81
 classification 287
 resistance to drug 82
 toxicity 305
 treatment of malignant glioma 80
N-Nitrosamide 14
Nitrosourea: synergistic effect with hyperthermia 83

Oak Ridge: Medical and Health Sciences Division 281
OK-432 157
Oligodendroglioma: treatment with radiation therapy 235–236
Oligodendroglioma-astrocytoma: heterogeneous cell populations 69
Optic glioma: incidence rate 8
Osteoporosis 126, 131, 182, 295
Osteosarcoma 22
 treatment with methotrexate 309
Oxygen
 hyperbaric: treatment of hypoxic tumor cells 212–213, 232
 photosensitizing agent 209, 231
Oxygen enhancement ratio 232

Pair production 195
Papavarine: treatment of C6 glioma 74
Papilloma: Shope virus-induced 274–275
PCB (see Procarbazine)
PCNU 54, 81
 cross resistance 83
 structural formula 82
 treatment of avian sarcoma virus-induced astrocytoma 51, 52
 treatment of ependymoblastoma group of tumors 48, 49
 treatment of glioma 52, 53
 treatment of 9L gliosarcoma 49
 treatment of L1210 leukemia 46, 47
Phenobarbital 182
Phenylketonuria 275
Phenytoin 371
Photoelectric effect 194
Podophyllotoxin
 classification 287
 penetrates central nervous system slightly 291

Podophyllotoxin (*continued*)
  toxicity 293
Podophyllum peltatum 86
Polycyclic aromatic hydrocarbons: induction of tumors 33
Prednisolone: chemical configuration 170
Prednisone
  biologic half-life 170, 171
  plasma half-life 171
Probenecid 293
Procarbazine 54, 84–85, 88, 91
  classification 287
  myopathy 327–328
  penetrates central nervous system freely 291
  structural formula 84
  toxicity 292, 293, 294, 305
  treatment of avian sarcoma virus-induced astrocytoma 51, 52
  treatment of ependymoblastoma group of tumors 48, 49
  treatment of glioma 52, 53
  treatment of L1210 leukemia 46, 47
Prolactinoma
  biology 126, 127
  clinical manifestations 125–126
  indications for operative and nonoperative management 130–131
  laboratory evaluation 127–128
  pathology 126–127
  prediction of postoperative result 129–130
  pregnancy 132
  radiographic diagnosis 128–129
  secretes prolactin in proportion to size 127
  treatment with radiation therapy 133
  treatment with transsphenoidal microsurgery 132
Prophylhydroxybenzoate 320

Radiation
  interaction with matter 196, 198
  mode of cell killing 198–201
  oxygen effect 201
  relative biological effectiveness 202–203
  tissue tolerance 203–205, 238–242
  tolerance of central nervous system 205–207
  treatment of brain metastasis 357–367
Radiation Therapy Oncology Group 227–229, 234
Rat
  F344 157

Fischer 344 36, 39, 41, 42, 43
Wistar-Furth 38
Reproductive death: definition 198
Reticulum cell sarcoma: increased risk by renal transplant patients 21
Retinoblastoma
  genetic factors 10
  incidence rate 8
  relationship to osteosarcoma 22
Rhabdomyosarcoma: measles encephalitis 325
Rous sarcoma virus 40

Sarcoma
  37 274
  180 31, 274
Selective recall bias: problem in studies 23
Semustine (*see* Methyl-CCNU)
Simian sarcoma virus 40
Spirohydantoin 80
  cross resistance 83
  in vitro assay 261
Spirohydantoin mustard 81
Stalk section effect 128
Stanford 224
State University of Iowa Hospital 9
Streptozotocin 81
Syndrome
  apathy 307
  Charcot-Marie-Tooth 326
  Fahr's 313, 322
  glioma-polyposis 397
  Korsakoff's 305
  multiple harmatoma 12
  Nelson's 124, 145
  nevoid basal cell carcinoma 12
  Sipple's 12
  steroid withdrawal 356
  Sturge-Weber 10, 313
  Thys 302
  Turcot's 12
  Wermer's 12
Systemic cancer: neurological complications 341–342

6TG: treatment of L1210 leukemia 46, 47
Therapeutic gain factor 202
Thio-TEPA
  classification 287
  toxicity 307
Thymosin 157

Tinea capitis 14, 15, 20
Toxoplasma encephalitis 324
Toxoplasmosis, acquired 323
Triazinate 90
Tuberous sclerosis 10, 11, 397
Tumor (*see also* Cancer; Carcinoma; specific type of tumor)
  avian sarcoma virus-induced 36
  brain
    compared with systemic cancer 65–66
    immunocompetence of patients 152–154
  central nervous system
    classification 2
    genetic factors 10–12
    morbidity 3–8
    mortality 3
    survival 8–10
  glomus: genetic factors 10
  in vitro assay 247
    biological considerations 249, 256–258
    cellular immune assays 254–256, 263–265
    cerebrospinal fluid tissue culture 248–249, 258, 259
    chemotherapeutic microcytotoxicity assays 258–261
    chemotherapy 250–253
    humoral immunology 253–254
    immunologic microcytotoxicity assay 261–262
    problems 265
    tissue culture 248
  mammary 68–69
  metabolic classification 274, 275–277
  nervous system
    etiology 13–22
    multicentric 13
  pituitary: occurs more frequently in blacks 8
Walker 274

United States Brain Tumor Study Group 226
University of California at San Francisco 118, 121, 125, 126, 127, 129, 130, 134, 139, 235
  General Clinical Research Center 138
  Neuroradiology Unit 129
  Reproductive Endocrinology Unit 132
University of North Carolina 159, 161
University of Rochester 38
University of Tennessee 277
  Center for the Health Sciences 279
University of Washington 232

Vidarbine 323
Vinblastine
  classification 287
  penetrates central nervous system slightly 291
  toxicity 292, 322
Vincristine 87–88, 91
  classification 287
  enhances radiation damage 208
  myelopathy following intrathecal chemotherapy 316
  penetrates central nervous system slightly 291
  structural formula 87
  toxicity 304, 311, 315, 319, 326–328
  treatment of ependymoblastoma group of tumors 47, 48
Vinyl chloride 17, 21
Virginia Commonwealth University 279
VM-26 86–87
  structural formula 87

Washington, D.C.: population-based study of residents 8
Weibull model 390, 392
Wernicke's encephalopathy 314, 322
WR-2721: radioprotector 211–212